Sewage Management

Edited by Başak Kılıç Taşeli

Published in London, United Kingdom

Sewage Management
http://dx.doi.org/10.5772/intechopen.104202
Edited by Başak Kılıç Taşeli

Contributors
Amir Moradinejad, Ye Duk Choi, Jun Yeon Lee, Jung Gone Joung, Kyu Mun Han, Zoo Ho Jang, Mok Young Lee, Kweon Jung, Maipady Ramu Dhanraj, A. Ganesha, Jose A. Muñoz Hernandez, Luis Eduardo Leguizamon Castellanos, Helmer Muñoz Hernandez, Ali Aghababai Beni, Arezoo Ghadi, Mohammad Saleh Samie Adel, Sayed Asaad Abdollahi, Kiarash Jalalvandi, Mojgan Zaeimdar, Vahid Hassani, Nityanand Singh Maurya, Amit Kumar, Rajanee Kant Yadav, Astha Kumari, Abhishek Kumar, Farzana Malik, Muhammad Yousuf, Zeeshan Umer, Wajid Malik, Abdul Majid, Sajid Iqbal, Heena Tabassum, Allika Sailaja, Huda Afreen, Minal Wani, Mihail Luca, Suchitra Rakesh, Radhakrishnan Vandana, Ahmed Abou-Shady, Heba El-Araby, Jesús Cisneros-Aguirre, Maria Afonso-Correa, Mounia Achak, Edvina Lamy, Noureddine Barka, Sonika Saxena, Amanpreet Kaur, Aditi Agarwal, Sudipti Arora

First published in London, United Kingdom, 2023 by IntechOpen
IntechOpen is the global imprint of INTECHOPEN LIMITED, registered in England and Wales, registration number: 11086078, 5 Princes Gate Court, London, SW7 2QJ, United Kingdom
Printed in Croatia

British Library Cataloguing-in-Publication Data
A catalogue record for this book is available from the British Library

Additional hard and PDF copies can be obtained from orders@intechopen.com

Sewage Management
Edited by Başak Kılıç Taşeli
p. cm.

This title is part of the Sustainable Development Book Series, Volume 5
Topic: Urban Planning and Environmental Management
Series Editor: Usha Iyer-Raniga
Topic Editors: Christoph Lüthi and Rui Alexandre Castanho

Print ISBN 978-1-83768-535-6
Online ISBN 978-1-83768-536-3
eBook (PDF) ISBN 978-1-83768-537-0
ISSN 2753-6580

IntechOpen Book Series

Sustainable Development

Volume 5

Aims and Scope of the Series

Transforming our World: the 2030 Agenda for Sustainable Development endorsed by United Nations and 193 Member States, came into effect on Jan 1, 2016, to guide decision making and actions to the year 2030 and beyond. Central to this Agenda are 17 Goals, 169 associated targets and over 230 indicators that are reviewed annually. The vision envisaged in the implementation of the SDGs is centered on the five Ps: People, Planet, Prosperity, Peace and Partnership. This call for renewed focused efforts ensure we have a safe and healthy planet for current and future generations.

This Series focuses on covering research and applied research involving the five Ps through the following topics:

1. Sustainable Economy and Fair Society that relates to SDG 1 on No Poverty, SDG 2 on Zero Hunger, SDG 8 on Decent Work and Economic Growth, SDG 10 on Reduced Inequalities, SDG 12 on Responsible Consumption and Production, and SDG 17 Partnership for the Goals

2. Health and Wellbeing focusing on SDG 3 on Good Health and Wellbeing and SDG 6 on Clean Water and Sanitation

3. Inclusivity and Social Equality involving SDG 4 on Quality Education, SDG 5 on Gender Equality, and SDG 16 on Peace, Justice and Strong Institutions

4. Climate Change and Environmental Sustainability comprising SDG 13 on Climate Action, SDG 14 on Life Below Water, and SDG 15 on Life on Land

5. Urban Planning and Environmental Management embracing SDG 7 on Affordable Clean Energy, SDG 9 on Industry, Innovation and Infrastructure, and SDG 11 on Sustainable Cities and Communities.

The series also seeks to support the use of cross cutting SDGs, as many of the goals listed above, targets and indicators are all interconnected to impact our lives and the decisions we make on a daily basis, making them impossible to tie to a single topic.

Meet the Series Editor

Usha Iyer-Raniga is a professor in the School of Property and Construction Management at RMIT University. Usha co-leads the One Planet Network's Sustainable Buildings and Construction Programme (SBC), a United Nations 10 Year Framework of Programmes on Sustainable Consumption and Production (UN 10FYP SCP) aligned with Sustainable Development Goal 12. The work also directly impacts SDG 11 on Sustainable Cities and Communities. She completed her undergraduate degree as an architect before obtaining her Masters degree from Canada and her Doctorate in Australia. Usha has been a keynote speaker as well as an invited speaker at national and international conferences, seminars and workshops. Her teaching experience includes teaching in Asian countries. She has advised Austrade, APEC, national, state and local governments. She serves as a reviewer and a member of the scientific committee for national and international refereed journals and refereed conferences. She is on the editorial board for refereed journals and has worked on Special Issues. Usha has served and continues to serve on the Boards of several not-for-profit organisations and she has also served as panel judge for a number of awards including the Premiers Sustainability Award in Victoria and the International Green Gown Awards. Usha has published over 100 publications, including research and consulting reports. Her publications cover a wide range of scientific and technical research publications that include edited books, book chapters, refereed journals, refereed conference papers and reports for local, state and federal government clients. She has also produced podcasts for various organisations and participated in media interviews. She has received state, national and international funding worth over USD $25 million. Usha has been awarded the Quarterly Franklin Membership by London Journals Press (UK). Her biography has been included in the Marquis Who's Who in the World® 2018, 2016 (33rd Edition), along with approximately 55,000 of the most accomplished men and women from around the world, including luminaries as U.N. Secretary-General Ban Ki-moon. In 2017, Usha was awarded the Marquis Who's Who Lifetime Achiever Award.

Meet the Volume Editor

Başak Kılıç Taşeli is a professor in the Environmental Engineering Department at Giresun University in Turkey. She has an MSc in soil pollution and a Ph.D. in environmental sciences, both from Middle East Technical University. Among the graduate-level sustainability courses she has taught are Environment, Energy and Sustainability, Biofuels, Biomass Conversion Systems, Biogas Production and Usage, and Sustainable Operation of Treatment Plants. She has taught undergraduate-level courses in Water Supply, Wastewater Disposal, Environmental Ecology, Sludge Treatment, Unit Operations, and Industrial Wastewater Treatment. She has 28 years of experience in protected area management, environmental management, renewable energies, and zero-waste and carbon footprint quantification of wastewater treatment plants and industries. She has worked for more than ten years on EU-funded Erasmus+ projects, coordinating several projects as a promoter and partner.

Contents

Preface

Over the past half-century, the world has experienced a significant economic boom and rapid urbanization, resulting in the uplifting of millions of individuals from poverty. Nevertheless, this progress has come at the cost of escalating resource consumption and environmental decline. The connection between resource utilization and environmental effects is undeniable. This is mainly due to the linear utilization of resources, which results in environmental consequences at every stage, culminating in various forms of waste.

Sewage management refers to the collection, treatment, and disposal of the millions of gallons of wastewater generated every day by households, industries, institutions, commercial establishments and public facilities. This critical aspect of modern society is often overlooked or taken for granted. Sewage is a complex mixture of macro and micropollutants, as well as pollutants from municipal solid waste and industrial wastewater. The first step in sewage management is the collection of wastewater through a system of pipes, pumps, and storage tanks. The collected wastewater is then transported to a treatment plant where it undergoes various processes such as screening, sedimentation, and biological treatment to remove solids, organic matter, and other pollutants. After treatment, the wastewater may be discharged into a water body, such as a river or ocean, or reused for irrigation, industrial processes, or other non-potable uses. Poorly managed sewage can lead to the spread of waterborne diseases, contamination of groundwater and surface water, and ecological damage to aquatic ecosystems. Effective sewage management requires a comprehensive approach that considers a range of factors such as technical feasibility, social acceptability, and economic viability. By adopting innovative and sustainable approaches, we can promote resource recovery and reuse, protect public health and the environment, and create a more sustainable future.

This book on sewage management aims to provide a comprehensive overview of the various aspects of this complex topic, drawing on knowledge from engineering, economics, and management. It covers the basics of wastewater treatment and disposal, the technologies and processes involved in sewage treatment, and the legal and regulatory frameworks that govern sewage management. The book also looks at the emerging trends and challenges facing sewage management, such as climate change, population growth, and urbanization. It includes case studies of successful sewage management practices from around the world, highlighting best practices and innovative solutions. There has recently been an increasing focus on sustainable and innovative sewage management approaches that promote resource recovery and reuse. For example, some sewage treatment plants now use anaerobic digestion to convert organic matter in wastewater into biogas, which can be used to generate electricity or heat. The treated wastewater can also be used for irrigation, aquaculture, or other non-potable uses.

The goal of this book is to educate and inform professionals, policymakers, students, and the general public about the importance of sewage management and the various strategies and technologies available to manage sewage effectively. It is the editor's hope that this book will serve as a valuable resource for anyone seeking to better understand this critical topic. The editor expresses sincere thanks to the authors for their contributions.

Başak Kılıç Taşeli
Professor,
Department of Environmental Engineering,
Giresun University,
Giresun, Turkey

Section 1

Treatment of Sewage

Chapter 1

Treatment Technologies and Guidelines Set for Water Reuse

Ahmed Abou-Shady and Heba El-Araby

Abstract

Water reuse is considered a practice that is currently embraced worldwide owing to the exacerbated water crisis, which is the result of several factors such as the increasing world population, urbanization, industrial sector, global climate change, limited water resources, and agricultural activities. Water reuse is not used intensively only in arid and semi-arid regions, which are characterized by limited water supply but can also be applied in countries that possess sufficient water resources (e.g., Brazil and Canada are implementing policies for water reuse). This chapter discusses the treatment technologies proposed for water reuse and presents some recent guidelines set for water reuse. Treatment technologies typically have three main processes: primary, secondary, and tertiary. There are several set guidelines worldwide for water reuse, however, a universal standard guideline to facilitate the reuse of reclaimed water has not been established. No federal regulations for reusing recycled water have been established in the United States; however, several individual states and territories have established specific regulations to manage reclaimed water for various purposes, including agricultural irrigation, animal watering, and crop production.

Keywords: water reuse, water treatment, guidelines, water crisis, process safety

1. Introduction

Globally, a big amount of potable water is being seawater, whereas less than 3% can be considered safe for use. This 3% of potable water exists in groundwater which accounts for 20% (requires energy for extraction and pumping) and glacial ice (79%), accordingly, the available amount of potable water that is suitable for direct use is account for less than 1% [1, 2]. In China, water crisis can be observed in two phases, scarcity and deterioration, in which two-thirds of Chinese cities suffer from water deficiency, river pollution, and lake eutrophication [3]. At present, the global economy withdraws approximately 4000 km^3 of water per year from natural resources, among which 45% is the discharged wastewater that cannot be handled in the currently available wastewater treatment facilities (only 11% of the total discharge is being treated through different treatment processes). The wastewater may be comprised agricultural runoff (56%), industrial effluents (28%), and household water in an urban area (14%) [4]. At present, half of the world's natural water bodies are severely contaminated, and by 2030, it will be imperative to reduce the proportion of untreated wastewater by half, according to the SGD agenda 2030 [5].

The scenario of water reuse is considered ancient as human civilizations themselves (e.g., several civilizations, including Egypt, Mesopotamia, and Crete civilizations utilized sewage (domestic wastewater) for agricultural irrigation from the beginning of the Bronze Age (approximately 3200–1100 BC). Afterward, Greek and Roman civilizations adapted water reuse during 1000 BC–330 AD [6].

Almost all continents at present such as Europe, Australia, Africa, Asia, and North America embrace the notion of potable reuse [7]. The amount of water being reused differs from one country to another (e.g., 46% in California, 7% in Japan, 32% in Asia, 75% in Israel, and 44% in Florida). Reclaimed water is being reused for environmental applications in northern Europe (51%). Also, water reuse is account for 44% in southern Europe for agricultural irrigation, 25% in Tunisia, and 25% in Spain for agriculture. In Singapore, approximately 500 Mm^3/year of treated wastewater is being reused to fulfill its water demands and by 2060, this amount is expected to be increased to 55% [5, 8].

Although some countries have sufficient water resources (e.g., Canada and Brazil), the arid regions may suffer in the future owing to the limited water supply. This is particularly true for expanding cities, with the situation being exacerbated by decreasing glaciers and depleting water sources due to climate change [9, 10]. The agricultural sector consumes a huge percentage of the global freshwater (70% of the withdrawal and 90% of the consumption), and the water consumption of this sector can grow in the future, as 56% of the globally irrigated crops experiences extremely high water stress [5, 11]. Approximately 12% of the globally irrigated land (36 million ha) irrigates with some urban wastewater, of which only 15% is reclaimed water [6].

Water crises are more evident in megacities, such as London, where water infrastructure and population growth pose severe challenges. Similar issues and incidents have been reported in Mexico City and Tokyo. In these areas, water reuse is implemented at two scales (large and small). At the large scale, reclaimed water is dedicated to drinking water (DW) (e.g., Texas, Orange County, and California), whereas at the small scale, reclaimed water is used for flushing toilets, cleaning streets, and irrigating urban areas [12]. A recently emerging term, "One Water," should be embraced worldwide; the term is used to promote the ideology that all water has value and should be managed in a sustainable, inclusive, and integrated way. Water reuse may be considered one of the available solution, to ensure sustainable water use and address food insecurity. The "One Water" approach may involve the following provisions: (1) all water types, from raw source water (for DW treatment plants) to water flushed down toilets and drains, should be viewed through the lens of source water protection, and (2) different types of water pollutants must be treated to establish appropriate standards for their intended downstream applications [11].

The main aim of this chapter is to provide an overview of treatment technologies and guidelines set for water reuse as is considered an important factor to overcome the future water crisis.

2. Recently proposed wastewater treatment technologies for water reuse purposes

2.1 Primary treatment (PT)

Sewage water is pretreated to remove grease, grit, and gross solids, which are considered an obstacle for the subsequent stages of treatment. Afterward, PT is conducted to settle and remove either inorganic or organic suspended solids using settlers and septic tanks [6]. The PT can be exploited to provide water for the controlled irrigation of forestland and parks, as long as the safety precautions are fulfilled. The PT can reduce 30–40% of the organic load and pathogens (**Figure 1**) [5].

2.2 Secondary treatment (ST)

The ST is conducted to remove or degrade soluble biodegradable organics via biological processes (e.g., aerobic or anaerobic processes using bacteria and protozoa). In this treatment, nutrients such as nitrogen and phosphorous may be also removed. The ST involves activated sludge, aerated lagoons, oxidation ditches, trickling filters, and constructed wetlands (CWs) [6]. The application of ST reduces the organic load and pathogens by ~95% and provides disinfection in some cases. This stage of treatment is convenient and can provide suitable water for the irrigation of trees (e.g., in olive orchards and vineyards), as long as there is no direct contact with the crops (**Figure 1**) [5].

2.3 Tertiary treatment (TT)

The TT primarily involves coagulation, flocculation, sedimentation, filtration, and UV treatment, aiming to purify the outlet discharge before ultimate reuse. Notably, TTs also remove nutrients and residual suspended matter via suitable filtration as well as microorganisms and provide disinfection through the application of UV radiation, ozone, and chlorine. Membrane filtration (micro-, nano-, ultra-, and reverse osmosis (RO)), activated carbon, and filtration/percolation are typically a part of the TT; however, their applications are not widespread in developing countries [6]. The TT is fully capable of reducing the organic load by 99%, and UV disinfection completely removes the pathogens. Therefore, TT may be convenient for all types of edible crops (**Figure 1**) [5]. In Spain, 1.4% (63,000 m^3) of the TT water produced from Lloret de Mar City (northeastern Mediterranean coast) was used for irrigation, and the rest was discharged into the sea [13]. Integrating different wastewater treatment technologies can also produce treated water suitable for special purposes. In the

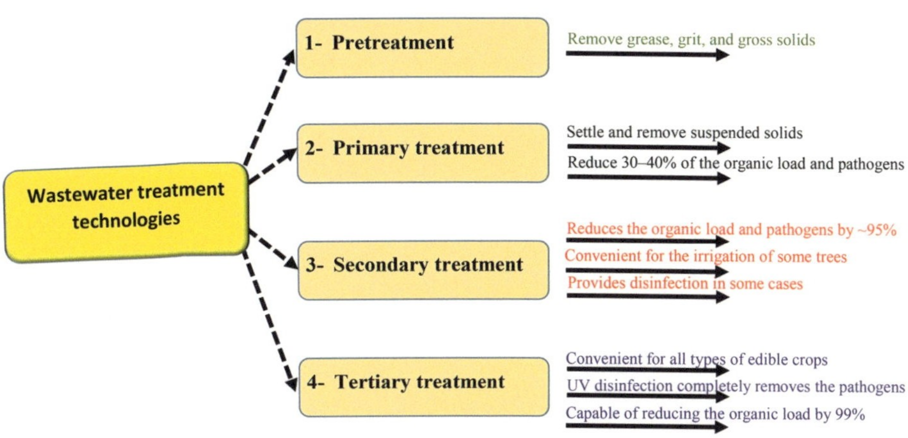

Figure 1.
Wastewater treatment technologies.

following paragraph, we have summarized the recent integrated approaches proposed for wastewater reuse.

The advanced wastewater treatment plant (AWTP) may be used for potable water production based on the wastewater effluent in terms of the feed. For example, AWTP was construed to serve Australian Antarctic Division's Davis Station at the Selfs Point Wastewater Treatment Plant, Hobart, Australia. This AWTP comprises seven barriers: ozone, ceramic microfiltration, biologically activated carbon, RO, UV radiation, calcite dissolution and chlorination, and activated sludge [14]. Moreover, the Groundwater Replenishment Scheme in Australia is operated by the Western Water Corporation; they manage the outlet discharge of the Beenyup Wastewater Treatment Plant by applying three advanced processes, ultrafiltration (UF), RO, and UV disinfection, to recharge the Yarragadee and Leederville aquifers using special recharge bores. The same concept was used in other areas to augment surface water and lake reservoirs [15]. The Old Ford Water Recycling Plant in London collects and treats wastewater from a combined sewer (Northern Outfall Sewer) using a membrane bioreactor, which is then treated by granular activated carbon following which sodium hypochlorite is used to produce reclaimed water suitable for irrigation and toilet flushing [12]. The Aqueous Phase Reforming approach has been recommended for removing the total organic carbon and micro pollutants (e.g., carbamazepine, caffeine, ibuprofen, and diclofenac) from sewage at high percentages (90%) with advantages of H_2 and CH_4 production (via the valorization of organic matter) [8]. Onsite chlorination has been proposed for vertical flow CWs, together with a small-scale solar-driven system, resulting in the reduction of total coliforms and *Escherichia coli* to $\geq$5.1 and $\geq$ 4.6 counts, respectively [16]. A review of the best available technologies and treatment trains in the EU countries, to overcome the deficiency of water supply through urban wastewater reuse, is published in the literature [17].

In the north of Spain, the performance of five WWTP was investigated based on the presence of pathogenic, intestinal protozoa, and nematode eggs. The fifth WWTP does not contain any primary or pretreatment stage; however, wastewater is directly allowed to be treated in an Imhoff tank. In the fifth WWTP, either nematode eggs, Cryptosporidium spp., and Entamoeba spp. were detected, demonstrating high resistance to wastewater treatments. Moreover, the produced sludge contained Cryptosporidium spp., even after aerobic digestion [18].

Mulugeta et al. (2020) suggested that the installation of low-cost bio-sand filters in the decentralized municipal wastewater produces reclaimed water that complied with the WHO and USEPA guidelines [19]. The integrated suspended growth biological process and postozonation (O_3) decreased the organic compounds by 92.1% diesel oil, 97.4% chemical oxygen demand, and 97.9% MB dye. This combination also established the efficacy of integrated industrial and domestic wastewater treatments that is ultimately capable of producing reclaimed water appropriate for agricultural irrigation [20]. The RO technology may be replaced with the ozone-biological activated carbon approach, owing to the reduction of capital and operation and maintenance costs, as well as the comprehensive enhancement of the system [21]. The integrated biochar vertical flow and free-water surface CW system is an effective approach for removing pollutants in wastewater, which is inversely correlated to the hydraulic loading rate [22]. A hybrid pretreatment process was proposed before the desalination of the cooling tower water effluents that included vertical subsurface flow, open water, and horizontal subsurface flow CWs [23]. The process of diluted desalination (using UF–RO) was explored as a preferable economical method to conduct the RO of seawater, even when both the UF recovery rate and water flux complied with the least values of thresholds [24]. The integrated biological trickling filters and CW may be suitable for treating wastewaters derived from food industry and have several advantages: (a) tolerance to loading shocks, (b) simplicity of operation, (c) durability, (d) low capital and operation costs, and (e) high pollutant removal efficiencies [25]. The combination of osmotic membrane bioreactor and RO could be used to treat wastewater before its subsequent reuse [26]. Five scenarios, namely, RO, evaporation, crystallization, de-supersaturation, and precipitation, have been evaluated for treating petrochemical unit effluents, in which the energy consumption and chlorine-derived compounds used in the pretreatment are considered the most influential factors [27]. The integration of ozonation (oxidation) and biofiltration (adsorption and biological degradation) is also considered an effective approach to overcome several obstacles during the conventional water reuse treatment process, owing to its effectiveness in removing trace organic pollutants, byproduct precursors, biodegradable organic matter, and concerning substances [28]. Integrating membrane distillation and RO can purify RO-concentrated wastewater for potable water reuse with high recovery percentages; however, subsequent treatments, such as advanced oxidation (posttreatment), may be required to treat nitrosodimethylamine [29]. A UV-based advanced oxidation can be used to treat RO concentrates characterized by high content of dissolved organic matter; this can ultimately improve the sustainability of water reuse systems [30]. Notably, the use of RO was suggested to reclaim wastewater that is characterized by high salinity [31]. In general, the membrane distillation process has been proposed for potable water reuse following the coagulation and filtration pretreatment processes to avoid severe membrane fouling [32]. In a previous study [33], the feasibility of combining a vertical flow CW and membrane system for the treatment and reuse of decentralized gray and black water was highlighted. Additionally, a breakthrough dynamic-osmotic membrane bioreactor/nanofiltration (NF) hybrid system was proposed for the treatment and reuse of real municipal wastewater, in which membrane fouling was minimized while maintaining high water quality [34]. An adsorption technology was proposed as a promising TT to treat anodized industrial wastewater for reuse purposes [35]. The intensification of supercritical water oxidation through ion exchange with zeolite was proposed to treat landfill leachates for reuse purposes. The intensified process was conducted without using auxiliary substances or oxidants, thereby promoting this method as more eco-friendly and less expensive; however, further improvements are required regarding arsenic concentrations and ammonium-saturated clinoptilolite [36, 37]. A photocatalytic membrane reactor was proposed to treat produced water; the reactor efficiently decomposes and mineralizes organic pollutants, inactivates viruses, detoxifies heavy metals, and recovers valuable minerals [38]. An integrated biological and advanced oxidation process followed by microfiltration and ultrafiltration, suggested to treat laundry wastewater, is a promising and highly efficient combination process for water reuse [39]. Integrated adsorption, RO, and TiO_2/Fe_3O_4 photocatalytic oxidation were proposed for the reuse and recycling of aquatic center sewage comprising shower wastewater, whirlpool tub discharge, pool overflow, and sauna wastewater. This method yielded satisfactory results; however, the process could not remove organic matter (3.3 mg/L) [40]. Jain et al. (2021) indicated that the removal of silica (reactive and colloidal) using diatom biofilms developed from water generated from cooling towers of thermal power plants may save annually 1485 MLD of fresh water, in addition to the generation of value-added products (biogenic silica) [41]. In a previous study, the water reuse systems were optimized using an advanced control system for RO, in which more than Euro 10,000,000 could be saved per year in large RO facilities (>30.000 m^3/day) [42]. Additionally, a study combined the mechanisms of sedimentation and flocculation for the in-situ treatment of quartz- and chlorite-containing water at the Sijiaying Iron Ore Mine; approximately 65.10% grade and 86.50% recovery rate were achieved for iron concentrate under optimum conditions [43]. The denitrifying woodchip bioreactor approach was proposed for treating nitrate-rich water-containing aquaculture effluents. The tannins-lignin and total ammonia nitrogen concentrations were simultaneously increased with Cu and Zn concentrations. It was not recommended to immediately reuse the outflows following start-ups or restart after a dry period [44]. Integration of decolorization and desalination using *Escherichia fergusonii* was recently proposed for remediation of textile effluents for water reuse; the water was subsequently treated with rice husk-activated charcoal treatment to ensure the disinfection and detoxification. This method was not energy-intensive, in addition to being simple, economic, and a two-step process as a complete solution for textile effluent treatment [45]. The process of ultrafiltration was proposed for municipal wastewater reuse as a TT; in this process, the treated water was used only for nonpotable applications, according to the national and international guidelines [46]. The pond-in-pond wastewater treatment system, in which anaerobic and aerobic ponds are combined into a single pond, was also proposed for water reuse. This system is considered highly efficient, with low costs and low maintenance [47]. Xing et al. (2021) revealed that the integrated chlor(am)ine-UV oxidation and UF may be considered a promising alternative for efficient wastewater recycle and reuse [48]. Additionally, molecularly imprinted polymers (MIPs) can be used for removing various contaminants of emerging concern (CECs); however, the MIPs cannot be applied at a large scale and are limited to the lab/bench scale. However, MIPs can remove a wide range of CECs, such as diclofenac, atrazine (pesticide), ketoprofen and ibuprofen (nonsteroidal anti-inflammatory drugs [NSAIDs]), ciprofloxacin and sulfamethazine (antibiotics), triclosan and parabens (personal care products [PCPs]), and bisphenol A (plasticizer) [49].

3. Recent guidelines set for water reuse

3.1 Food and Agriculture Organization (FAO), World Health Organization (WHO), and United States Environmental Protection Agency (USEPA) guidelines

The guidelines for the safe use of reclaimed water in agricultural irrigation, according to FAO, WHO, and USEPA are discussed by Hacıfazlıoğlu et al. [50]; the guidelines comprise several parameters, such as electrical conductivity (EC), sodium absorption ratio (SAR), total dissolved solids (TDS), total nitrogen, total suspended solid (TSS), NO_3-N, PO_4-P, PAR, HCO_3, boron, chloride, sodium, free chloride, pH, and turbidity. These parameters were classified according to the degree of restrictions: none, slight–moderate, and severe. For example, reused water having EC values <0.7 mS/cm is considered to not have any harmful effects on the ecosystem, whereas EC values >3 mS/cm will cause severe effects. Additionally, there is a high correlation between the SAR values and EC, based on the three classes of restriction on soil permeability (none, slight–moderate, and severe) [50].

3.2 Food and drug administration (FDA) guidelines

The regulations set by the United States FDA as part of the Food Safety Modernization Act (FSMA) for water quality metrics were discussed by Rock et al. [51].

The regulations set by the FSMA are applied to the crops that are subject to raw agricultural commodities and do not receive commercial processing required to decrease the growth of microorganisms (to ensure public health). The FSMA has set standard processes, practices, and procedures that reduce the risk of serious health consequences or biological hazards in fresh raw crops (e.g., edible leafy greens) to decrease foodborne diseases resulting from the consumption of contaminated crops [51]. According to the Produce Safety Rule, farmers establish a microbial water quality profile (MWQP) for irrigation water sources (e.g., untreated surface and ground water) and are requested to perform annual surveys that can be used in the following years. The levels of generic *E. coli* are considered a basic parameter for the water quality profile in agricultural (pre-harvest) water. First, the MWQP must be performed, with not less than 20 water samples collected as near to the harvest period as possible, over not less than 2 years and not more than 4 and 5 years for surface water and ground water, respectively. The geometric mean (GM) or statistical threshold value (STV) is to be calculated from the collected samples using a minimum of five samples. The MWQP comprises both GM (126 CFU/100 mL) and STV (410 CFU/100 mL). The GM is an average that reflects the central tendency of the water source, whereas the STV represents the variations in water quality. Further details about the FDA FSMA regulations are presented in the literature [51].

3.3 State-level guidelines in the United States

In the United States, there are no established federal regulations for reusing recycled water. However, several individual states and territories (e.g., Arizona, California, Colorado, Florida, Virginia, Delaware, Massachusetts, New Jersey, Hawaii, North Carolina, Idaho, Minnesota, and Washington) have established specific regulations to manage reclaimed water for various purposes, including agricultural irrigation, animal watering, and crop production. Details about these state-level regulations are presented in the literature [51]. Notably, the regulations established by state standards for irrigation purposes to promote the use of recycled water to grow food crops are considered restrictive compared to the FSMA Agricultural Water metrics. This is indicated by the higher permitted concentrations of *E. coli*, total coliform, or fecal coliform bacteria, which are comparatively lower than those detailed in the FSMA metrics [51].

3.4 Europe (Spain, Italy, Greece, and Cyprus) guidelines

A comparison of guidelines for wastewater reuse in irrigational applications in Europe (Greece, Italy, Spain, and Cyprus) and the United States are discussed by Otter et al. [16]. These guidelines contain uncommon parameters for water quality, prescribing a limit for the number of pathogen indicators, to represent the effectiveness of the wastewater treatment plants in removing the nutrients, organic matter, and pathogens. Accordingly, the disinfection approach (e.g., ultraviolet [UV] radiation, membrane filtration, onsite chlorine generation system, and ozonation) is also a mandatory step to minimize public health risks resulting from potential exposure to reclaimed water [16]. In May 2020, a new regulation was proposed by European Union (EU) regarding the minimum quality requirements (MQR) for reclaimed water (EU, 2020/741) to be reused in the agriculture sector; this regulation will be implemented on June 26, 2023, in all the member states. There is also a growing concern about potential noncompliance situations regarding the MQR [52].

3.5 Egypt

The Egyptian government (Ministry of Housing, Utilities, and Urban Communities) has published the Egyptian Code, with two versions, for treated wastewater (TWW) reuse (code 501, 2005, and 2015). The 501 code in the 2005 version classifies the TWW into three categories, whereas four categories were proposed in (ECP 501, 2015) based on the treatment level. In each category, the Egyptian Code specifies the crops that can be cultivated. In general, the Egyptian Code prohibits the cultivation of raw vegetables, such as cucumber and tomatoes, using TWW [53].

4. Conclusions

The deficiency of water supply is considered a global issue and in the near future (2030; 160% of currently available resources) should be provided to overcome this threat. In the present chapter, we have discussed both treatment technologies and guidelines set for water reuse that may be considered one of the available solutions presented to overcome the future water crisis. The treatment technologies are divided into three main categories including pretreatment, PT, ST, and TT based on the main endeavors suggested for water reuse. The PT can reduce 30–40% of the organic load and pathogens, accordingly, it may be exploited to provide water for the controlled irrigation of forestland and parks, as long as the safety precautions are fulfilled. The ST is capable of reducing the organic load and pathogens by ~95% and provides disinfection in some cases. The ST is suitable water for the irrigation of trees (e.g., olive orchards and vineyards), as long as there is no direct contact with the crops. Lastly, The TT may be convenient for all types of edible crops owing to it being fully capable of reducing the organic load by 99%, and UV disinfection completely removes the pathogens. The standard guideline for the application of water reuse at a global scale is unavailable, whereas, several guidelines for water reuse were discussed in the present chapter including FAO, WHO, USEPA, FDA, State-level guidelines in the United States, Europe guidelines, and Egypt.

Acknowledgements

This work was funded by Science and Technology & Innovation Funding Authority (STIFA) known also as Science and Technology & Development Fund (STDF) (project number: 39369).

Conflict of interest

The authors declare no conflict of interest.

Author details

Ahmed Abou-Shady[1*] and Heba El-Araby[2]

1 Soil Physics and Chemistry Department, Water Resources and Desert Soils Division, Desert Research Center, Cairo, Egypt

2 Independent Researcher, Kafr El Sheikh, Egypt

*Address all correspondence to: aboushady@drc.gov.eg

References

[1] Abou-Shady A, El-Araby H. Electro-agric, a novel environmental engineering perspective to overcome the global water crisis via marginal water reuse. Natural Hazards Research. 2021;**1**(4):202-226. DOI: 10.1016/j.nhres.2021.10.004

[2] Abou-Shady A. Recycling of polluted wastewater for agriculture purpose using electrodialysis: Perspective for large scale application. Chemical Engineering Journal. 2017;**323**:1-18

[3] Wang H-C et al. A2O-MBR as an efficient and profitable unconventional water treatment and reuse technology: A practical study in a green building residential community. Resources, Conservation and Recycling. 2019;**150**:104418

[4] López-Morales CA, Rodríguez-Tapia L. On the economic analysis of wastewater treatment and reuse for designing strategies for water sustainability: Lessons from the Mexico Valley Basin. Resources, Conservation and Recycling. 2019;**140**:1-12

[5] Ricart S, Rico AM. Assessing technical and social driving factors of water reuse in agriculture: A review on risks, regulation and the yuck factor. Agricultural Water Management. 2019;**217**:426-439

[6] Ait-Mouheb N, Mayaux P-L, Mateo-Sagasta J, Hartani T, Molle B. Water reuse: a resource for Mediterranean agriculture. In: Water Resources in the Mediterranean Region. Elsevier; 2020. pp. 107-136

[7] Zhiteneva V, Carvajal G, Shehata O, Hübner U, Drewes JE. Quantitative microbial risk assessment of a non-membrane based indirect potable water reuse system using Bayesian networks. Science of The Total Environment. 2021;**780**:146462

[8] Faria D et al. Sewage treatment using aqueous phase reforming for reuse purpose. Journal of Water Process Engineering. 2020;**37**:101413

[9] Van Rossum T. Water reuse and recycling in Canada-history, current situation and future perspectives. Water Cycle. 2020;**1**:98-103

[10] Marangon BB, Silva TA, Calijuri ML, do Carmo Alves S, dos Santos VJ, de Sousa Oliveira AP. Reuse of treated municipal wastewater in productive activities in Brazil's semi-arid regions. Journal of Water Process Engineering. 2020;**37**:101483

[11] Sapkota AR. Water Reuse, Food Production and Public Health: Adopting Transdisciplinary, Systems-based Approaches to Achieve Water and Food Security in a Changing Climate. Elsevier; 2019

[12] Goodwin D, Raffin M, Jeffrey P, Smith HM. Collaboration on risk management: The governance of a non-potable water reuse scheme in London. Journal of Hydrology. 2019;**573**:1087-1095

[13] Santana MVE, Cornejo PK, Rodríguez-Roda I, Buttiglieri G, Corominas L. Holistic life cycle assessment of water reuse in a tourist-based community. Journal of Cleaner Production. 2019;**233**:743-752

[14] Scales PJ et al. A critical control point approach to the removal of chemicals of concern from water for reuse. Water Research. 2019;**160**:39-51

[15] Khan SJ, Anderson R. Potable reuse: Experiences in Australia. Current Opinion in Environmental Science & Health. 2018;**2**:55-60

[16] Otter P et al. Disinfection for decentralized wastewater reuse in rural areas through wetlands and solar driven onsite chlorination. Science of the Total Environment. 15 Jun 2020;**721**:137595. DOI: 10.1016/j.scitotenv.2020.137595

[17] Rizzo L et al. Best available technologies and treatment trains to address current challenges in urban wastewater reuse for irrigation of crops in EU countries. Science of the Total Environment. 2020;**710**:136312

[18] Benito M, Menacho C, Chueca P, Ormad MP, Goñi P. Seeking the reuse of effluents and sludge from conventional wastewater treatment plants: Analysis of the presence of intestinal protozoa and nematode eggs. Journal of Environmental Management. 2020;**261**:110268

[19] Mulugeta S, Helmreich B, Drewes JE, Nigussie A. Consequences of fluctuating depth of filter media on coliform removal performance and effluent reuse opportunities of a bio-sand filter in municipal wastewater treatment. Journal of Environmental Chemical Engineering. 2020;**8**(5):104135

[20] Egbuikwem PN, Mierzwa JC, Saroj DP. Evaluation of aerobic biological process with post-ozonation for treatment of mixed industrial and domestic wastewater for potential reuse in agriculture. Bioresource Technology. 2020;**318**:124200

[21] Sundaram V, Pagilla K, Guarin T, Li L, Marfil-Vega R, Bukhari Z. Extended field investigations of ozone-biofiltration advanced water treatment for potable reuse. Water Research. 2020;**172**:115513

[22] Nguyen XC et al. Combined biochar vertical flow and free-water surface constructed wetland system for dormitory sewage treatment and reuse. Science of the Total Environment. 2020;**713**:136404

[23] Wagner TV et al. Pilot-scale hybrid constructed wetlands for the treatment of cooling tower water prior to its desalination and reuse. Journal of Environmental Management. 2020;**271**:110972

[24] Pazouki P, Stewart RA, Bertone E, Helfer F, Ghaffour N. Life cycle cost of dilution desalination in off-grid locations: A study of water reuse integrated with seawater desalination technology. Desalination. 2020;**491**:114584

[25] Tekerlekopoulou AG, Economou CN, Tatoulis TI, Akratos CS, Vayenas DV. Wastewater treatment and water reuse in the food industry. In: The Interaction of Food Industry and Environment. 2020. pp. 245-280. DOI: 10.1016/b978-0-12-816449-5.00008-4

[26] da Silva JRP et al. Applicability of osmotic bioreactor using potassium pyrophosphate as draw solution combined with reverse osmosis for removal of pharmaceuticals and production of high quality reused water. Journal of Environmental Chemical Engineering. 2021;**9**(6):106487

[27] Sakamoto H, de Sá Teles BA, Kulay L. An eco-efficiency analysis of refinery effluent pretreatments for water reuse under a zero liquid discharge regime. Science of the Total Environment. 2021;**793**:148564

[28] Guarin TC, Pagilla KR. Microbial community in biofilters for water reuse applications: A critical review. Science of the Total Environment. 2021;**773**:145655

[29] Ngo MTT et al. Membrane distillation for achieving high water recovery for potable water reuse. Chemosphere. 2021;**288**(3):132610

[30] Mangalgiri K, Cheng Z, Cervantes S, Spencer S, Liu H. UV-based advanced oxidation of dissolved organic matter in reverse osmosis concentrate from a potable water reuse facility: A parallel-factor (PARAFAC) analysis approach. Water Research. 2021;**204**:117585

[31] Wei X, Sanders KT, Childress AE. Reclaiming wastewater with increasing salinity for potable water reuse: Water recovery and energy consumption during reverse osmosis desalination. Desalination. 2021;**520**:115316

[32] Jeong S, Song KG, Kim J, Shin J, Maeng SK, Park J. Feasibility of membrane distillation process for potable water reuse: A barrier for dissolved organic matters and pharmaceuticals. Journal of Hazardous Materials. 2021;**409**:124499

[33] Lakho FH et al. Decentralized grey and black water reuse by combining a vertical flow constructed wetland and membrane based potable water system: Full scale demonstration. Journal of Environmental Chemical Engineering. 2021;**9**(1):104688

[34] Cong Nguyen N et al. A breakthrough dynamic-osmotic membrane bioreactor/nanofiltration hybrid system for real municipal wastewater treatment and reuse. Bioresource Technology. 2021;**342**:125930. DOI: 10.1016/j.biortech.2021.125930

[35] Acosta-Herrera AA, Hernández-Montoya V, Castillo-Borja F, Pérez-Cruz MA, Montes-Morán MA, Cervantes FJ. Competitive adsorption of pollutants from anodizing wastewaters to promote water reuse. Journal of Environmental Management. 2021;**293**:112877. DOI: 10.1016/j.jenvman.2021.112877

[36] Scandelai APJ, Zotesso JP, Vicentini JCM, Cardozo Filho L, Tavares CRG. Intensification of supercritical water oxidation (ScWO) by ion exchange with zeolite for the reuse of landfill leachates. Science of the Total Environment. 2021;**794**:148584. DOI: 10.1016/j.scitotenv.2021.148584

[37] Scandelai APJ, Zotesso JP, Jegatheesan V, Cardozo-Filho L, Tavares CRG. Intensification of supercritical water oxidation (ScWO) process for landfill leachate treatment through ion exchange with zeolite. Waste Management. 2020;**101**:259-267. DOI: 10.1016/j.wasman.2019.10.005

[38] Chen L, Xu P, Wang H. Photocatalytic membrane reactors for produced water treatment and reuse: Fundamentals, affecting factors, rational design, and evaluation metrics. Journal of Hazardous Materials. 2022;**424**:127493. DOI: 10.1016/j.jhazmat.2021.127493

[39] Zoroufchi Benis K, Behnami A, Aghayani E, Farabi S, Pourakbar M. Water recovery and on-site reuse of laundry wastewater by a facile and cost-effective system: Combined biological and advanced oxidation process. Science of the Total Environment. 2021;**789**:148068. DOI: 10.1016/j.scitotenv.2021.148068

[40] Izadpanah M, Sarrafzadeh MH, Rezaei-DashtArzhandi M, Vojoudi H. Aquatic center sewage reclamation and water reuse, using an integrated system combining adsorption, RO membrane system, and TiO_2/Fe_3O_4 photocatalytic oxidation. Journal of Environmental

Chemical Engineering. 2021;**9**(1):104957. DOI: 10.1016/j.jece.2020.104957

[41] Jain R, Nigam H, Mathur M, Malik A, Arora UK. Towards green thermal power plants with blowdown water reuse and simultaneous biogenic nanostructures recovery from waste. Resources, Conservation and Recycling. 2021;**168**:105283. DOI: 10.1016/j.resconrec.2020.105283

[42] Galizia A et al. Advanced control system for reverse osmosis optimization in water reuse systems. Desalination. 2021;**518**:115284. DOI: 10.1016/j.desal.2021.115284

[43] Liang G, Zhao Q, Liu B, Du Z, Xia X. Treatment and reuse of process water with high suspended solids in low-grade iron ore dressing. Journal of Cleaner Production. 2021;**278**:123493. DOI: 10.1016/j.jclepro.2020.123493

[44] Lepine C, Christianson L, Soucek D, McIsaac G, Summerfelt S. Metal leaching and toxicity of denitrifying woodchip bioreactor outflow—Potential reuse application. Aquacultural Engineering. 2021;**93**:102129. DOI: 10.1016/j.aquaeng.2020.102129

[45] Santhana Raj D, Nagarajan SV, Raman T, Venkatachalam P, Parthasarathy M. Remediation of textile effluents for water reuse: Decolorization and desalination using *Escherichia fergusonii* followed by detoxification with activated charcoal. Journal of Environmental Management. 2021;**277**:111406. DOI: 10.1016/j.jenvman.2020.111406

[46] Yang J, Monnot M, Eljaddi T, Ercolei L, Simonian L, Moulin P. Ultrafiltration as tertiary treatment for municipal wastewater reuse. Separation and Purification Technology. 2021;**272**:118921. DOI: 10.1016/j.seppur.2021.118921

[47] Adhikari K, Fedler CB, Asadi A. 2-D modeling to understand the design configuration and flow dynamics of Pond-In-Pond (PIP) wastewater treatment system for reuse. Process Safety and Environment Protection. 2021;**153**:205-214. DOI: 10.1016/j.psep.2021.07.012

[48] Xing J, Du L, Quan X, Luo X, Snyder SA, Liang H. Combining chlor (am) ine-UV oxidation to ultrafiltration for potable water reuse: Promoted efficiency, membrane fouling control and mechanism. Journal of Membrane Science. 2021;**635**:119511

[49] Parlapiano M et al. Selective removal of contaminants of emerging concern (CECs) from urban water cycle via molecularly imprinted polymers (MIPs): Potential of upscaling and enabling reclaimed water reuse. Journal of Environmental Chemical Engineering. 2021;**9**(1):105051

[50] Hacıfazlıoğlu MC, Tomasini HR, Bertin L, Pek TÖ, Kabay N. Concentrate reduction in NF and RO desalination systems by membrane-in-series configurations-evaluation of product water for reuse in irrigation. Desalination. 2019;**466**:89-96

[51] Rock CM et al. Review of water quality criteria for water reuse and risk-based implications for irrigated produce under the FDA Food Safety Modernization Act, produce safety rule. Environmental Research. 2019;**172**:616-629

[52] Truchado P et al. New standards at European Union level on water reuse for agricultural irrigation: Are the Spanish wastewater treatment plants ready

to produce and distribute reclaimed water within the minimum quality requirements? International Journal of Food Microbiology. 2021;**356**:109352

[53] Abd-Elhamid HF, Daiem MMA, El-Gohary EH, Elnaga ZA. Safe reuse of treated wastewater for agriculture in EGYPT. In: Conf. Exceed. Expert Work. Water Effic. Cities, Marrakech, Morocco, November 3-8, 2017. 2017 [Online]. Available from: http://mena.exceed-swindon.org/wp-content/uploads/2015/03/0024-Abd-Elhamid-Manuscript-Marrakech-2018-02-27-final-pp.pdf

Chapter 2

Constructed Wetlands Process for Treating Sewage to Improve the Quantitative and Qualitative Management of Groundwater Resources

Mounia Achak, Noureddine Barka and Edvina Lamy

Abstract

Water scarcity limits access to safe water for drinking and communities face some form of water stress, which can be related to insufficient supplies or inadequate infrastructures. Climate change plays a crucial role in water stress worldwide, as rising temperatures lead to more unpredictable weather and extreme weather events. In face of this challenge, the need to seek an alternative to protect groundwater resources and to decrease the use of public water is imposed. Sewage management seems to be a significant treatment of removing contaminants and undesirable components from polluted waters and safely return it to environment for irrigation and other uses. For this consideration, many treatment technologies are discussed in the literature including biological, physical and chemical processes. Among biological processes principally used for the treatment of sewage figured constructed wetlands. Constructed wetland system is considered as an economic, efficient and environmentally friendly sewage treatment method, based on adsorption and retention of pollutants by substrates, sorption by plants, and decomposition by microorganisms. Therefore, the chapter of this book throws will light on the principal mechanisms responsible to organic matter, nitrogen and phosphorus removal in different types of constructed wetlands, and provides recommendations concerning the factors affecting pollutants removal performance of constructed wetlands from sewage.

Keywords: water scarcity, sewage characteristics, treatment, constructed wetlands process, pollutants removal, mechanism approach

1. Introduction

Many parts of the world suffer of water scarcity and insufficient water quality demanded by population and environment. Water stress effects can vary dramatically from one region to another; it can damage not only public health, economic development, and global commerce, but also can increase migrations and spark conflict.

IntechOpen

According to Food and Agriculture Organization (FAO), in the case of demand side, around 70% of the world's freshwater is used for agriculture, 19% for industrial uses and 11% for domestic uses [1]. In the case of supply side, the water sources include groundwater and surface water such as rivers lakes and reservoirs. Indeed, many causes can trace to the water stress including the increase of human withdrawals from surface or groundwater and the increased need of agricultural irrigation. The population growth who lead to economic development, change of lifestyle and consumption patterns, and the increase of agriculture production impact dramatically the planet's reserves, also the climate disruption as rising temperatures, floods and drought who can decrease water supply. Another important approach of water scarcity is identified by discharge of polluted wastewater generated from industry, agriculture and household activities causing deterioration of freshwater quality and infiltration of contaminant compounds in the groundwater [2, 3]. Around 80% of wastewater in the world is evacuated, largely untreated, in the lakes, rivers, oceans and environment [4]. In face of this challenge, many countries seek an alternative to protect water resources and to improve international collaboration on water practices, as well as to implement innovative and sustainable technologies in order to ameliorate the water quality by suggesting significant processes to reduce hazardous organic pollutants from polluted wastewater. Several approaches have been investigated to eliminate the contaminants from domestic sewage using physical, chemical and biological processes. Physical processes include evaporation, distillation, combustion, centrifugation, filtration, pyrolysis and membrane separation [5–7]. While chemical processes include coagulation-flocculation, electrocoagulation, oxidation and advanced oxidation and ion exchange [8–10]. Infiltration-percolation, aerobic, anaerobic and constructed wetland are included in biological processes [11–13].

The limitations of physical and chemical processes like expensiveness, high power consumption and production of toxic products, as well as the risk of secondary pollution, can affect the performance of these techniques for the treatment of sewage [14, 15]. Thus, biological processes are widely considered among the efficient techniques and have become the center of interest of several researchers in the pollutants removal of sewage. Nowadays, constructed wetlands have been attracting increasing attention compared to other treatment techniques. Constructed wetlands are a cost-effective and sustainable approach to treat not only sewage but also agriculture and industrial wastewater as olive mill wastewater, tannery wastewater, winery wastewater and petrochemical industry wastewater [16–19]. Constructed wetlands system offers many benefits, such as less expensive to construct, low operational and maintenance expenses, tolerate fluctuation in water flow, eliminate odors associated with wastewater, allow a high process stability during wastewater treatment and facilitate wastewater reuse and recycling.

Constructed wetlands are a natural wastewater treatment system that use natural wetland vegetation (aquatic plants), substrate (sand, soil, gravel, zeolite, pozzolan) and microorganisms for the removal of conventional pollution parameters such as metals, organic matter and nutrients (nitrogen and phosphorus) from polluted water [17, 20–23]. Aquatic plants are considered as a central of sewage treatment and they play a major role in the adsorbion, precipitation and degradation of organic compounds, as well as concentration of heavy metal from contaminated water. While substrates are also beneficial to the efficiency of the constructed wetlands, they provide storage for many contaminants by the adsorption process and they report support to many microorganisms responsible to degradation of organic matter in the favorable conditions by biological process. On the other hand, different designs and structures of constructed wetlands are investigated in the literature. Overall, we can distinguish three types of constructed wetlands (i) surface flow constructed wetlands (SF CWs), (ii) horizontal subsurface flow constructed wetlands (HSF CWs) and (iii) vertical flow constructed wetlands (VF CWs). The selection of the effective constructed wetlands depends on several parameters such as load and type of pollutants, material of substrates, local available vegetation, climatic conditions, nature of polluted water, and hydraulic loading rate (HLR) and hydraulic retention time (HRT).

In this chapter, the treatment of sewage using constructed wetlands as approach to improve the quantitative and qualitative management of groundwater resources will be discussed. For this purpose, selection of favorable constructed wetlands design, optimization of operation parameters and factors affecting remediation processes such as substrates, aquatic plants type and availability of microorganisms can improve the performance of sewage treatment. Finally, the recommendations and future possibilities in this field will also be discussed.

2. Characteristics and composition of sewage

Sewage is classified mainly into two categories, namely, domestic and industrial wastewater. Domestic wastewater is derived from human activities in households such as bath, laundry, dishwashing, garbage disposal and human waste, mainly feces and urine [24]. The variation in the characteristics of domestic wastewater depends on several factors such as the water use, quality and type of water supply, nature and condition of sewerage system and population habits. Compared to industrial wastewater, domestic wastewater usually contains low content of pollutants but even small amounts of contaminants can be responsible of environment pollution. Industrial wastewater is generated from various industrial processes, namely, the water released from battery, chemical and pharmaceutical manufacturing, agricultural and mine activities, paper and fiber plants, refining and petrochemical operations and other industrial activities [25]. The volume, flow and load of wastewater is closely linked to the type of industries and industrial establishment.

Generally, the characteristics and composition of sewage depend on the nature of wastewater discharged and its source. Sewage contains a high content of organic compounds, which may be in dissolved, suspension and colloidal state. These compounds may be toxic, resist to biological degradation, can damage sewers and other structures, and affect the condition operation of the wastewater treatment plant. Sewage also may contains some heavy metals provided through industrial discharges. These compounds limit the treatment by biological process and their disposal in stream and land affect the human and aquatic life. On the other hand, various types of microorganisms may be identified in sewage. Some of these are pathogens and are harmful to the human and animal life.

Many heavy metals are identified in sewage, such as Lead (Pb), Zinc (Zn), Mercury (Hg), Nickel (Ni), Cadmium (Cd), Copper (Cu), Chromium (Cr) and Arsenic (As) released by paint and dye manufacturing, textile, pharmaceutical, paper and fine chemical industries. These metals are non-biodegradable and can be carcinogenic, their easily adsorption to suspended particles in water threat dramatically ecosystems and their infiltration in the groundwater can present a risk for living organisms. Solids contained in sewage may be classified into dissolved solids, suspended solids and colloidal solids. Suspended solids and dissolved colored material and reduce water clarity by creating an opaque, hazy or muddy appearance. Their remove can be

Industrial wastewaters	pH	BOD_5 (mg/L)	COD (mg/L)	NH^{4+}-N, (mg/L)	NO_3^--N, mg/L	PO_4^{3-} (mg/L)	Phosphorus (mg/L)	TSS (mg/L)	Chlorides (mg/L)	Total phenols (mg/L)	Cr-Ca-K (mg/L)	References
Domestic	7.6	78.4	ND	27.4	0.4	3.1	ND	75.1	ND	ND	ND	[29]
Pulp and paper	7.41	1928	4103	47.52	ND	ND	105	612	ND	ND	ND	[30]
Textile	9	1010	3200	ND	ND	ND	ND	2185	213	ND	ND	[31]
Olive oil mill	5.06	ND	7022	1960	0.64	0.40	4200	2070	1420	1345	ND-0.06-2.11	[32]
Coke plant	8.28	80.60	692.11	454.95	49.30	ND	ND	1122.65	ND	92.82	ND	[33]
Distillery	7.5–8	8000	45,000	ND	ND	ND	ND	40.700	7997	7202	ND	[34]
Winery	7.15	2950	4283	170.6	ND	7.0	ND	580	ND	ND	ND	[35]
Brewery	2–12	1000–37,500	1900–50,000	19–22	ND	ND	5–100	560–3000	ND	ND	ND	[36]
Tannery	7.6	ND	17.618	ND	ND	ND	ND	20.498	ND	ND	44-372-50	[37]
Slaughterhouse	7.28	10,172	16,910	1520	ND	ND	25.5	7267	ND	ND	ND	[38]
Meat processing	7.1	1966.7	2250	50.37	80.8	ND	9.6	1120.7	ND	ND	ND	[39]
Cigarette industry	6.9–7.4	516–540	1120–1245	ND	ND	ND	ND	998–112	460	ND	ND	[40]
Fish market	7.1	ND	1079	66.7	ND	ND	56.0	ND	ND	ND	ND	[41]

Table 1.
Examples of concentrations of major pollutants in various industrial wastewater reported in the literature.

assured by sedimentation if their size comparatively large or by filtration process. Generally, solids are an important factor for sewage treatment processes. Sewage also contains high load of organic pollutants, namely, phenolic compounds, dioxins and dibenzofurans, polycyclic aromatic hydrocarbons, and organochlorine pesticides [26]. Organic pollutants are highly resistant to environmental degradation via physical, chemical and biological processes. They have long half-lives in soil, water, and air and their toxicity can pose a threat to human health and environment. High content of nitrogen and phosphorus are also one of the major contaminants present in sewage. The principal nitrogen elements in sewage are proteins, amines, amino acids, and urea. Indeed, the ammonia nitrogen in sewage results from the bio-degradation of organic matter present in the water. In domestic wastewater, human urine contains approximately more than 95% of total nitrogen, 90% of total phosphorus and 50% of the total COD [27]. Phosphorus is contributing to sewage from laundry detergent, human waste and other household cleaning products with 0.30, 0.60 and 0.10 kg of phosphorus per capita/year, respectively [28]. It considered as an essential element for the biological method. With an adequate concentration, phosphorus can support aerobic biological wastewater treatment. **Table 1** summarizes the concentration of major pollutants resulting from different industrial sector.

3. Constructed wetlands as sewage management pattern

3.1 Free water surface constructed wetlands

Free water surface constructed wetlands (FWS CWs) are defined as wetland systems comprise shallow basins or channels, with a sealed bottom to prevent wastewater to infiltrate in the groundwater. FWS CWs are characterized by the presence of dense aquatic plants covering more than 50% of surface, emerged in 20–40 cm of water into 20–30 cm of rooting soil [42]. In FWS CWs, water flows through over vegetated soil surface from an inlet to an outlet point (**Figure 1**). This operation allows the physical, chemical and biological processes to take place in order to remove and degrade the various contaminants. However, in some cases, standing water may increase the possibility of mosquito breeding, or water may completely lost by evapotranspiration especially in the hot region, which affect the treatment process.

Based on the literature, FWS CWs are efficient in removal of organic pollutants through bio-degradation and settling of colloidal particles. Suspended solids removal is assured by sedimentation, filtration and aggregation mechanisms in function of particles size and structure [43]. The smaller and lighter particles may settle out through the dense wetland plants, while largest and heaviest particles may settle out in

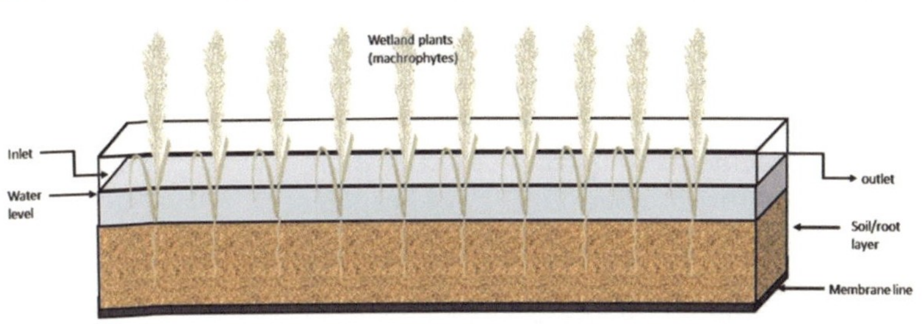

Figure 1.
Free water surface constructed wetlands (FWS CWs).

the inlet open water zone. Indeed, the wetland plant tissue play a major role in suspended solids removal by reducing wind speed which supports sedimentation of suspended solids and prevents re-suspension [44]. On the other hand, many studies indicated that the high suspended solid concentrations can cause clogging of the soil and effects negatively the treatment process. To avoid the excess accumulation of solids, FWS CWs facilities should be coupled with a pretreatment stage, namely septic tank, lagoons, settling basins or compost filter [45].

Nitrogen is generally removed from wastewater through nitrification/denitrification mechanisms. Before nitrification step, organic-nitrogen is converted to ammonia by hydrolysis process, ammonia is oxidized to nitrate by nitrifying bacteria under aerobic conditions. Nitrate is reduced to free nitrogen or nitrous oxide under anaerobic conditions by denitrification process. Many parameters may influence nitrification-denitrification processes such as carbon source, temperature, pH, dissolved oxygen availability and nitrite accumulation. In CWs, nitrogen is effectively removed primarily by nitrification/denitrification, and ammonia volatilization under higher pH values due to algal grow. However, the presence of the ammonia in the atmosphere can pollute aquatic and terrestrial environments [46]. Indeed, in FWS CWs, the growth of algae is very limited, due to the presence of emergent wetland plants, which cover completely the surface water and as consequence limit algal photosynthesis by preventing light to penetrate into the water column. Denitrification can be increased by availability of carbon provided from decaying plant biomass. The supply of dissolved oxygen in FWS CWs is assured by the air-water interface and the plant roots, which release oxygen into the environment media, creating the favorable conditions for nitrification process. While in the soil layer below the water, the dissolved oxygen is practically non-existent. A viable solution to this behavior is to report an extended aeration to achieve nitrification process and provide biological treatment for the removal of bio-degradation pollutants. Air may be supplied by diffusion or mechanical in required conditions to maintain the aerobic biological process.

FWS CWs are efficient in removal of organic pollutants trough aerobic and anaerobic processes. The aerobic process is carried out under redox conditions in the water columns, while anaerobic process is realized by fermentation or biomethanation in the litter layer near the bed bottom. The decomposition performance of organic pollutants is assessed by the balance between organic load and available oxygen. Other pathways for organic matter removal are identified, namely, photochemical reactions, uptake by plants and metabolization [47, 48]. Removal of phosphorus in CWs is assured by adsorption, complexation-precipitation and plant uptake if the biomass is harvested, otherwise, the phosphorus releases in the system and its concentration increase in wastewater. The complexation-precipitation mechanism is highly dependent on the origin and chemical composition of the substrate used. Soil rich in reactive minerals such as Fe, Al or calcareous materials that favorite Ca phosphate precipitation are very significant in phosphorus removal. In FWS CWs the elimination of phosphorus is relatively with a slow rate, due to little contact between water column and the soil containing mineral elements for precipitation.

Based on the literature studies, CWs have been generally designed to treat municipal and domestic wastewater. Recently, CWs are strongly applied to other type of wastewater as industrial wastewater. Among effluents treated by FWS CW are petrochemical wastewater, pulp mill wastewater, tannery wastewater, aquaculture wastewater, distillery wastewater, brewery wastewater, meat processing wastewater, abattoir wastewater, seafood wastewater and olive mill wastewater. Goulet and

Sérodes [49] reported that the treatment of abattoir wastewater using FWS CWs planted with *Typha sp*. and consisted of storage tank (750 m^3) and two cells a total surface area of 1420 m^2 allowed removal efficiencies amounted to 95%, 85%, 66%, 54% and 74% for TSS, BOD_5, TKN, NH_4^+-N and TP respectively. Two FWS CWs occupied an area of 45.5 m^2 for each, planted with *Phragmites australis* and filled with gravel were used to treat diluted olive mill wastewater. The results showed that the bed without recirculation allowed a reduction of 80%, 83%, 80%, 78% and 74% for TSS, COD, total phosphorus, NH_4^+-N and phenols respectively. While the reduction using bed with recirculation acceded 90%, 98%, 85%, 55% and 87% respectively [50]. In another study, the treatment of petrochemical wastewater by 15.5 m^2 FWS CW planted with *P. australis*, *Typha angustifolia* and *Typha latifolia*, and HLR ranged between 1.3 and 1.6 cm d^{-1} was investigated. The average annual removal of COD, BOD_5, TN and total phosphorus from pretreated wastewater reached 54%, 59%, 22% and 43%, respectively [51]. Allen et al. [52] presented the results from a full-scale FWS CW (8.19 ha) designed to treat increasing amounts of pre-treated domestic wastewater. The system received an annual average loading rate of 947 kg/year BOD_5, 19,644 kg/year TN, 31039 kg/year NH_4-N, 18140 kg/year TKN and 807 kg/year total phosphorus and removal rate of 8%, 72%, 73%, 78% and −246% respectively. The study of Bydalek et al. [53] aimed to assess microplastics fate in FWS CW following an oxidation process. The FWS CW has an operational surface of approximately 8000 m^2 and planted with *Schoenoplectus lacustris* and *Typha angustifolia*. The results showed that over 95% of microplastics were retained within the first 20% of the FWS CW and allowed an aerial removal rate exceed 4000 microplastic m^{-2} d^{-1}.

3.2 Horizontal sub-surface flow constructed wetlands

Horizontal sub-surface flow constructed wetland (HSF CW) systems are filled with gravel or soil and usually planted with common reeds [54]. The depth of the substrate varies between 30 and 80 cm and it is most often selected based on many parameters such as low patterns, effluent charge and aquatic plant types as a nutrient source required for removal process [55]. Generally, the media mostly used include gravel, sand, soil and compost. In HSF CWs, the wastewater is fed in at the inlet and flows horizontally along bed media below substrate surface through its pores and plant roots. The treated wastewater is collected in outlet zone from the opposite side of the system before leaving level control arrangement at the outlet, thus keeping a large contact between water and substrate (**Figure 2**). Consequently, the availability

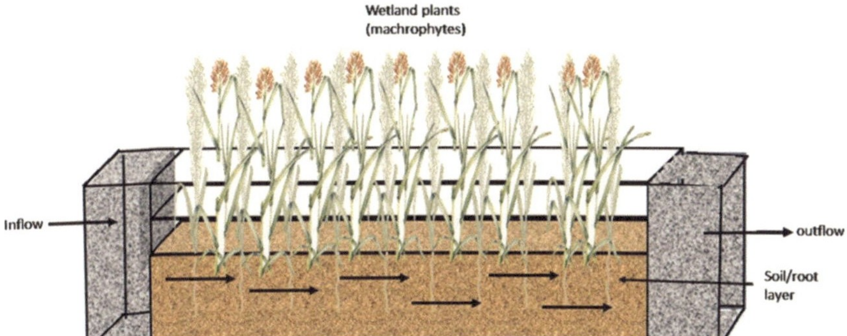

Figure 2.
Horizontal sub-surface flow constructed wetlands (HSF CW).

of oxygen in the substrate for organic matter oxidation is limited despite the creation of some aerobic zones by plants by transportation of oxygen from the atmosphere to the roots through the plant stems. HSF CWs included many processes and mechanisms for removal of contaminant elements and compounds from sewage such as sedimentation, filtration, and aerobic and anaerobic microbial processes for organic matter remove, nitrification/denitrification for nitrogen elements remove, and sorption and precipitation for phosphorus remove.

Organic compounds in HSF CWs are very effectively degraded by aerobic and anaerobic microbial processes as well as by filtration and sedimentation. The aerobic process is carried out in plant roots and surface waters where oxygen can be supplied from atmosphere, while anaerobic process is occurred in the soils with microorganisms adapted to each condition. The aerobic microorganisms consume oxygen to degrade organic contaminants to produce biomass for microorganism. The methane is produced through the degradation of organic matter by anaerobic bacteria. The presence of plant roots favors the development of biofilm, which increase the organic matter decomposition. The amount decomposed of organics is related dramatically to the availability of dissolved oxygen in HSF CWs. On the other hand, plant biomass are capable to storage organic carbon thus making constructed wetlands as an alternative approach to remove organic matter. Suspended solids are removed in HSF CWs by gravity sedimentation, flocculation/settling of colloidal particulates and adsorption. Gravity sedimentation occurred when particle sediment independently without contact with other particles. While, flocculation/settling involves the interacting of particles in order to form large flocculants result from charge imbalances on the surface of particles. The settling rapidly take place when a new particle as larger flocs are formed. Filtration and adsorption onto gravel and plant media play an important part in suspended solids removal. The surface of gravel, stem, and root plants are coated by biomass film, which can absorb colloidal and soluble matter. The solids adsorbed may be metabolized, and then converted to biomass or gases.

As viewed in the literature for nitrogen removal, regardless of removal pathways including plant uptake, ammonification, plant root and substrate adsorption, volatilization in forms of ammonia, fixation by converting nitrogen gas to organic nitrogen, and transformation into nitrogen gas, the nitrification-denitrification is usually considered as the principal mechanism to reduce nitrogen amount in constructed wetlands. Nitrification is carried out by converting ammonia and ammonium to nitrite and then nitrite to nitrate. Many parameters influence nitrification process such as temperature, pH, alkalinity, dissolved oxygen concentration, retention time, and organic load. According to Vymazal et al. [55], the optimum temperature for nitrification is generally considered from 30°C to 40 °C, pH values range from 6.6 to 8.8, and dissolved oxygen and alkalinity amounts reach 4.3 mg of oxygen and 8.8 mg/mg of ammonia oxidized respectively. Denitrification process is the process in which microorganisms reduce nitrate to nitrite and nitrite to nitrogen gas. The nitrogen gas produced is in the form of nitric oxide (NO), nitrous oxide (N_2O) or nitrogen gas (N_2). Environmental conditions that affect the efficiency of denitrification include nitrate concentration, anoxic conditions, presence of organic matter as well as pH, temperature, alkalinity and the effects of trace metals. Denitrification rate decreases as dissolved oxygen increases and can occur between 5 and 30°C, and under optimum pH values between 7.0 and 8.5. In HSF CWs the denitrification process is favorable due to availability of carbon source contained in raw sewage. However, the nitrification process is limited by the absence of oxygen in these systems. The alimentation process of HSF CWs maintain filtration bed submerged continually by water, consequently the presence of oxygen is limited and, therefore, the removal of nitrogen is less effective. The major phosphorus removal mechanisms in wastewater wetlands are physical/chemical (adsorption, absorption, precipitation) and biological processes. The main mean for phosphorus removal in constructed wetland is plant absorption. The absorption is carried out through leaves and root plants, and is increased at the beginning of the plants growing season. Many aquatic plants were showed their efficiency in the storage of phosphorus such as *Iris pseudacorus*, *Panicum virgatum*, *Canna sp*., *Oenanthe javanica*, *Myriophyllum aquaticum*, etc. [56, 57]. However, the die of plant parts and the beginning of their decay generates the release of plant matter above ground in the water and the secretion of decaying roots into the soil and, therefore, the contamination of groundwater. Adsorption and precipitation is another process to removal phosphorus amounts. Different cations present in the substrate such as Fe^{3+}, Ca^{2+}, Al^{3+} can be combined with orthophosphate and forme insoluble phosphates. In addition, phosphates charged negatively can react by anions-cations exchange with substrate elements. Various substrates such as steel slag and oyster shell approved their efficiency to removal phosphorus from sewage [58]. Although the removal of phosphorus can takes place by microorganisms absorption due to their availability and quick multiplication, microorganisms are unable to storage a large contents of phosphorus.

Several studies applied HSF Cws as a promote technology to treat industrial wastewater. Chapple et al. [59] study focused on the reduction of dissolved hydrocarbons (Diesel Range Organics-DRO- C_{10}-C_{40}) from oil refinery using four pilot wetlands. Two filled with soil and others filled with gravel (300 m^2 each), and planted with *P. australis*. The results showed that with the mean inflow DRO concentration of 410 mg/L, all beds presented an efficiency removal, which exceed 99%. Choudhary et al. [60] studied 5.25 m^2 HSF CW filled with gravel and planted with *Canna indica* to treat chlorinated resin and fatty acids (RFAs) from a paper mill wastewater. The system achieved 92% removal of 9,10,12,13-tetrachlorostearic acid and 96% removal of 9,10-dichlorostearic acid. The authors concluded that at hydraulic retention time of 5.9 days, the most probable mechanism for this removal is microbial decomposition in the plant roots as well as adsorbtion/absorption. To treat a primary treated sewage, Shukla et al. [61] study three HF CWs (35 m^2 each) filled sequentially and supplied with gravel media. CW_1 (unplanted), CW_2 (planted with *Typha latifolia*) and CW_3 (planted with *Typha latifolia* and *Commelina benghalensis*). The CWs were aerated and operated in continuous mode at an average hydraulic loading rate of 250 L/h with different hydraulic retention time (HRT) 12, 24, 36 and 48 h. The authors reported that among the constructed wetlands used, CW_3 was the best performer reducing 79, 77, 79, 79 and 78% of BOD, COD, N-NO_3, N-NH_4 and phosphate respectively in 48 HRT.

3.3 Vertical sub-surface flow constructed wetlands

Vertical sub-surface flow constructed wetlands (VSF CWs) are a flat bed filled by graded gravel topped with sand layers and planted with macrophytes usually is common reeds. Wastewater is poured onto the bed surface from above. The water is draining vertically down by gravity through the porous media to the bottom of the bed where is collected by a drainage pipe (**Figure 3**). With this mode of operation, the bed inside is aerated by pushing out the trapped air, thus increasing aeration. On the other hand, the aeration may be enhanced by insertion of aeration pipes, and employing a wet-dry cycle of operation. This way of feeding plays an important part

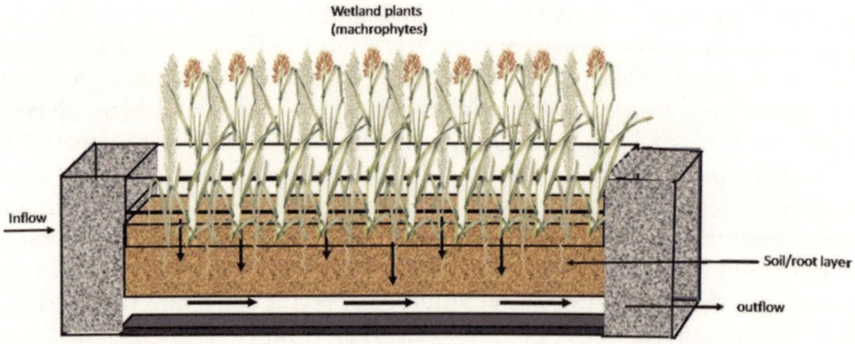

Figure 3.
Vertical sub-surface flow constructed wetlands (VSF CW).

in the provide of aerobic conditions, since increase of oxygen transfer within the filter enhance the oxidation of ammonia nitrogen by nitrification and decomposition of organic contaminants. The crucial difference between HSF CWs and VSF CWs is not only the direction of the flow path, but also the availability of the oxygen in the bed.

In VSF CWs, the mechanisms removal of pollutants follow the same approach applied in HSF CWs. The main chemical-physical and biological processes mentioned are filtration, sedimentation, sorption, chemical oxidation, evaporation as well as aerobic/anaerobic degradation, plants adsorption, phytodegradation and phytoevaporation. In addition, due to better aeration of the bed, VSF CWS are very effective in the removal of organic contaminants (COD, BOD_5) and ammonia nitrogen. Concerning phosphorus, the removal rate remains limited. For this purpose, different filter materials were investigated in order to improve the efficiency of phosphorus removal such as fragmented Moleanos limestone, bauxite, and zeolite, a mixture of river sand and dolomite and wollastonite [62, 63].

VSF CWS are applied for treatment of sewage from different sources, namely, olive mill wastewater [17, 64], laundry wastewater [65], aquaculture wastewater [66], textile wastewater [67] and olive pomace leachate [68].

Achak et al. [17] showed the potential application of experimental system composed of sand filter and VSF CW to achieve nutrient and COD removal from olive mill wastewater. VSF CW consists of a tank (1 m^3) filled with gravel and soil, and planted with a mixture of aquatic plants (*Phragmites australis*, *Typha latifolia* and *Arundo donax*). The presence of aquatic plants was more efficient in removing of nutrients and organic load. The average elimination of experimental system in terms of flow was 62.48% for NTK, 90.43% for NH_4^+, 77.25% for NO_3^-, 98.51% for PO_4^{3-}, 97.53% for PT and 99.05% for COD. The same wastewater was treated by Herouvim et al. [64] using three identical series with four pilot scale VSF CWs filled with various porous media such as gravel, sand and cobble, with several sizes. Two series of pilot scale units were planted with common red and the third is as control unit (unplanted). The authors concluded that COD, phenol and TKN removal seems to be significantly higher in the planted series, while orthophosphate removal shows no significant differences among the three series. The purpose of the study reported by Sotiropoulou et al. [65] aims to indicate the significant improvement in the overall quality of laundry wastewater due to the use of the VSF CW. The VSF unit had a length of 64 cm and a diameter of 20 cm, filled with sand/gravel of different gradations, and planted with *Zantedeschia aethiopica*. Results showed an important decrease of microfibers concentration, COD and turbidity of 98, 93 and 94% respectively from laundry wastewater when a hydraulic load retention of 63.7 mm/d was applied. The treatment of aquaculture wastewater by a system composed of an intermittent sand and VF CWs with a total surface area of 9.5 m^2 was studied by Behrends et al. [66]. At 5.5 days of hydraulic retention time, the inflow concentrations of BOD_5, COD, TN, ammonium and TP of 771 mg/L, 3609 mg/L, 67.4 mg/L, 0.22 mg/L and 40.5 mg/L respectively were amounted to rate removal of 99.5% for BOD_5, 99.1% for COD, 95.5% for TN, 85.9% for ammonia and 84.4% for TP. Davies et al. [67] used VF CWs to remove an azo dye acid orange (AO7) from textile wastewater. VF CWs planted with *Phragmites australis* received an organic loading rates varied between 21 and 105 g COD/m^2 d and reduced to 11–67 g COD/m^2 d. The rates removal of COD, total oxygen carbon and AO7 amounted to 64%, 71% and 74%, respectively. The authors concluded that *Phragmites australis* not simply had a principal role in AO7 removal but also can degrade aromatic amines released during AO7 degradation. The combination of electrochemical oxidation for 360 min at 20 A and VSF CWs planted with *Phragmites australis* at retention time of 3 days and hydraulic organic loading between 5 and 15 g COD/m d was studied by Grafias et al. [68] for treating olive pomace leachate fed intermittently at organic loadings between 5 and 15 g COD/m d and a residence time of 3 d. The treatment by VF CWs followed by electrochemical oxidation enhance rates removal from 86 to 95% for COD and from 77 to 94% for color. However, the reverse approach yielded only 81% COD and 58% color removals.

4. Conclusion and recommendations

Constructed wetland is a natural wastewater system that have been proven sustainable, eco-friendly and low-cost technology compared with many other wastewater treatment technologies. Constructed wetlands consist on aquatic plants, soils, bed media, water, and microorganisms and encompass many processes for eliminating various wastewater pollutants such as biological process (microbial oxidation), physical process (sedimentation, precipitation, adsorption, filtration, and absorption) as well as chemical process (ion exchange and chelation). Many different designs and structures of constructed wetlands are experimented to remove the maximum contaminants from different wastewater sources including FWS CWs, HSF CWs and VSF CWs. Each system as it has many advantages it also has disadvantages. FWS CWs are characterized by low operating costs, can be built with locally available materials, do not use chemical elements for treatment process and provide a high reduction of BOD and suspended solids. However, FWS CWs require expertise for system construction and a large land area for setup; also, they are not tolerant to cold climates. Although HSF CWs necessity low operation and maintenance cost, and allow a high reduction of BOD and suspended solids, but require large land area, allow a low remove of nitrogen and phosphorus (absence of oxygen), and can present a clogging risk. Contrary to FWS CWs and HSF CWs, VSF CWs are a high oxygen transfer capacity, require less land area, report a good nitrification (presence of oxygen) and present less clogging. Nevertheless, VSF CWs require frequent maintenance and pre-treatment of wastewater to prevent clogging.

The increase of water treatment efficiency is mainly based on the optimization of all operation parameters factors affecting processes application such as vegetation, wastewater sources, substrate media, feeding mode, and hydraulic loading rate (HLR) and hydraulic retention time (HRT).

- Vegetation is one of the important parameters in CW systems and could influence the depollution performance of wetlands. The performance of plant species principally depends on plant root characteristics. Since the roots are spread broadly and uniformly in the constructed wetlands filter bed, the decontamination is performed. On the other hand, the selection of vegetation with high biomass productivity while the transpiration needs is low (Higher Water Use Efficiency index), is a key factor in selection of constructed wetland species.

- Wastewater can be provided of many industries. Its concentration of organic matter, suspended solid, nitrogen, phosphorus and other contaminants is very high. The selection of constructed wetlands type to use for depollution of sewage is based on the BOD/COD ratio. If the ration is greater than 0.5, the sewage is biologically biodegradable such as wastewaters from dairies, breweries, food industry, abattoirs or starch and yeast production. While, BOD/COD ratio is lower than 0.5, the low level of biodegradability is reported such as pulp and paper wastewater an olive mill wastewater. In order to keep the efficient constructed wetlands operation, its use proceeded by preliminary treatment is recommended to remove various contaminants and reduce the potential risk of system clogging.

- Substrate media is one the main factors makes the use of constructed wetlands benefit since media can attribute to the decontamination of wastewater. Sand, soil and gravel are the current conventional materials of porous media. To enhance the removal efficiency of different types of constructed wetlands, alternative materials are used including (i) natural substrates which are directly used without any pretreatment such as zeolite, maifanite, pozzolane and limestone; (ii) agricultural by-products and industrial wastes are residue materials produced in the processes of many industries and agricultural operation such as quartz sand, fly ash, steel slag, oyster shell, sawdust, banana peel and coconut endocarp; (iii) synthetic substrates are the materials synthesized in the laboratory which their physical-chemical properties are modified including bio-ceramic, activated carbon and biochar.

- Feeding mode of wastewater is one of the main parameters, which influence constructed wetlands treatment processes. There are three modes, including continuous, batch and intermittent. The effect of each mode is dramatically related to transfer and diffusion of oxygen in constructed wetlands. The removal efficiency of pollutants in batch mode is better than in continuous mode, since the hydraulic retention time in batch mode is higher. While intermittent mode can provide a high removal of nitrogen and organic matter.

- HLR and HRT are the dominant factors which significantly affect the treatment processes in constructed wetlands and control flow rate in the system. Long HRT in constructed wetlands allows enough contact time between microorganisms and pollutants for organic matter degradation and nitrogen removal. While, larger HLR does not allow a sufficient contact time and leads to faster transfer of wastewater trough media that affect the removal efficiency of pollutants.

Acknowledgements

The authors gratefully acknowledge the financial support provided by research funds from the region Hauts-de-France and French National Center for Scientific Research (No. 20-HDF1-0004-RI).

Author details

Mounia Achak[1,2]*, Noureddine Barka[3] and Edvina Lamy[4]

1 Science Engineer Laboratory for Energy, National School of Applied Sciences, Chouaïb Doukkali University, El Jadida, Morocco

2 Chemical and Biochemical Sciences, Green Process Engineering, CBS, Mohammed VI Polytechnic University, Ben Guerir, Morocco

3 Multidisciplinary Research and Innovation Laboratory, FP Khouribga, Sultan Moulay Slimane University of Beni Mellal, Khouribga, Morocco

4 Integrated Transformations of Renewable Matter (TIMR), Sorbonne University, University of Technology of Compiègne, UTC/ESCOM, Compiègne, France

*Address all correspondence to: achak_mounia@yahoo.fr

References

[1] Steflova M, Koop S, Elelman R, Vinyoles J, Van Leeuwen K. Governing non-potable water-reuse to alleviate water stress : The case of Sabadell. Water. 2018;**10**:739

[2] Van Vliet MTH, Flörke M, Wada Y. Quality matters for water scarcity. Nature Geoscience. 2017;**10**:800-802

[3] UN-Water. Policy Brief on Water Quality. Geneva, Switzerland: UN-Water; 2011

[4] Dutta MA, Hanifa MA, Nadeema F, Bhatti HN. A review of advances in engineered composite materials popular for wastewater treatment. Journal of Environmental Chemical Engineering. 2020;**5**:104073

[5] Fane A, Tang C, Wang R. Membrane technology for water: Microfiltration, ultrafiltration, nanofiltration, and reverse osmosis treatise. Water Science. 2011;**113**:301-335

[6] Shahraki S, Miri M, Motahari-Nezhad M. Experimental analysis of pyrolysis of sewage sludge. Energy Sources, Part A: Recovery, Utilization, and Environmental Effects. 2018;**40**: 2037-2043

[7] Shi W, Liu C, Ding D, Lei Z, Yang Z, Yang Y, et al. Immobilization of heavy metals in sewage sludge by using subcritical water technology. Bioresource Technology. 2013;**137**:18-24

[8] Hidaka T, Tsuno H, Kishimoto N, Nakamoto M. Advanced treatment of sewage with pre-coagulation and bio-filtration reactor. Environmental Science. 2001;**685**:147-156

[9] Kanjo Y, Somiya I, Tsuno H, Kagayama M, Kinoshita I, Imashiro M, et al. Advanced treatment of municipal sewage by pre-coagulation and biofilm process. Japan Sewage Works Association. 2000;**37**:131-145

[10] Devlin TR, Kowalski MS, Pagaduan E, Zhang X, Wei V, Oleszkiewicz JA. Electrocoagulation of wastewater using aluminum, iron, and magnesium electrodes. Journal of Hazardous Materials. 2019;**368**:862-868

[11] Li WW, Yu HQ. Advances in energy-producing anaerobic biotechnologies for municipal wastewater treatment. Engineering. 2016;**2**:438-446

[12] Stazi V, Tomei MC. Enhancing anaerobic treatment of domestic wastewater: State of the art, innovative technologies and future perspectives. Science of the Total Environment. 2018; **635**:78-91

[13] Yang J, van Lier JB, Li J, Guo J, Fang F. Integrated anaerobic and algal bioreactors: A promising conceptual alternative approach for conventional sewage treatment. Bioresource Technology. 2022;**343**:126115

[14] Pathak A, Dastida MG, Sreekrishnan TR. Bioleaching of heavy metals from sewage: A review. Journal of Environmental Management. 2009;**90**: 2343-2353

[15] Sharma R, Vymazal J, Malaviya P. Application of floating treatment wetlands for stormwater runoff: A critical review of the recent developments with emphasis on heavy metals and nutrient removal. Science of the Total Environment. 2021;**777**:146044

[16] Wang Y, Cai Z, Sheng S, Pan F, Chen F, Fu J. Comprehensive evaluation of substrate materials for contaminants

removal in constructed wetlands. Science of the Total Environment. 2020; **701**:134736

[17] Achak M, Boumya W, Ouazzani N, Mandi L. Preliminary evaluation of constructed wetlands for nutrients removal from olive mill wastewater (OMW) after passing through a sand filter. Ecological Engineering. 2019;**136**: 141-151

[18] Akratos CS, Tatoulis TI, Tekerlekopoulou AG. Biotreatment of winery wastewater using a hybrid system combining biological trickling filters and constructed wetlands. Applied Sciences. 2020;**10**:619

[19] Jain M, Majumder A, Gupta AK, Ghosal PS. Application of a new baffled horizontal flow constructed wetland-filter unit (BHFCW-FU) for treatment and reuse of petrochemical industry wastewater. Journal of Environmental Management. 2023;**1**:116443

[20] Ilyas H, Hullebusch ED. Performance comparison of different constructed wetlands designs for the removal of personal care products. International Journal of Environmental Research and Public Health. 2020;**17**: 3091

[21] Vystavna Y, Frkova Z, Marchand L, Vergeles Y, Stolberg F. Removal efficiency of pharmaceuticals in a full scale constructed wetland in East Ukraine. Ecological Engineering. 2017; **108**:50-58

[22] Malaviya P, Singh A. Constructed wetlands for management of urban stormwater runoff. Critical Reviews in Environmental Science and Technology. 2012;**42**:2153-2214

[23] Cooper PF, Job GD, Green MB, Shutes RBE. Reed Beds and Constructed Wetland for Wastewater Treatment. UK: WRc Swindon; 1996

[24] Larsen TA, Maurer M. Source separation and decentralization. In: Wilderer P, editor. Treatise on Water Science. Vol. 4. Oxford: Academic Press; 2011. pp. 203-229

[25] Munter R. Industrial Wastewater Characteristics. Sweden: The Baltic University Programme (BUP); 2003. pp. 185-194

[26] Shumbula P, Maswanganyi C, Shumbula N. Type, sources, methods and treatment of organic pollutants in wastewater. In: Rashed MN, editor. Persistent Organic Pollutants (POPs)—Monitoring, Impact and Treatment. London, UK: Intechopen; 2021. DOI: 10.5772/intechopen.101347

[27] Hu M, Fan B, Wang H, Qu B, Zhu S. Constructing the ecological sanitation: A review on technology and methods—A review. Journal of Cleaner Production. 2016;**125**:1-21

[28] Parsons SA, Smith JA. Phosphorus removal and recovery from municipal wastewaters. Elements. 2008;**4**:109-112

[29] Ung H, Leu BT, Tran HTH, Nguyen LN, Nghiem LD, Hoang NB, et al. Horizontal combining flowform cascade with constructed wetland to enhance domestic wastewater treatment. Environmental Technology & Innovation. 2022;**27**:102537

[30] Jagaba AH, Kutty SRM, Baloo L, Birniwa AH, Lawal IM, Aliyu MK, et al. Combined treatment of domestic and pulp and paper industry wastewater in a rice straw embedded activated sludge bioreactor to achieve sustainable development goals. Case Studies in Chemical and Environmental Engineering. 2022;**6**:100261

[31] Patel S, Rajor A, Jain BP, Patel P. Performance evaluation of effluent treatment plant of textile wet processing industry: A case study of Narol textile cluster, Ahmedabad, Gujarat. International Journal of Engineering Science and Innovative Technology. 2013;**2**:4

[32] Achak M, Hafidi A, Ouazzani N, Sayadi S, Mandi L. Low cost biosorbent "banana peel" for the removal of phenolic compounds from olive mill wastewater: Kinetic and equilibrium studies. Journal of Hazardous Materials. 2009;**166**:117-125

[33] Ghose MK. Complete physico-chemical treatment of coke plant effluent. Water Research. 2002;**36**: 1127-1134

[34] Mohana S, Desai C, Madamwar D. Biodegradation and decolorization of anaerobically treated distillery spent wash by a novel bacterial consortium. Bioresource Technology. 2007;**98**: 333-339

[35] Serrano L, de la Varga D, Ruiz I, Soto M. Winery wastewater treatment in a hybrid constructed wetland. Ecological Engineering. 2011;**37**:744-753

[36] Kandasamy S, Devarayan K, Bhuvanendran N, Bo Z, He Z, Narayanan M. Accelerating the production of bio-oil from hydrothermal liquefaction of microalgae via recycled biochar-supported catalysts. Journal of Environmental Chemical Engineering. 2021;**9**:105321

[37] Espinoza-Quinones FR, Fornari MMT, Módenes AN, Palácio SM, da Silva FG Jr, Szymanski N, et al. Pollutant removalfrom tannery effluent by electrocoagulation. Chemical Engineering Journal. 2009;**151**: 59-65

[38] Loganath R, Mazumder D. Performance study on organic carbon, total nitrogen, suspended solids removal and biogas production in hybrid UASB reactor treating real slaughterhouse wastewater. Journal of Environmental Chemical Engineering. 2018;**6**: 3474-3484

[39] Nowrouzi M, Abyar H, Rostami A. Cost coupled removal efficiency analyses of activated sludge technologies to achieve the cost-effective wastewater treatment system in the meat processing units. Journal of Environmental Management. 2021;**283**:111991

[40] Bejankiwar RS. Electrochemical treatment of cigarette industry wastewater: Feasibility study. Water Research. 2002;**36**:4386-4390

[41] Nocetti M, Maine MA, Hadad HR, Mufarrege MM, Di Luca GA, Sánchez GC. Selection of macrophytes and substrates to be used in horizontal subsurface flow wetlands for the treatment of a cheese factory wastewater. Science of the Total Environment. 2020;**745**:141100

[42] Vymazal J. Constructed wetlands for wastewater treatment. Water. 2010;**2**: 530-549

[43] Vymazal J. Constructed wetlands for treatment of industrial wastewaters: A review. Ecological Engineering. 2014;**73**: 724-751

[44] Pettecrew EL, Kalff J. Water flow and clay retention in submerged macrophyte bed. Canadian Journal of Fisheries and Aquatic Sciences. 1992;**49**:2483-2489

[45] Cronk JK. Constructed wetlands to treat wastewater from dairy and swine operations: A review. Agriculture, Ecosystems and Environment. 1996;**58**: 97-114

[46] Asman WAH. Emission and deposition of ammonia and ammonium. Nova Acta Leopoldina. 1994;**228**:263-297

[47] Lyu T, Zhang L, Xu X, Arias CA, Brix H, Carvalho PN. Removal of the pesticide tebuconazole in constructed wetlands: Design comparison, influencing factors and modelling. Environmental Pollution. 2018;**233**:71-80

[48] Jain M, Majumder A, Ghosal PS, Gupta AK. A review on treatment of petroleum refinery and petrochemical plant wastewater: A special emphasis on constructed wetlands. Journal of Environmental Management. 2020;**272**: 301-4797

[49] Goulet R, Sérodes J, Principles and Actual Efficiency of Constructed Wetlands Field Trip Guide during 2000 INTECOL Wetland Conference; Québec, Canada 2000

[50] Kapellakis IE, Paranychianakis NV, Tsagarakis P, Angelalis AN. Treatment of olive mill wastewater in constructed wetlands. Water. 2012;**4**:260-271

[51] Czudar A, Gyulai I, Keresztúri P, Csatári I, Serra-Páka S, Lakatos G. Removal of organic material and plant nutrients in a constructed wetland for petrochemical wastewater treatment. Studia Universitatis Vasile Goldis. Seria Stiintele Vietii. 2011;**21**:109-114

[52] Allen JD, Farrell M, Huang J, Reynolds C, Rupasinghe M, Mosley LM. Long-term water quality response to increased hydraulic loadings in a field-scale free water surface constructed wetland treating domestic effluent. Journal of Environmental Management. 2022;**311**:114858

[53] Bydalek F, Ifayemi, Reynolds L, Barden R, Kasprzyk-Hordern B, Wenk J. Microplastic dynamics in a free water surface constructed wetland. Science of the Total Environment. 2023;**858**:160113

[54] Vymazal J. Horizontal sub-surface flow and hybrid constructed wetlands systems for wastewater treatment. Ecological Engineering. 2005;**25**:478-490

[55] Vymazal J. Removal of nutrients in various types of constructed wetlands. Science of the Total Environment. 2006; **380**:48-65

[56] Keizer-Vlek HE, Verdonschot PF, Verdonschot RC, Dekkers D. The contribution of plant uptake to nutrient removal by floating treatment wetlands. Ecological Engineering. 2014;**73**:684-690

[57] Luo P, Liu F, Liu X, Wu X, Yao R, Chen L, et al. Phosphorus removal from lagoon-pretreated swine wastewater by pilot-scale surface flow constructed wetlands planted with myriophyllum aquaticum. Science of the Total Environment. 2017;**576**:490-497

[58] Yang C, Zhang X, Tang Y, Jiang Y, Xie S, Zhang Y, et al. Selection and optimization of the substrate in constructed wetland: A review. Journal of Water Process Engineering. 2022;**49**: 103140

[59] Chapple M, Cooper PF, Cooper DJ, Revitt D. Pilot trials of a constructed wetland system for reducing the dissolved hydrocarbon in the runoff from a decommissioned oil refinery. In: 8th International Conference on Wetland Systems for Water Pollution Control. Arusha, Tanzania. 16 to 19 September, 2002

[60] Choudhary AK, Kumar S, Sharma C. Removal of chlorinated resin and fatty acids from paper mill wastewater through constructed wetland. World Academy of Science, Engineering and Technology. 2010;**80**:67-71

[61] Shukla R, Gupta D, Singh G. Performance of horizontal flow constructed wetland for secondary treatment of domestic wastewater in a remote tribal area of Central India. Sustainable Environment Research. 2021;**31**:13

[62] Vohla C, Kõiv M, Bavor H, Chazarenc F, Mander U. Filter materials for phosphorus removal from wastewater in treatment wetlands—A review. Ecological Engineering. 2011;**37**: 70-89

[63] Wang D, Jin X, Zhang H, Zhang X, Zeng H, You S, et al. Research progress on phosphorus removal in substrates of constructed wetlands. In: 3rd International Conference on Bioinformatics and Biomedical Engineering. Beijing, China. 2009. pp. 1-6. DOI: 10.1109/ICBBE.2009.5163343

[64] Herouvim E, Akratos CS, Tekerlekopoulou A, Vayenas DV. Treatment of olive mill wastewater in pilot-scale vertical flow constructed wetlands. Ecological Engineering. 2011; **37**:931-939

[65] Sotiropoulou M, Stefanatou A, Schiza S, Petousi I, Stasinakis AS, Fountoulakis MS. Removal of microfiber in vertical flow constructed wetlands treating greywater. Science of the Total Environment. 2023;**858**:159723

[66] Behrends LL, Houke L, Bailey E, Brown D. Reciprocating subsurface flow constructed wetlands for treating high-strength aquaculture wastewater. In: Means JL, Hinchee RE, editors. Wetlands and Remediation. Columbus, Ohio: Battelle Press; 2000. pp. 317-324

[67] Davies LC, Carias CC, Novais JM, Martins-Dias S. Phytoremediation of textile effluents containing azo dye by using *Phragmites australis* in a vertical flow intermittent feeding constructed wetland. Ecological Engineering. 2005; **25**:594-605

[68] Grafias P, Xekoukoulotakis NP, Mantzavinos D, Diamadopoulos E. Pilot treatment of olive pomace leachate by vertical-flow constructed wetland and electrochemical oxidation an efficient hybrid process. Water Research. 2010; **44**:2773-2780

Chapter 3

Problems of Centralized Depuration Systems

Jesús Cisneros-Aguirre and Maria Afonso-Correa

Abstract

Sewage management produces one of the worse impacts on our environment. The current technology applied is obsolete, which results in a huge public spent on installation and maintenance, with very negative consequences on the health of people and environment. The administration and the water companies try to hide these consequences, but the impacts are everyday more and more evident. This situation blocks any development of new technology that can solve the problem in a few years, changing the management, with strict control of every cubic meter of treated water and mud produced, saving an enormous quantity of money from public administration and avoiding a huge negative sanitary and environmental impact. New technologies can change the centralized depuration for decentralized depuration, avoiding the current problems, with a certificate control and saving between 80 and 90% of public inversion, and with the possibility to reuse the mud and treated water in place.

Keywords: pollution, sewage, sewerage management, maintenance, hiding problems, bad quality treatment

1. Introduction

Wastewater management is one of the best indicators of the low health and environmental awareness prevalent at different levels of society today. Except for some examples of efficient treatment systems implemented, most of the world uses highly inadequate technology that requires centralized management of wastewater treatment. This produces a series of associated problems well known to public administrations, which put the health of the public at risk and lead to a huge environmental impact that is reducing the quality of inland aquatic systems and coastal waters.

The vast majority of wastewater treatment plants currently being built and operated around the world are based on technology that has not seen an appreciable development in more than 100 years. This produces purification systems with a low capacity to eliminate contaminants from water as well as numerous associated problems that make it necessary to move treatment plants away from urban centers due to the danger to human health they pose [1–3].

The need to move treatment plants away from wastewater production areas makes it necessary to create an entire sewage infrastructure, with pumping stations and kilometers of pipes; and this represents one of the main problems in wastewater management throughout the world. Only by taking into account the losses from this sewage network can we get an idea of the scale of the problem of this management.

The best example demonstrating the precariousness of purification technology in a standard wastewater management system is the obligatory marine outfall in coastal areas, or discharge pipes to aquifers, lakes, and rivers in inland environments; these provide little confidence in being able to manage wastewater properly. They are totally necessary because purification processes under normal conditions produce an effluent with a high concentration of pollutants, which makes it very difficult to achieve the health guarantees for its reuse; so, the most widely used solution is to discharge it into the aquatic environment.

Sludge produced from purification is another of the major problems associated with these systems since the high concentrations of pollutants it contains exceed the capacity of the system and make it extremely dangerous to manage properly. Most of the time, this sludge ends up in the environment: discharged through outfalls, accumulated in landfills, dumped into the sea from ships, or unfortunately scattered over crop fields [4–8].

All of this bleak panorama—which can be summarized only briefly in this chapter—costs an enormous amount of public money, both for initial investment and for maintenance and management of the systems in our cities that treat wastewater; yet this water mostly ends up in the environment.

2. The danger of urban wastewater

Increasingly universally accepted, urban wastewater represents one of the main problems for coastal ecology. These are liquid residues that are not taken into sufficient consideration, as it is assumed they only contain organic matter and its degradation products.

As a brief introduction to the danger of urban wastewater, it should be highlighted that, in addition to organic matter, it contains all the toxic products from different industries, for example, heavy metals, chemical products, and micro and macro plastics. These are primarily responsible for over half of hydrocarbon discharges into the marine environment; this applies to the so-called emerging and dangerous compounds, which the public administration has recently begun to take into account as compounds to monitor [4–7, 9–12].

Another of the contents of these wastewaters are pharmaceutical products, which when dumped uncontrollably into the sewer, end up in all aquatic systems, whether inland or coastal [13].

It should be remembered that wastewater also contains the entire sum of pathogenic microorganisms for humans, which when cultivated with sublethal doses of antibiotics in the sewage system and in the bioreactors of purification systems, are an ideal culture medium for the creation of drug-resistant pathogenic microorganisms, which are inevitably released into the environment [14–17].

3. Types of discharge into the environment

Having outlined the danger that urban wastewater poses, not only for the environment but for public health, what follows is a description of the routes of entry into the environment of this dangerous waste.

3.1 Discharges from treatment plants

It should be pointed out that this section describes how wastewater is generally managed throughout the world; however, there are a few places where the management is better than in others and may be acceptable, but this is after huge investment in civil works and maintenance that most countries cannot or will not address.

There is not enough space in this summary to address Membrane Biological Reactor (MBR) systems or plant filters, but their results are not very satisfactory either. So, this introduction will focus on traditional purification systems that consist of several chain treatments, where the most common is a so-called primary treatment followed by a secondary one, which is a combination of filtering, biological reactor (bacteria culture), and sedimentation process.

In general, pollutant removal from currently used treatment plants operates at low efficiency, despite their enormous size, initial investment, and maintenance costs, leaving the projects with very little capacity. A typical situation is due to the large civil works required which take several years on average, the plant is normally undersized when it is opened because of the population increase.

One of the main problems of these plants is their extreme fragility, which cannot withstand changes in effluent conditions. The quality of the treated water is affected by changes in the concentrations of initial wastewater, such as organic matter, suspended sediments, nutrients, and numerous parameters, which make management a continuous headache; even the weather can modify the treatment process [18].

Consequently, their management is very complex and depends on the ability of the management team. However, they are not always adequately trained in maintaining a culture of living organisms (biological reactor), and engineers are generally overwhelmed since these do not depend on precise parameters of pressure, flow, pumping, and temperatures, as happens in desalination plants. Here, the recommended plant conditions can be achieved quite well, but the growth of living organisms depends on their interrelationship, biogeochemical cycles, stratification, denitrification, foaming, etc. [19].

One of the most critical parameters that need to be controlled, especially in coastal places, is the salinity of the effluent. Small levels of salinity affect drastically the biological reactor, the core of the treatment [20].

All these changes and variations in the effluent can produce a total collapse of the system. This is when it is necessary to empty tanks to the environment and restart the cultivation. This period can last several weeks, and lead to even worse quality of treated water during this time.

The sewage system does not separate rainwater from wastewater, which overwhelms the purification system in heavy rain, and then has to be directly discharged without purifying. Of all the wastewater management deficiencies, this one-off wastewater malfunction is the only one that is admitted within the management control system, and yet there are rarely any statistics on it or independent assessment of the impact that occurs in these cases [21].

The different public administrations in charge of controlling water treatment, whether private or public, do not use to admit these problems and try to hide them. Therefore, it is not so easy to have reliable statistics on the frequency with which systems are overcome; how long they take to restart; how much is sent to submarine outfall or directly into the environment and under what conditions, and for how long. It is typically admitted that discharges occur at night, as detected by the bad odors suffered in our cities, near the points of discharge.

The biggest problem of this treatment system, in addition to achieving quite poor purification results after the secondary treatment, is that the composition of the resulting effluent is tremendously variable [22].

This great variability makes it very difficult to design an adequate disinfection system or tertiary treatment. As the dosage of hypochlorite or ultraviolet radiation intensity, for example, should depend on the concentration of bacteria; and the same happens with filtration systems, which find it impossible with this great unpredictability.

In addition, these discharges from treatment plants also include products added in plant management processes, such as substances to promote decantation, cleaning of pipes, disinfectants, and other chemicals, all of which end up dumped into the environment. For instance, the impact of hypochlorite when it reacts with organic matter is well known.

Figure 1 shows the result of the water treatment plant in the Mar Menor (Murcia, Spain) with the best traditional technology (activated sludge, prolonged aeration, coagulation processes, flocculation, ring filter, and ultraviolet treatment) with tertiary treatment. However, what cannot be seen in the figure is the bad smell from the treatment plant discharge, which is proof of the serious difficulties in meeting the minimum purified water parameters.

In addition to the bad smell from the same discharge, several dead fish, apparently, mullets (*Mugil cephalus*) of about 15 or 20 cm could be seen, with other smaller, unidentified ones, marked with red circles in **Figure 2**. This gives an idea of the impact that this discharge has when it reaches the coastal lagoon of Mar Menor.

The light foam accumulating on the edges of the discharge is a sign of a significant concentration of surfactants. The toxicity of these compounds for the life of water

Figure 1.
The outlet from the Torre Pacheco treatment plant into the Rambla de Albion, which ends up in the nearby lagoon after passing through the flowmeter located downstream. The bad smell of decomposing organic matter dominates the area (Mar Menor, Murcia, Spain).

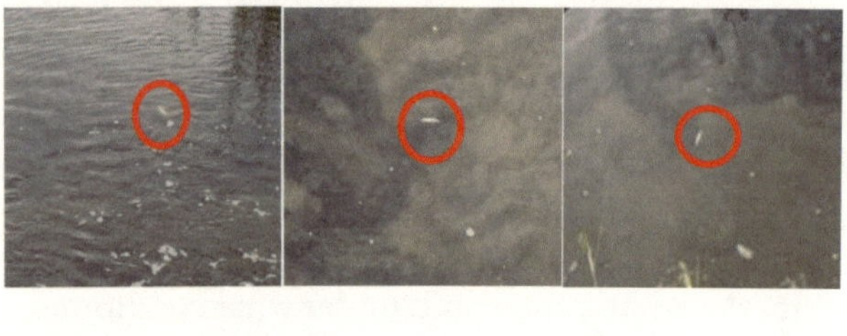

Figure 2.
View of the dead fish at the treatment plant outlet from **Figure 1***, marked with red circles for easier identification; and at the entrance to the treatment plant from which the emanation of foul smells was already evident. This is all indicative of the limitations of the technology.*

courses is well known, as they disturb gaseous exchange between the water surface and atmosphere, among other conditions [23].

This is only a small example of the most common day-by-day management, but it is also very indicative of the weakness of this technology.

3.2 Discharges from the sewerage and cesspools

The risk of treatment plants means they must be located away from population centers, which implies a huge sewerage system is required. This kilometric net is almost impossible to control, leading to a large number of problems, including an important number of leaks into the soil and aquifer.

This enormous net of pipes needs pumping stations to impulse the effluent to the treatment plants. Pumping increases the pressure within the tubes and elevates drastically the leaks to the soil, also the turbulence washes the sediments accumulation, producing a variability of the effluent quality, which the treatment plants do not support easily. Because of the bad smells effluent pumping produce, this maneuver used to be developed during the night.

It must be taken into account that it is estimated that between 8% and more than 50% of the sewerage effluent is lost by leaks during its travel to the treatment plants, which represents an enormous quantity of pollution and sanitary risk that gives it an idea about the size of the problem [24–28].

There are numerous works to replace the drinking water system, also due to the fact that the danger for human health as a result of these losses in sewerage used to be great, if one takes into account that the two pipes usually run parallel and losses from the wastewater pipes can seep into drinking water pipes, create a serious health problem over the world. Taking into account that the standard consumes of water per person in Europe is about 250 liters per day, which approximately means leaks of about 20 liters per day per person of wastewater to the aquifer, applying the lower rate of leaks (8%), is easy to reckon the size of the problem [29].

Figure 3 shows an example of how the sewage reaches the sand itself and its low tightness, producing a continuous discharge that quickly reaches the shore. These losses provide a continuous nutrient flow that feeds the primary producers of any aquatic system.

Fixing this problem is a very complicated matter because the budgets to carry out this work are sky-high and require a large amount of time. In addition, the high cost of sewerage means many homes are not connected to the general network of pipes and are forced to use filtering cesspools. The municipalities are realizing the problem, but there is not enough budget to solve it and the long deadlines needed to place them at a dead end.

Figure 3.
Sewage leaks make a significant contribution to nutrients, in addition to numerous more dangerous pollutants. This can be seen in this figure and is directly related to the algae growth on the shore of the beaches. (Mar Menor, Murcia, Spain).

The sewage system contaminates not only when the wastewater leaks into aquifers and the shoreline but also when salt water penetrates pipelines in coastal places. Together with the discharges from pumps in flood-prone areas in buildings, this increases the salinity of water reaching the purification plants, as explained above, which hinders the purification processes and makes the quality and purification capacity drop even more, as we discuss in the previous chapter.

The sewage system also collects rainwater, which causes the entire treatment system to break down, producing a general discharge throughout the hydraulic basin made up of sewage system wastewater, washing of the sediments accumulated in the sewer system, discharges from the blocked treatment plants, and rainwater, which produces a significant polluting effluent. Then, it is necessary to understand that the majority of human settlements are a continuous source of a big amount of wastewater to the soil, which finished in continental and coastal waters.

The processes of water movement in porous media and the amount of effluent spilled by the sewerage and filtering cesspools are difficult to follow and any approximation must be carried out carefully using a series of data that, usually, do not exist. It is very difficult to find the hydraulic model of wastewater spread in the aquifer, where speed, quality, and flow, for example, could be estimated [30–32].

However, the filtering will retain mainly larger particles or groups in the form of flocs, they can also retain fats and oils that will be adsorbed to the soil particles; similar to the effect sought in sand filters from sewage treatment plants [33].

In the case at hand, it should be stated that highly soluble compounds will pass through this filtrate to reach the aquifer, while those that precipitate, and form crystals or flocs will be retained.

Is typical to the behavior of nitrogen that due to it elevated solubility moves almost freely with the effluent through the soil and aquifer, arriving to the coastal areas, lagoons, rivers, etc., producing a continuous increment of nitrogen in those places.

As will be seen in the section on the effects of pollution, the arrival of soluble phosphorus compounds is one of the novelties that has changed the response of the lagoon ecosystem in Europe; producing the possibility of new growth that was not possible until 2015–2016, due to the change in the formulations of laundry and dishwashing detergents [34, 35].

The phosphorus compounds that were used prior to the European Union regulation of 2012 formed precipitate crystals that were relatively easily retained in the ground and most phosphorus did not reach the coastal areas or continental waters. However, the high solubility of current phosphorus compounds has led to this being an important pathway for the change in speciation.

3.3 Direct discharges

As we have already discussed above, purification systems are often too small, very fragile, and with a low capacity to remove pollutants, then a standard sewage management organization requires direct discharge points to evacuate raw effluent before reaches the treatment system, and partially treated effluent with poor quality, by spillways or submarine outfalls.

Spillways are normally used for this purpose, which is discharge points that in theory should be used to evacuate excess water in the event of floods. These spillways use to be operating all the time when the system is too small, at most, a bar screen is usually used, more to prevent the piping infrastructure from clogging than for the purpose of reducing environmental impact. However, even an adequately sized system that in theory could absorb all the effluent, commonly produces effluent quality not suitable for purification, then needs to be evacuated as well.

Storm tanks have an interesting use in these operations. In principle, they are meant floodwater but are used to accumulate effluent during the day that can then be discharged at night, or pumped into the treatment plant, depending on the effluent quality.

Clearly, these storm tanks are usually placed at the head of the start of marine outfalls or in spillway pipes, since they need a discharge point to manage this accumulation of wastewater that often has significant odor problems. They are usually found in the middle of cities, promenades, squares, or recreational spaces with an admirable combination of recklessness and lack of professionalism.

4. Connivance of the public administrations

This state of affairs is only possible due to collusion between wastewater managers and public administrations that should closely monitor the entire system, aware of the enormous problems, but they simply do not carry out this function.

The complicity between administrations and water managers begins with drafting legislation that is extremely benign and condescending to managing companies. The list of flagrant cases of laxness in the legislation of various countries around the world would fill this introductory summary.

To give an idea of the legislative problem, it should suffice to point out that until 1999 the European Union authorized and recommended the discharge of treated wastewater into the sea together with the sludge. In other words, after a multimillion investment with equally high maintenance costs, continuous discomfort for the public due to unpleasant odors, insects, rodents, pests, noise from pumps at night, the facilitation of disease transmission, and exposing the public to contact with the dangerous effluent, in the end, the treated water is discharged into the sea, with the sludge as well. This process occurs in other parts of the world, and it seems to have been the general solution for many years, continuing to be so today due to minimal control from the administration [36, 37].

The only way to control the water management system would be to monitor each process of management and movement of water and sludge, so that the impact the wastewater is having can be at least approximately evaluated. This would be important not only for the environment, but also to see the impact on the proliferation of resistant bacteria, the return of diseases that had already been overcome, and the incidence of parasites, or malnutrition problems in affected populations as a result of these diseases from this wastewater mismanagement [8].

The cost of such a control system, for example, with adequate sensors, would be negligible compared to the amount of money already spent in this sector on the construction of conventional treatment plants, the pipes to carry wastewater to these points, and the construction of pumping stations, not to mention their maintenance [38].

The results of a sensor system could be offered to the public, so that management was transparent. For example, users could know and limit their use of the coast, by not fishing when dumping in a submarine outfall occurs; or safe bathing areas or freshwater courses affected by a spill could be properly segregated. But anybody with a small experience in this sector believes that this transparency will be real in the next 20 years.

Other examples of the ineffective control of the public administrations are submarine outfalls, which are lucky to be visited once a year, with totally irrelevant surface sampling taken. Also, the specific sampling in treatment plants is carried out in accordance with the operators, and both samplings are carried out in full agreement with the operating companies [39].

The increasing privatization of wastewater management together with this poor control by public administrations, augment the still huge problems we have. Private management means that the purification quality is varied to save money, probably the most common way is to reduce energy consumption in the process. Firstly, as much effluent as possible is removed, by direct discharges before it reaches the treatment plant, and the biological tank aeration flow rates are decreased, constituting practically the biggest expense a traditional treatment plant has. This is very easy to see in sewage treatment plants with agitator aeration, with a reduction in stirring speed during the night leading to an award for the plant manager for the energy savings achieved. This decrease in the already low purification result goes unnoticed, since everything goes through submarine outfall, which is not controlled, much less at night.

5. Sanitation and environmental impact

To understand the impact on the environment of this wastewater management, a short introduction about the evolution of effluents is needed. There is a greater emphasis on marine processes since they are less known and perhaps more difficult to understand than continental water systems.

Produced through submarine outfalls, spillways, or streams that discharge on the shore or a few meters below sea level, the maximum depth should be around 50 meters depth. This chapter will describe the fundamental processes that control the development of these effluents and will summarize their impact, which is very important in the phenomena being studied.

Marine and especially coastal processes are very complicated to differentiate as established academic subjects. However, in this chapter, we will try to differentiate between physical, chemical, and biological processes that occur after the direct discharge of wastewater into the sea. It is undoubtedly necessary to repeat some concepts in various sections, as there may be various points to be discussed.

5.1 Physical evolution of wastewater discharges

5.1.1 Evolution in the water column

From the moment it leaves the pipe, the discharge encounters certain conditions of depth, density, salinity, nutrients, marine life, temperature, waves, wind, and coastal currents, which establish their mixing and movement processes with respect to the point of discharge and coast. This determines the environmental impact on the coast and in the area near the discharge point.

The evolution of the discharge on the water column is part of what is technically called the Near Field, where the physical processes of evolution are dominated by the discharge system design [40–42].

From the point of view of the movement of effluent, when leaving the pipe, it will feel several forces, for example, density difference, with the discharge coming up against water of greater density since wastewater is less dense, giving it a tendency to float toward the surface. It is also found in an environment of slow-moving water, as the speeds of marine currents are very weak near the bottom; the discharge normally has a higher speed and will run into the seawater until it is stopped (**Figure 4**).

Figure 4.
The view of a spillway on the surface of the water. These discharge points are well known to seagulls, for which they are often an easy source of food.

This decrease in speed causes the turbulence to decrease, so that the particles that were kept in suspension with this turbulent energy within the effluent begin to precipitate as the effluent loses speed.

At the outlet to the sea, there will be a sum of these two forces, one floating and the other as a jet in the direction of the outlet, which will lead to a result as determined by the Froude number for the jet:

$$Fr = \frac{\mathrm{U}}{\sqrt{\mathrm{g}'\mathrm{D}}} \tag{1}$$

where U is the exit velocity g' is a modified gravity that takes into account the density difference $\left(g' = g\left(\frac{\rho 1 - \rho 2}{\rho 1}\right)\right)$, and D is the pipe diameter.

Figure 5 shows a couple of examples of the exit from the final mouth of a discharge through an underwater outlet without a diffuser section, where the distance traveled by the effluent from the mouth depends on the exit speed, where the tendency to rise from the bottom is dependent on the difference in density of the effluent and seawater.

At this moment, which is represented in the photos, the exit velocity and the difference in density cause the discharge to travel a distance in the jet mode, so that

Figure 5.
Pictures show a coupled of examples of physical behavior of two different submarine outflows. For the administration both are considered as depurated wastewater without any environmental impact, despite the awful visual appearance.

after losing the exit velocity, its evolution is in the form of a plume, where the forces of buoyancy dominate the movement, producing those volutes that grow and incorporate water from the environment into the effluent.

In this ascent, the phenomenon of dilution occurs through very specific mixing processes that form increasingly larger eddies or volutes, according to the interface between the effluent and seawater. In a limited mixture of the effluent with water with less concentration of contaminants, causing the difference between the two waters to decrease.

During the ascent, important sedimentation phenomena take place due to the change in conditions. This sedimentation is enhanced by the effect of aggregates, a mixture of different compounds (bacteria, organic matter, flocculant products, detergents, and other pollutants) that form small flocs that are ingested by fish and other filtering organisms. This is perhaps the most direct process of damage to the environment and to people's health, since these fish incorporate contaminants directly into their tissues in a rapid bioaccumulation process [43–45].

The formation of flocs that gives rise to the phenomenon of "Marine Snow" increases sedimentation, and also produces a very negative effect on the organisms that are attached to the bottom or depend on it to live. The high sedimentation in the areas impacted by the submarine discharge points covers the bottom and suffocates the benthic organisms in rocky or sandy systems. However, the ones that suffer the most from this abnormal sedimentation process are the filter feeders, whose mechanisms for capturing food are obstructed and are not able to survive in these conditions.

An indirect effect of sedimentation is the anoxia produced on the seabed where it is deposited. The marine snow flocs transport a large amount of organic matter mixed with bacteria to the bottom, which is consumed by them that need oxygen for their degradation. The amount of oxygen is not sufficient and the processes of oxygenation of the sediment stop with this layer of sediments, creating a layer with a high oxygen deficit that eliminates all aerobic life [46, 47].

Sedimentation depends as well on the hydrodynamic conditions of the area: the higher the current, the less sedimentation there is in each area since turbulence keeps the sediments in suspension. Flocs in this form will travel in the water column and settle when they encounter appropriate low turbulence conditions. They will be concentrated at these points and will be resuspended when the dynamics increase.

This happens in storms arrive or due to high-speed boats in small depths, which put a large amount of organic matter back into the water column along with the pollutants that have settled during the calm period.

5.1.2 Evolution on surface

When the discharge finally reaches the surface, it is still very buoyant and accumulates against the surface due to this thrust of the effluent density difference. This accumulation creates a gravity current that causes the spill to spread out horizontally and accelerate toward the surface currents that dominate the area, usually driven by the wind (video 1).

At this point, the dilution of the effluent depends on the natural conditions of the environment and is independent of the discharge type; the process that takes place from this moment on is called Far Field [40–42].

So, a surface "cloud" is created that remains continuously stationary, with slight side movements, which can be pushed in different directions only by energy processes, such as the wind, currents, or movement of boats in shallow places.

Figure 6 shows how the tidal ellipse gives shape to the spill on the surface, which remains stationary, with slight movements around the spill point, with the tail of this case being dragged by the coastal drift. In this case, it is to the right of the photo, which corresponds to the southern component. The slick remains stable in this situation for the duration of the spill and gradually loses its contamination tail.

Figure 7 shows the output of a numerical simulation model for the dispersion of pollutants from the discharge at a depth of 42 m from an underwater outfall (corresponding to the spill in **Figure 6** and the left picture in **Figure 5**).

Figure 7 shows various images of a numerical model for the scenario shown in **Figure 5** left and 6, where the discharge finds a denser medium once it leaves the tube and rises to the surface, due to the difference in buoyancy. On the surface, it accumulates and gives rise to the typical spill slick from underwater pipelines.

Figure 6.
View of the slick of a discharge from a submarine outlet on the coast of the gran Canaria capital. The discharge stays on the surface in an elliptical shape, stationary at the mercy of the coastal currents, wind, and waves.

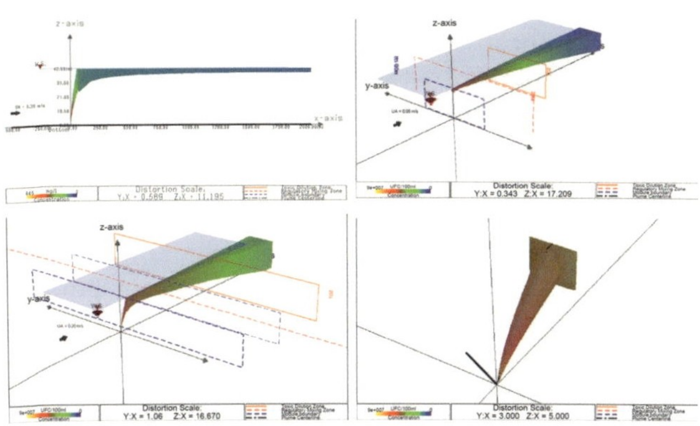

Figure 7.
Numerical simulations were performed by a pollutant dispersion model of the discharge in **Figure 6**, *produced by a submarine outlet at a depth of 42 m. firstly, the ascent to the surface can be verified, with accumulation when it reaches this point and a progression on the surface up to 2000 m away. The color code indicates the evolution of the concentration.*

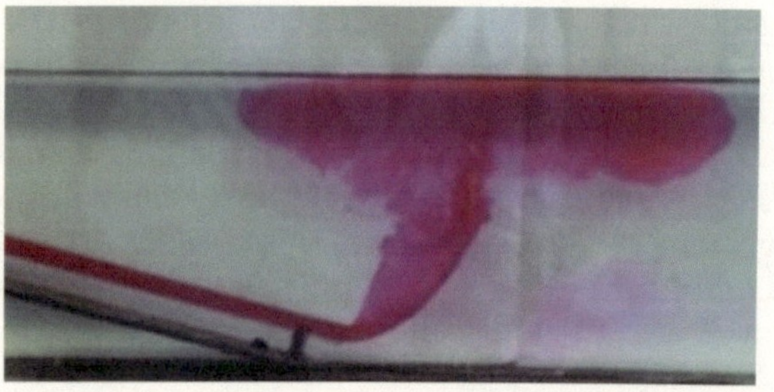

Figure 8.
View of an experiment carried out in the fluid mechanics' laboratory of the physics department, Las Palmas university, gran Canaria. It is a scale model of an underwater outlet discharge. An experiment was conducted to determine the distribution of effluent with a lower density than the receiving medium and thus calibrate a numerical model of contaminant distribution (video 1).

In coastal lagoons and close seas like the Mediterranean and Baltic, tidal currents are negligible, and the slick will grow around the outcrop point, which will be dragged by the coastal drift current, although the wind is mainly responsible for the movement.

As a summary of the dynamic evolution of the discharge when it comes out of the mouth of a submarine drainage conduit, initially, it rises to the surface where significant sedimentation and dilution take place, which continues until it reaches the surface. During this ascent, the dynamics of the currents lift this plume to the seepage point. Here, there is a process of concentration of the discharge pushed by the buoyancy against the surface of the water, more or less above the outlet (**Figure 8**) (video 2).

This is why the odor produced by gases from decomposing organic matter is easily detectable, especially in a shallow system, such as the Mar Menor and other coastal lagoons along the world, since the initial dilution, that is, in the Near Field, is reduced and the effluent appears on the surface with practically the same properties as that circulating through the sewage system, where hydrogen sulfide clearly stands out from other gases in the putrefaction of organic matter.

These areas need to be more controlled by the administrations because is a very high dangerous places that are attractive by coastal users, like fishermen, because of the fish concentration, sailboat racers (especially kits with small boats like an optimist) because is a point to turn, or to berth, because use to have a big buoy to mark the end of outfall point.

Figure 8 shows a scale model experiment of the behavior of dilution and evolution in the water column and on the surface of wastewater discharge. This is one of the actions necessary to calibrate a numerical model of contaminant dispersion, for the case when the discharge is carried by a discharge pipe without a diffuser system.

As already explained initial process is dominated by the effluent outlet velocity and the difference in buoyancy between the receiving medium and discharge, or the water of the sea and the effluent of a wastewater discharge.

However, the starting situation can change when there is stratification due to heavy rain flood processes, estuary, or river presence when the freshwater from the continental streams enters the sea and produces a surface layer of water less dense than that of the sea.

In this case, wastewater discharge poses a greater risk, since it can be trapped between these two water layers, leading to a totally underwater discharge situation.

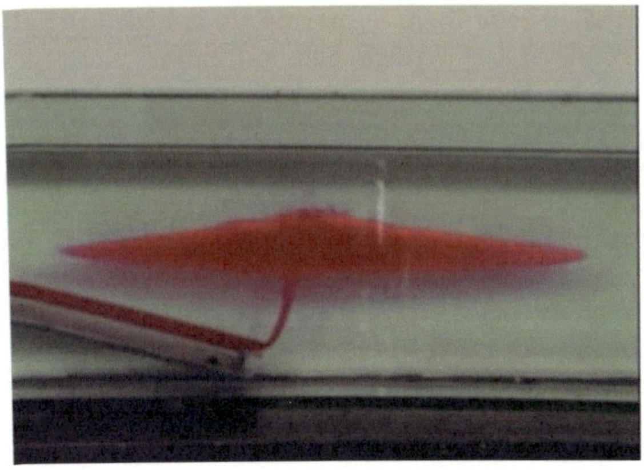

Figure 9.
View of a video frame of a physical simulation of a single point discharge in a receiving medium with strong stratification. The discharge rises until it finds the corresponding density and evolves between the two layers of water due to the difference in density from the point of accumulation where the effluent arrives (video 2).

This discharge makes no contact with the surface of the water, that is, there is no possibility of oxygen entering from the atmosphere. It cannot undergo the energetic mixing processes, produced by wind and waves at the surface (See **Figure 9**) (videos 1, 2).

Figure 9 shows a laboratory simulation using a physical scale model, where the effluent discharged between two layers of water is seen to develop. The spill evolves between the two layers, almost exclusively by density difference processes and with limited dilution, produced solely by friction between the layers of water.

This process is very likely to be taking place in coastal lagoons when there are discharges under these conditions of strong stratification after a rain flood of water, where it is very easy for it to produce large general oxygen deficits in the two layers, since it tends to sequester oxygen from both, further impeding oxygenation of the bottom layer.

The effluent can travel in this way, without mixing, for long distances and seep out at any point, either on the shoreline or propelled to the surface by powerful storms or vessels, which causes it to reach the surface creating a serious foul smell nuisance.

5.2 Chemical evolution of wastewater discharge

When urban wastewater discharges, it meets water from marine coastal areas, which has a different chemical composition. In this encounter, in addition to the mixing processes already mentioned, chemical changes are produced in the composition of the mixing effluent.

Perhaps the most important change is that of acidity or pH. Seawater has a basic pH, above 8, while wastewater maintains a pH of around 7, depending on the chemical compounds diluted in it [48].

This change in pH favors a process previously established as marine snow, which is the product of the transformation of species that are dissolved at acidic pH (around 7), which when they encounter a basic pH become insoluble compounds. The most common example is the change from oxides (soluble) to hydroxides (insoluble). Inorganic contaminants, such as heavy metals and trace elements, experience this change the most.

Acidity modification affects the ecosystem because, while the dilution occurs and since the discharge is continuous, it creates a differential gradient of acidity that compromises the normal development of marine life, which is dependent on Ph, that is, it prevents, for example, the growth of shells and tissues made of calcium carbonate, which poses a danger to the ecosystem and especially to organisms with a shell or exoskeleton.

Figure 10 shows a clear example of sedimentation and turbidity processes, while in **Figure 11**, a detail of the consequences of this sedimentation on the seabed is shown. The flocs most visible in this photo are the light-colored ones, but the abundance of dark flocs is much higher and they appear in the photo as dark turbidity.

This high organic load provides this effluent with a high content of nitrogen, phosphorus, and other compounds, which are responsible for the problems that these effluents produce when they are discharged into the environment. They cause growth above usual levels of macro or microalgae and bacteria of species in coastal waters.

A large part of organic contaminants are associated with these flocs (**Figure 10**) and remain in suspension. They are ingested by marine organisms or precipitate, as described above. Many dissolve in the water, where they are absorbed by the respiratory organs (gills), incorporated into the bloodstream directly, and quickly injected into the food chain.

Another material dumped into the sea by wastewater is microplastic, mainly due to the discharge of water from industry and washing synthetic textiles, but also due to the discharge of numerous hygiene products. Plastics not only increase turbidity and leave annoying-colored grains on beaches; they also chemically degrade and release highly toxic products into the environment that add to the series of pollutants outlined in this study.

For example, bisphenols, vinyl, and carbamates, among others, are compounds (monomers) of high toxicity and are released into the environment from the degradation of the polymers in objects made of plastic materials [49–56].

Some of the monomers that act chemically with very toxic biological effects are called endocrine disruptors because when organisms assimilate them in their tissues, they modify their metabolic and endocrine processes, as their structure is similar enough to replace them; they are all highly active carcinogens [57].

Chemical components of wastewater have been shown to have a catalytic effect on the populations of opportunistic competitor organisms. They stand out among all the vitamins and hormones, and their kinetics are described in the scientific literature as being responsible for bacterial emergence [58].

Figure 10.
Photograph of a spill on the coast where you can see the fall of marine snow, in the form of white dots, which transports a lot of pollution to the seabed. You can also see the turbidity that obscures the photo in broad daylight and at a relatively shallow depth. (Sta Cruz de Tenerife, Canary Islands, Spain).

Figure 11.
Photograph of a coastal seabed affected by an underwater discharge. You can see the dark color of muddy sediment, which indicates sedimentation and anaerobic degradation processes. In addition, you can see the microplastic particles that cover this sludge where a higher life is not found.

These organic substances are part of a group called persistent compounds since, in rapid purification cycles, they do not degrade and even less during their passage through the sewage system, poor technology treatment plants, and of course, in untreated discharge. In fact, these persistent molecules end up degrading into smaller compounds (CO2, N2, and H2O) after at least 96 hours and, during this long period, they can be incorporated into the tissues of marine organisms as the complete compound or a degradation product [59].

Over time, the remaining nutrients end up mineralizing, which can lead to the growth of sulfate-reducing or methanogenic bacteria in increasingly tropical seas that are being related to local changes in the pH and sulfur content of these seabeds [60].

All these processes are accentuated in coastal lagoons, or places where accumulation conditions are greater and renewal rates are low.

Thus, in a short time, these areas can no longer support life, as shown in **Figure 11**, of the bed in the vicinity of an underwater discharge pipeline. The dark-colored bottom confirms the existence of these processes with oxygen deficit; the fact that it is mud instead of sand, also indicates large fine particulate sedimentation, which covers an initial bed of light sand, located just underneath, which leaves the upper sediment lifeless.

This sediment devoid of oxygen and with a large amount of partially degraded organic matter produces anaerobic processes that can be divided into two broad categories:

a. Closed and reducing or hydrogenating anaerobic environments, in which hydrogen is produced and digested by hydrogenotrophic bacteria. Organic acid and polysaccharide chains break down to CH4 and CO2. Purines and amino acids break down to NH3 or NH4+ ammonium anions [61].

b. Nonreducing open anaerobic environment, in which the hydrogen produced escapes. In these cases, the organic chains degrade to acetic acid:

$$CH_3-\mathrm{COOH}+HR_2\mathrm{R}_3\mathrm{N}=CH_3-\mathrm{CONR}_2\mathrm{R}_3\,(\text{Toxic Amide})+\mathrm{H_2O}.$$

The seabed is usually type (b) since it is an open and nonreducing system, which produces short-chain acids that, in turn, react with amines to produce highly toxic amides [62].

The generation of these toxins also finds an absorbent medium through which they can percolate, as if they were leachate, eliminating species capable of photosynthesis on that bed. They can be resuspended, however, by especially energetic storms or much more efficiently by high-speed sports boats with high-power motors and propellers near the bottom, which put new toxins into the water column until they settle elsewhere or are ingested by marine life. Thus, there is continuous pollution from different sources, from which it has no ability to recover its balance; so-called chronic action.

The layer of small, brightly colored, microplastic particles on the seabed, as seen in **Figure 11**, can end up having absorbent spongy properties when degraded, constituting ideal niches for the stabilization of bacteria provided by fecal water, in particular *Clostridium prefigens*, which causes skin and soft tissue infections in humans and fish. Proof of this is the excoriations, infections, and ulcers that fish have that live in the environments of the discharge pipe [63, 64].

In addition, there is the dangerous genus of *Cryptosporidium* protozoa that live in wastewater and are one of the most dangerous parasites worldwide, and are very difficult to eradicate with common disinfection systems. These are just two examples of the dangers to which users of coastal areas polluted by sewage are exposed, and they are cited here so that sewage receives attention as a toxic product, with increased control over these effluents [65, 66].

In the same way that these microorganisms can affect humans, they also affect these phenomena of accumulation of pathogens and other pests in fisheries, and the rest of marine organisms.

A clear example is the increase of ciguatera (which produces an accumulation of ciguatera toxin in the fatty parts of the fish), which is related to the opportunistic growth of populations of bacteria, such as dinoflagellates and some species of cyanobacteria that produce hepatotoxins and neurotoxins, in contaminated environments. Wastewater especially favors this growth, and it is found in coastal waters with increasing frequency [67].

Many of these plastic granules are ingested by living beings so, although in theory, they are inert chemically, they are vectors of diseases spread by bacteria. Even if the granule is subsequently returned or defecated, the bacteria will affect the fish, mollusk, or mammalian host [52–56].

And wastewater impacts in more ways. One of the clearest that gives an idea of the enormous damage caused in coastal waters is the oxygen consumption in the water column, an environment with a limited amount of oxygen.

Seawater has on average about 6 mg/l of dissolved oxygen. This is consumed by bacteria that degrade organic matter, and so can cause anoxia situations that produce degradation of organic matter with more harmful by-products.

A good measure of the impact is to calculate how much volume of water will be affected by the discharge only in Ref. to the dissolved oxygen to be consumed by the degradation of organic matter [68].

A flow of wastewater of one cubic meter, with an average BOD5 of 360 mg/l has an impact on approximately 60 m^3 of seawater, which will be left without oxygen to be able to degrade organic matter.

5.3 Biological evolution of wastewater discharges

As briefly described in the previous section, the discharge carries a large number of polluting compounds, in addition to appearing in various forms and undergoing different processes, such as sedimentation and dilution.

The sedimentation process and formation of flocs generally called marine snow, evolves in two fundamental ways within the ecosystem. These flocs are either ingested by fish before reaching the bottom, or they are deposited and otherwise impact the coastal ecosystem.

During the descent of this particulate material, the fish are strongly attracted by these flocs since they are made up of organic matter of terrestrial origin, upon which bacterial colonies are developing; providing energy to move and proteins to grow without making any significant effort.

In addition to fish, there is a type of organism that is especially affected by these particles before they reach the bottom, which specifically concerns protected species that normally inhabit all marine ecosystems; these are filtering organisms [69, 70].

Filter feeders capture these particles, which also have a negative effect on their body, assimilating contaminants in a similar way, but at a faster rate than fish. If the amount of sediment or the size of the flocs is greater than the dimensions or capacity of the filtering system, these organisms have their feeding mechanisms obstructed and disappear from the areas affected by the discharges.

To understand the global impact of this process, it is enough to remember that coral reefs disappear due to this same problem, but other filtering organisms, such as oysters, will also be affected [71–73].

The enormous increase in suspended particles produced by a discharge pipe decreases the transparency of the water, and sunlight has more difficulty reaching the bottom. It is not difficult to understand that sunlight is essential for the development of another of the most important systems in the coastal waters, such as the Cymodocea nodosa meadows, or any seabed plant. However, in this area, they also have a negative influence on the development of the coastal rocky areas, since the algae that form the basis of this habitat need sunlight to develop, reducing their habitat depth as the turbidity caused by suspended sediments increases [74–76].

Once it reaches the bottom, the sediment can be ingested by detritivore organisms that incorporate the pollutants. Normally, the amount of sediment is very large, and they end up accumulating rapidly, exceeding the capacity of the detritivores [77, 78].

However, the marine ecosystem goes against the intention of the discharges which is to dilute the concentration of pollutants, through a process called bioaccumulation.

Bioaccumulation is a process by which marine organisms living in a contaminated environment concentrate in their tissues a greater number of pollutants, than in the medium, increasing its concentration at each step in the food chain [79, 80].

Wastewater discharges also produce numerous indirect effects in the coastal areas, one of the most recognized is their responsibility in the eutrophication process because they are by far the largest contributors of both N and P in many places. This process of fertilization produces an increase in different opportunistic vegetables in water [81–83].

The abnormal and widespread growth of plants is evidence of this excess of nutrients. The specific abundance of species and their growth also depend on other factors, such as the dynamic regime, salinity, and availability of consistent substrate.

Figure 12.
The problems suffered by the area due to an overabundance of nutrients are shown. In this case, they are taken advantage of by opportunistic macroalgae specializing in assimilating nutrients in enormous quantities and growing rapidly. Mar Menor (Murcia, Spain).

The larger these algal growths, the more pollution the coast is suffering from, and in the images, it can be seen that these coastal algae are extensive and enormously widespread (see **Figure 12**). These abnormal growths of algae alter the entire ecosystem, where opportunistic species dominate over specialists, changing the ecosystem from its baseline status [84, 85].

In an environment with the anthropomorphic input of nutrients, the formation of persistent foam is common, especially in large storm events or due to the beating of waves on the shores and motorboats. The beating of the water forms this foam from the emulsification of biological products associated with the growth and degradation of organic matter together with water. In **Figure 13**, you can see several places on the coast of the Mar Menor lagoon where this foam is formed [86, 87].

The biological evolution of the discharge must include a reference to the very high concentration of microorganisms that are part of this waste, *Escherichia coli* values, of the order of 10^8–10^9 CFU/100 ml (1,000,000,000–100,000,000 formation units of colonies/in 100 ml of sample), are normal values in wastewater [88, 89].

But these bacteria are an indirect reference to the number of other species of microorganisms transported that are dangerous for the environment and human health. A large part of them are pathogens, which were briefly discussed in the chapter on chemical impacts [63–66].

When the waste effluent comes into contact with the sea and this mixture is produced, the microorganisms find aggressive conditions, that is, high salinity, temperature differences, and when they reach the surface, great sunstroke.

These changes make it difficult for organisms from an environment of fresh water and darkness to progress. An indicative parameter for the Mediterranean Sea says that in 2 hours the concentration of bacteria has been reduced by 90%, which seems an

Figure 13.
Brown-colored persistent foam on the coast of the Mar Menor lagoon. (Spain).

important decrease; but not enough, considering the starting concentrations. As the mortality of microorganisms seems high, this process has led to incorrect speaking of the great purification capacity of the sea and associating this behavior in general with any pollutant (hydrocarbons, nutrients, and all the rest of chemical pollutants); this aspect is important because it is the basis of all legislation and wastewater management [90, 91].

5.4 Summary of the consequences of wastewater discharge

Everything described above shows that wastewater must be treated as a highly dangerous waste, which requires precise and extremely controlled management.

The controls that the legislation requires for a discharge pipeline are practically nonexistent and very inefficient. Thus, an underwater discharge pipeline or spillway becomes a dangerous point of contamination almost impossible to evaluate. The real impact it produces on the environment can only be approximated, in the face of a general lack of reliable and certified data.

Common management of discharges both for effluents and sludges is to concentrate them at night, which can be ease detected by the bad smells in our capitals or coastal population centers around the underwater discharge pipe or in the vicinity of the pipes that lead to their start on the coast.

Night discharges are especially serious for coastal areas, especially in lagoon or shallow and non-ventilated places, since, at that moment, the aquatic plants stop producing oxygen and begin to consume it to fulfill their respiratory metabolism. This produces a dangerous situation where oxygen consumption is generalized, and anoxic zones can quickly form if the movement of water is limited or is not capable of spreading the discharge quickly.

The first thing to mention is that the coastal lagoons are enclosed systems with limited connection to the coastal sea, so the dilution of pollutants is less, and they accumulate for a longer time. To this is added the low depth that makes the volume of waterless and so the water column can be quickly affected in its entirety. Also, the low depth means that the benthic system is accustomed to continuous lighting that promotes plant growth processes both on the bed and specializes in these conditions. However, at the same time, they are likely to produce exponential growth of phytoplankton when the right circumstances occur, which blocks the arrival of light to the bottom of the lagoon, in a system that is not prepared for low levels of lighting.

Sewage discharges produce a chronic impact because are continuous and it alters any marine area, which the ecological system is incapable of coping with, and which conditions it absolutely; thereby reducing the capacity of normal development of the ecosystem, no matter how small the discharge.

When considering dilution or dispersion or assimilation of discharge into the sea, as the final situation that solves the contamination problem, this is nothing more than the injection of pollutants into the marine environment. The incidence is direct and instantly affects organisms that in one way or another suffer the consequences.

Plastic may be a good example to understand the problem of pollution. It can be verified that it does not disappear but breaks into small pieces until it becomes part of the trophic web, so we are now aware that we are eating plastic even though it disappears from our sight in the sea. The same happens with pollutants, which become part of the food chain and ultimately end up on our tables as well as having a major impact on the coastal ecosystem.

It must be taken into account that treatment plants and the large movements of wastewater in sewers maintain cultures of all the pathogenic organisms that humans have plus sublethal doses of antibiotics, that is, it is the perfect breeding ground to produce resistant microscopic species which are discharged since the purification systems are not capable of retaining these microorganisms before discharge.

Having submarine discharges does not reduce pollution, it only dilutes the concentration of the discharge. This dilution is counteracted by the bioaccumulation effect of the contaminants, which are quickly incorporated and concentrated in the food chain; consequently, affecting all steps in the ecosystem and returning the concentrated contamination to our table in the form of fish.

The poor state of coastal ecosystems means that organisms are conditioned by pollution, having to use a significant part of their energy to defend themselves against this aggression, thus reducing their ability to grow, develop or, for example, fight against diseases and parasites. Parasites, especially microparasites, represent an increasingly worrying problem and the most worrying are those that prey on fish and can transmit themselves to us, such as Ciguatera or Anisakis.

This weakness means that the environment cannot recover from the changes we are experiencing, such as global warming, overfishing (both professional and recreational), coastal works, hydrocarbon spills, coastal erosion, and invasive species, are all part of the panorama of our bleak coastal waters.

This has led to an immense increase in pollution on the coast, which has been accumulating for decades, decreasing water quality dramatically. Therefore, today's biodiversity is very poor, and the marine ecosystem is affected entirely. The system is so affected that it is changing the balance at important levels, and the idea that nitrogen is no longer a limiting nutrient for the growth of marine plants is finally beginning to be generalized, and it is increasingly accepted that phosphorus has also become the limiting nutrient in this environment.

Urban wastewater represents by far the greatest impact on the marine environment in general and on the coastal areas in particular. Both the management and the technology involved do not allow the control of the entire production or, of course, the elimination of pollutants in the effluent from our cities and urban settlements.

The list of pollutants that these effluents carry includes all the chemical products made in our society, plus a large amount of organic matter and chemical compounds from sewage treatment plants, which increase their toxic potential for the environment and, above all human health.

This effluent is a very dangerous toxic waste for people's health and must be handled with great care. It also contains and concentrates all the organisms that cause diseases in human beings, some enhanced by its passage through the sewage system and by the treatments in sewage treatment plants.

6. Alternatives to centralized management of sewages

The alternative is decentralized treatment, based on having treatment plants arranged along the sewerage network to treat wastewater at the place it is produced and to make water available for reuse.

This system must necessarily be based on treatment plants that do not produce the usual problems, such as bad odors, noise, and proliferation of rodents and insects, and which can be installed even inside buildings in optimal health conditions.

The advantages of this new technology are great and largely solve the current purification problems, beginning with a change in sewage management, which will be used only for the transport of rainwater; so preventing the discharge of raw wastewater into the aquifer. The plants can achieve high purified water quality for complete reuse in the production area, without resorting to marine outfalls and discharges from pumping stations.

To manage this multiplication of treatment plants, sensor-based control systems connected to software that manages each part of the plant can be used. After calibration, they respond to changes in the variables monitored by the treatment plant automatically. The data can then be sent to a central station, where alarms and maintenance notices are managed and recorded to keep accurate statistics of all the water treated.

Plants can produce a totally degraded sludge that can be used without any hazard to health in electrical energy production to reduce consumption; hence, avoiding one of the biggest problems of the current purification of wastewater.

The investment necessary for the installation of these new technologies would save between 70% and 90% of the initial investment necessary for installing purification systems in places where these have not been installed. The same range of investment savings could be achieved when introducing these systems in cities where the traditional centralized purification systems are already installed. This is because only the reparation of the entire sewerage or the creation of a separate sewage network involves an investment of the order of two orders of magnitude superior to the installation of compact plants that do not need to use this network.

Energy consumption is higher in these plants but is between 1.8 and 2 kWh/m^3 of treated water, which is not much higher than the consumption of traditional treatment plants where, in addition to the treatment cost, between 0.5 and 1 kWh, the cost of pumping wastewater from several kilometers to the treatment plant must be added.

The higher cost of energy consumption is offset by savings in management since the need for personnel, chemical products, sewage maintenance, and sludge purification treatment would be less; producing purified water that could be sold with all health guarantees.

When considering costs, the pollution parameter must also be considered; that is, **how much does it cost to pollute**? A system may consume slightly less energy, which could be compensated by maintenance savings, but where is the pollution produced? Clearly, a wastewater management system that produces a lot of pollution should be removed from the choice of systems. However, which is not something that is shared by the majority of water managers, who simply do not take this effect into account and continue as if nothing has happened, seeking to compare electricity consumption and justifying that the change is difficult, preventing even a debate or any objective evaluation.

These systems of small, high-efficiency, and low-price plants open up the possibility for minor municipalities or isolated communities to have a wastewater treatment solution, which until now has not been addressed with guarantees; preventing health and ecological hazards distributed throughout the world that has not even been evaluated. Only in Spain, there are thousands of municipalities without any type of purification, because the installation of alternative systems is not allowed.

There are enough examples for comparison; however, wastewater management in Venice, Italy is a good example. The difficulties in the area, the awareness of the water authorities and the conjunction of the private initiative led by applied research to solve problems have produced a decentralized system that is an example to follow. Over 100 treatment plants have been installed in the entire historic center of the island, specifically designed for each institution (hospitals and medical centers, chemical laboratories, grouped houses, hotels, sewage sludge treatment, mechanical workshops, theaters, universities, naval construction areas, etc.) and managed by eight people in 24-hour shifts.

There is also the Giniginamar plant, on the island of Fuerteventura, Canary Islands, which has been operating since 2008 for a coastal town of 200 inhabitants (smaller plants are more difficult to manage than large plants), where water has been reused in gardens and not discharged into the environment *via* the old filtering cesspools that were the solution adopted up to that moment. This has led to savings in the initial investment of several million euros compared with a standard solution. Despite having all the problems of a small, coastal system, with salinization of the effluent, seasonal variations in tourists, discharges from the chemical toilets of around 50 caravans, and all kinds of other discharges, including from a familiar craft cheese factory, output values are stable and, without changing membranes since 2008, the concentration of bacteria (*Escherichia coli*) has been zero during all the time and rest values under the limits of European water reuse legislation.

A typical bad case is a problem that has arisen in recent years in the Mar Menor lagoon, on the southeastern Mediterranean coast in Murcia, Spain. Here, water discharges with different degrees of purification along with other minor contributions of nutrients from agricultural and livestock activity have produced a spectacular growth of phytoplankton that has revealed the over-exploitation of tourism in the area, which caters to hundreds of thousands of people in the summer season.

Discharges that go unnoticed in other parts of the Mediterranean coast are more evident in a lagoon with these features. The poor state of the sewage system and the deficiencies of the technology explained above, together with the delicacy of the lagoon system, which is also visible to all and under the bright lights of the media have given rise to an enormous controversy where economic interests are intermingled with political ones, leading to a problem with no solution until contaminant inputs are addressed.

The high investment required and the enormous time it takes to improve the traditional purification system have given rise to a stalemate situation, where agriculture has received the full burden of responsibility for the degradation of the lagoon environment. This is because, as in the rest of the cases, the public administration is not capable of assuming responsibility for the real wastewater management situation. Since implementing only, the leaks of wastewater due to a sewage fault in the 1980s, without much care and in a hurry, both the contributions of organic matter and nutrients far exceed any contribution from other sources, together with other hazardous pollutants a pattern repeated all over the world.

7. Conclusions

Current wastewater management shows the failure of all sectors of our society. Beginning with a public administration that has no choice but to depend on the large water companies, with huge wastewater management businesses; as well as the hundreds of research centers that are incapable, not only of offering effective alternative solutions to the current management but of warning about the hazards entailed. In the end, it is clear that there is not the slightest awareness about health protection or, of course, about caring for the environment.

The increased concentration in the world population makes separating the contact between wastewater and the public much more complicated. There is increasing contact with wastewater, which reaches the dams from which we take drinking water, coasts, rivers, and lakes where we bathe, fish, and drink water, as well as aquifers and wetlands.

The saddest part of the situation is that the technology is available, capable of closing the water cycle for continuous reuse, with reduced initial investment and maintenance savings; managing to prevent the current health and environmental problem we suffer [92].

New relevant technology in our society is currently developed by small private companies, sometimes with associations of certain scientists from research centers. They have sufficient knowledge of reality and are promoting tangible alternatives to the great social problems we are facing at the moment. However, these developments are systematically blocked, and it is very difficult to generalize their use.

For improvement to be possible, we need independent, transparent public administrations associated with research institutes that are close to the real problems of our society.

The solution to the problem of water pollution is there, as with other contaminating sectors. We need to start thinking as a human species since our quality of life depends on it.

Videos

Video 1 https://www.youtube.com/watch?v=k3KTXtMGoIw.
Video 2 https://www.youtube.com/watch?v=F7k-xn11mBA&t=223s.

Author details

Jesús Cisneros-Aguirre[1]* and Maria Afonso-Correa[2]

1 University of Las Palmas GC/PONTHO Eng. Ltd, Las Palmas GC, Canary Islands, Spain

2 PONTHO Engineering. Ltd, Las Palmas GC, Canary Islands, Spain

*Address all correspondence to: jesus.cisneros@ulpgc.es

References

[1] Ardern E, Lockett WT. Experiments on the oxidation of sewage without the aid of filters, Part I. Journal of Society Chemical Indigenous. 1914a;**33**:523

[2] Ardern E, Lockett WT. Experiments on the oxidation of sewage without the aid of filters, Part II. Journal of Society Chemical Indigenous. 1914b;**33**:1122

[3] Ardern E, Lockett WT. Experiments on the oxidation of sewage without the aid of filters, Part III. Journal of Society Chemical Indigenous. 1915;**34**:937

[4] THirata T, Kawamura K, Sonoki S, Hirata K, Kaneko M, Taguchi K. Clostridium Perfringens, as an indicator microorganism for the evaluation of the effect of wastewater and sludge treatment systems. Water Science Technology. 1991;**24**:367-372. DOI: 10.2166/wst.1991.0092

[5] Tao J et al. Composition of waste sludge from municipal wastewater treatment plant. Procedia Environmental Sciences. 2012;**12**:964-971

[6] Lake DL, Kirk PWW, Lester JN. Heavy metal solids association in sewage sludges. Water Research. 1989;**23**(3):285-291

[7] Moreira R, Sousa JP, Canhoto C. Biological testing of a digested sewage sludge and derived composts. Bioresource Technology. 2008;**99**(17): 8382-8389

[8] Adegoke AA, Amoah ID, Stenström TA, Verbyla ME, Mihelcic JR. Epidemiological evidence and health risks associated with agricultural reuse of partially treated and untreated wastewater: A review. Frontiers in Public Health. 2018;**6**:337. DOI: 10.3389/fpubh.2018.00337

[9] Thompson RC, Swan SH, Moore CJ, Saal FS. Our plastic age. Philosophical Transaction on Royal Society. 2009;**364**(1526):1973-1976. DOI: 10.1098/rstb.2009.0054

[10] Deblonde T, Cossu-Leguille C, Hartemann P. Emerging pollutants in wastewater: A review of the literature. International Journal of Hygiene and Environmental Health. 2011;**214**(6):442-448. DOI: 10.1016/j.ijheh.2011.08.002. Available from: https://www.sciencedirect.com/science/article/pii/S1438463911001325

[11] Zhou Y, Meng J, Zhang M, Chen S, He B, Zhao H, et al. Which type of pollutants need to be controlled with priority in wastewater treatment plants: Traditional or emerging pollutants. Environment International. 2019;**131**:104982. DOI: 10.1016/j.envint.2019.104982. Available from: https://www.sciencedirect.com/science/article/pii/S0160412019311699

[12] Ivan Muñoz M, Gómez J, Molina-Díaz A, Huijbregts MAJ, Fernández-Alba AR, García-Calvo E. Ranking potential impacts of priority and emerging pollutants in urban wastewater through life cycle impact assessment. Chemosphere. 2008;**74**(1):37-44. DOI: 10.1016/j.chemosphere.2008.09.029. Available from: https://www.sciencedirect.com/science/article/pii/S0045653508011879

[13] Chen Y, Vymazal J, Březinová T, Koželuh M, Kule L, Huang J, et al. Occurrence, removal and environmental risk assessment of pharmaceuticals and personal care products in rural wastewater treatment wetlands. Science of The Total Environment, Volumes. 2016;**566-567**:1660-1669. DOI: 10.1016/j.

scitotenv.2016.06.069. Available from: https://www.sciencedirect.com/science/article/pii/S0048969716312426

[14] Hembach N, Alexander J, Hiller C, et al. Dissemination prevention of antibiotic resistant and facultative pathogenic bacteria by ultrafiltration and ozone treatment at an urban wastewater treatment plant. Scientific Reports. 2019;**9**:12843. DOI: 10.1038/s41598-019-49263-1

[15] Kim S, Aga DS. Potential ecological and human health impacts of antibiotics and antibiotic-resistant Bacteria from wastewater treatment plants. Journal of Toxicology and Environmental Health. 2007;**10**(8):559-573. DOI: 10.1080/15287390600975137

[16] Karkman A, Do TT, Walsh F, Virta MPJ. Antibiotic-resistance genes in waste water. Trends in Microbiology. 2018;**26**(3):220-228. DOI: 10.1016/j.tim.2017.09.005. Available from: https://www.sciencedirect.com/science/article/pii/S0966842X1730210X

[17] Rizzo L, Manaia C, Merlin C, Schwartz T, Dagot C, Ploy MC, et al. Urban wastewater treatment plants as hotspots for antibiotic resistant bacteria and genes spread into the environment: A review. Science of the Total Environment. 2013;**447**:345-360. DOI: 10.1016/j.scitotenv.2013.01.032. Available from: https://www.sciencedirect.com/science/article/pii/S0048969713000429

[18] Moghaddam AA. Prediction of effluent quality parameters of a wastewater treatment plant using a supervised committee fuzzy logic model. Journal of Cleaner Production. 2018;**180**:539-549. DOI: 10.1016/j.jclepro.2018.01.139. Available from: https://www.sciencedirect.com/science/article/pii/S0959652618301616

[19] Sanna K, Kettunen RH, Rintala JA. Occurrence and removal of organic pollutants in sewages and landfill leachates. Science of The Total Environment. 2003;**301**:1-12. DOI: 10.1016/S0048-9697(02)00302-9. Available from: https://www.sciencedirect.com/science/article/pii/S0048969702003029

[20] Linarić M, Markić M, Sipos L. High salinity wastewater treatment. Water Science Technology. 2013;**68**(6):1400-1405. DOI: 10.2166/wst.2013.376

[21] Lisle J, Smith JJ, Edwards DD, McFeters GA. Occurrence of microbial indicators and *Clostridium perfringens* in wastewater, water column samples, sediments, drinking water, and Weddell Seal Feces Collected at McMurdo Station, Antarctica. ASM Journals, Applied and Environmental Microbiology. 2013;**70**(12)

[22] Spellman FR. Wastewater treatment operations. In: Handbook of Water and Wastewater Treatment Plant Operations. 2nd ed. Boca Raton: CRC Press, T&F eBooks; 2008. p. 872. DOI: 10.1201/9781420075311

[23] Ying G-G. Fate, behavior and effects of surfactants and their degradation products in the environment. Environment International. 2006;**32**(3):417-431. Available from: https://www.sciencedirect.com/science/article/pii/S0160412005001625. DOI: 10.1016/j.envint.2005.07.004

[24] Reynolds J, J.H. and Barrett, M.H. A review of the effects of sewer leakage on groundwater quality. Water Environment Journal. 2003;**17**:34-39. DOI: 10.1111/j.1747-6593.2003.tb00428.xH

[25] Bishop PK, Misstear BD, White M, Harding NJ. Impacts of sewers on groundwater quality. Water

Environment Journal. 1998;**12**:216-223. DOI: 10.1111/j.1747-6593.1998.tb00176.x

[26] Wolf L, Held I, Eiswirth M, Hötzl H. Impact of leaky sewers on groundwater quality. Acta Hydrochimica et Hydrobiologica. 2004;**32**:361-373. DOI: 10.1002/aheh.200400538

[27] Rutsch M, Rieckermann J, Krebs P. Quantification of sewer leakage: A review. Water Science Technology; **54**:135-144. DOI: 10.2166/wst.2006.616

[28] Sojobi AO, Zayed T. Impact of sewer overflow on public health: A comprehensive scientometric analysis and systematic review. Environmental Research. 2022;**203**:111609. DOI: 10.1016/j.envres.2021.111609

[29] Paul M, Wolf L, Fund K, Held I, Winter J, Eiswirth M, et al. Microbiological condition of urban groundwater in the vicinity of leaky sewer systems. Acta Hydrochimica et Hydrobiologica. 2004;**32**:351-360. DOI: 10.1002/aheh.200400539

[30] Karpf C, Krebs P. Modelling of groundwater infiltration into sewer systems. Urban Water Journal. 2013;**10**(4):221-229. DOI: 10.1080/1573062X.2012.724077

[31] Rasmussen P, Kidmose J, Kallesøe AJ, et al. Evaluation of adaptation measures to counteract rising groundwater levels in urban areas in response to climate change. Hydrogeology Journal. 2022. DOI: 10.1007/s10040-022-02573-7

[32] Tscheikner-Gratl F, Caradot N, Cherqui F, Leitão JP, Ahmadi M, Langeveld JG, et al. Sewer asset management – State of the art and research needs. Urban Water Journal. 2019;**16**(9):662-675. DOI: 10.1080/1573062X.2020.1713382

[33] Yavuz Corapcioglu M, Haridas A. Microbial transport in soils and groundwater: A numerical model. Advances in Water Resources. 1985;**8**(4):188-200. DOI: 10.1016/0309-1708(85)900636. Available from: https://www.sciencedirect.com/science/article/pii/0309170885900636

[34] REGULATION (EU) No 259/2012 OF THE EUROPEAN PARLIAMENT AND OF THE COUNCIL of 14 March 2012

[35] REGULATION (EC) No. 648/2004 of the European Parliament and of the Council of 31 March 2004

[36] REGULATION (CEE) N° 91/271/CEE

[37] REGULATION (CE) 2000/60/CE. Available from: https://www.boe.es/eli/es/rd/2012/09/07/1290

[38] Hoar C, McClary-Gutierrez J, Wolfe MK, Bivins A, Bibby K, Silverman AI, et al. Looking forward: The role of academic researchers in building sustainable wastewater surveillance programs. Environmental Health Perspectives. 2022;**130**(12):125002. DOI: 10.1289/EHP11519

[39] SPANISH REGULATION. Discharges from inland to sea. Orden de 13 Julio de 1993. Sobre vertidos de tierra a mar.

[40] Zhao L, Chen Z, Lee K. Modelling the dispersion of wastewater discharges from offshore outfalls: A review. Environmental Reviews. 2022;**19**:107-120. DOI: 10.1139/a10-025

[41] Morelissen R, Vlijm R, Hwang I, Doneker RL, Ramachandran AS. Hydrodynamic modelling of large-scale cooling water outfalls with a dynamically coupled near-field–far-field modelling system. Journal of Applied Water Engineering and Research. 2016;**4**(2):138-151. DOI: 10.1080/23249676.2015.1099480

[42] Roberts PJW. Sea Outfalls. In: Singh VP, Hager WH, editors. Environmental Hydraulics. Dordrecht: Springer; 1996. DOI: 10.1007/978-94-015-8664-1_3

[43] Alldredge AL, Timothy J, et al. Occurrence and mechanisms of formation of a dramatic thin layer of marine snow in a shallow Pacific fjord. MEPS. 2002;**233**:1-12. DOI: 10.3354/meps233001

[44] Tansel B. Morphology, composition and aggregation mechanisms of soft bioflocs in marine snow and activated sludge: A comparative review. Journal of Environmental Management, Volume. 2018;**205**:231-243. DOI: 10.1016/j.jenvman.2017.09.082

[45] Passow U, Ziervogel K, Asper V, Diercks A. Marine Snow Formation in the Aftermath of the Deepwater Horizon Oil Spill in the Gulf of Mexico. In: Environmental Research Letters. Vol. 7, No. 3. IOP Publishing Ltd; 2012. DOI: 10.1088/1748-9326/7/3/035301

[46] Barnes RO, Goldberg ED. Methane production and consumption in anoxic marine sediments. Geology. 1976;**4**(5):297-300. DOI: 10.1130/0091-7613(1976)4<297:MPACIA>2.0.CO;2

[47] Burdige DJ. The kinetics of organic matter mineralization in anoxic marine sediments. Source: Journal of Marine Research. 1991;**49**(4):727-761. DOI: 10.1357/002224091784995710

[48] Jason M, Spencer H, Harvey BP. Ocean acidification impacts on coastal ecosystem services due to habitat degradation. Emerging Topical Life Science. 2019;**3**(2):197-206. DOI: 10.1042/ETLS20180117

[49] Lee H-B, Peart TE. Bisphenol A contamination in canadian municipal and industrial wastewater and sludge samples. Water Quality Research Journal. 2000;**35**(2):283-298. DOI: 10.2166/wqrj.2000.018

[50] Pan Y, Liu Y, Wu D, Shen C, Ma C, Li F, et al. Application of Fenton pre-oxidation, Ca-induced coagulation, and sludge reclamation for enhanced treatment of ultra-high concentration poly(vinyl alcohol) wastewater. Journal of Hazardous Materials. 2020;**389**:121866. DOI: 10.1016/j.jhazmat.2019.121866

[51] Nakamura J, Kasuga T, Sakka Y. Preparation of carbamate-containing vaterite particles for strontium removal in wastewater treatment. Journal of Asian Ceramic Societies. 2017;**5**:364369. DOI: 10.1016/j.jascer.2017.06.011

[52] Barnes DKA, Galgani F, Thompson RC, Barlaz M. Accumulation and fragmentation of plastic debris in global environments. Philosophical Transactions of the Royal Society B. 2009;**364**:1985-1998. DOI: 10.1098/rstb.2008.0205

[53] Harshvardhan K, Jha B. Biodegradation of low-density polyethylene by marine Bacteria from pelagic waters, Arabian Sea, India. Marine Pollution Bulletin. 2013;**77**(1-2):100-106. DOI: 10.1016/j.marpolbul.2013.10.025

[54] Chamas A, Moon H, Zheng J, Yang Q, Tabassum T, Jang JH, et al. Degradation rates of plastics in the environment. ACS Sustainable Chemistry & Engineering. 2020;**8**(9):3494-3511. DOI: 10.1021/acssuschemeng.9b06635

[55] Pathak VM. Navneet review on the current status of polymer degradation: A microbial approach. Bioresources Bioprocess. 2017;**4**:15. DOI: 10.1186/s40643-017-0145-9

[56] Webb HK, Arnott J, Crawford RJ, Ivanova EP. Plastic Degradation and

Its Environmental Implications with Special Reference to Poly(ethylene terephthalate). Polymers. 2013;**5**:1-18. DOI: 10.3390/polym5010001

[57] Kasprzyk-Hordern B, Dinsdale RM, Guwy AJ. The removal of pharmaceuticals, personal care products, endocrine disruptors and illicit drugs during wastewater treatment and its impact on the quality of receiving waters. Water Research. 2009;**43**(2):363-380. DOI: 10.1016/j.watres.2008.10.047

[58] Selimoğlu F, Öbek E, Karataş F, Arslan EI, Tatar ŞY. Determination of amounts of some vitamin B groups in domestic wastewater treatment plants. Turkish Journal of Science & Technology. 2015;**10**(2):1-5. Available from: https://dergipark.org.tr

[59] Mohapatra DP, Brar SK, Tyagi RD, Picard P, Surampalli RY. Analysis and advanced oxidation treatment of a persistent pharmaceutical compound in wastewater and wastewater sludge-carbamazepine. Science of the Total Environment. 2014;**470-471**:58-75. DOI: 10.1016/j.scitotenv.2013.09.034

[60] Kosma CI, Lambropoulou DA, Albanis TA. Investigation of PPCPs in wastewater treatment plants in Greece: Occurrence, removal and environmental risk assessment. Sci Total Environ. 1 Jan 2014;**466-467**:421-438. DOI: 10.1016/j.scitotenv.2013.07.044. Epub 2013 Aug 7. PMID: 23933429

[61] Rachel M, Tfaily MM, Rich VI, Keller JK, Bridgham SD, Zalman CM, et al. Hydrogenation of organic matter as a terminal electron sink sustains high CO2:CH4 production ratios during anaerobic decomposition. Organic Geochemistry. 2017;**112**:22-32. DOI: 10.1016/j.orggeochem.2017.06.011

[62] Roth F, Broman E, Sun X, et al. Methane emissions offset atmospheric carbon dioxide uptake in coastal macroalgae, mixed vegetation and sediment ecosystems. Nature Communications. 2023;**14**:42. DOI: 10.1038/s41467-022-35673-9

[63] Nikaeen M et al. Occurrence of Clostridium difficile in two types of wastewater treatment plants. Journal of the Formosan Medical Association. 2015;**114**:663-665. DOI: 10.1016/j.jfma.2014.12.005. Available from: https://www.sciencedirect.com/science/article/pii/S0929664615000248

[64] Lisle J et al. Occurrence of microbial indicators and *Clostridium perfringens* in wastewater, water column samples, sediments, drinking water, and weddell seal feces collected at McMurdo Station, Antarctica. ASM Journals, Applied and Environmental Microbiology. 2015;**2015**:70

[65] Widmer G, Carmena D, Kváč M, Chalmers RM, Kissinger JC, Xiao L, et al. Update on Cryptosporidium spp.: Highlights from the Seventh International Giardia and Cryptosporidium Conference. Parasite. 2020;**27**(14):32167464. DOI: 10.1051/parasite/2020011

[66] Rose JB, Cifrino A, Madore MS, Gerba CP, Sterling CR, Arrowood MJ. Detection of cryptosporidium from wastewater and freshwater environments. Water Science Technology. 1986;**18**(10):233-239. DOI: 10.2166/wst.1986.0134

[67] Friedman MA, Lora E, et al. Ciguatera fish poisoning: Treatment, prevention and management. Marine Drugs. 2008;**6**(3):456-479. DOI: 10.3390/md6030456

[68] Montgomery HAC, Thom NS, Cockburn A. Determination of dissolved oxygen by the Winkler method and the

solubility of oxygen in pure water and sea water. Journal of Applied Chemistry. 1964;**14**:280-296. DOI: 10.1002/jctb.5010140704

[69] Magaud H, Migeon B, Morfin P, Garric J, Vindimian E. Modelling fish mortality due to urban storm run-off: Interacting effects of hypoxia and un-ionized ammonia. Water Research. 1997;**31**(2):211-218. DOI: 10.1016/S0043-1354(96)00255-2

[70] McDermott, Deirdre J, et al. Metal contamination of flatfish around a large submarine outfall. Journal (Water Pollution Control Federation). 1976;**48**(8):1913-1918. Available from: http://www.jstor.org/stable/25039961 [Accessed 30 Jan 2023]

[71] Roth F et al. Impacts of a high-discharge submarine sewage outfall on water quality in the coastal zone of Salvador (Bahia, Brazil). Marine Pollution Bulletin. 2016;**106**(1-2):43-48

[72] Pastorok RA, Bilyard GR. Effects of sewage pollution on coral-reef communities. Oldendorf. 1985;**21**(1):175-189

[73] Shuval H. Estimating the global burden of thalassogenic diseases: Human infectious diseases caused by wastewater pollution of the marine environment. Journal of Water and Health. 2003;**1**(2):53-64

[74] Tuya F et al. Decadal changes in the structure of Cymodocea nodosa seagrass meadows: Natural vs. human influences, Estuarine. Coastal and Shelf Science. 2014;**137**:41-49. DOI: 10.1016/j.ecss.2013.11.026

[75] Papathanasiou V. Sotiris Orfanidis, anthropogenic eutrophication affects the body size of Cymodocea nodosa in the North Aegean Sea: A long-term, scale based approach. Marine Pollution Bulletin. 2018;**134**:38-48. DOI: 10.1016/j.marpolbul.2017.12.009

[76] Moreno D, Aguilera PA, Castro H. Assessment of the conservation status of seagrass (Posidonia oceanica) meadows: Implications for monitoring strategy and the decision-making process. Biological Conservation. 2001;**102**(3):325-332. DOI: 10.1016/S0006-3207(01)00080-5

[77] Spies RB et al. Stable isotope ratios and contaminant concentrations in a sewage-distorted food web. Marine Ecology Progress Series. 1989;**1989**:157-170

[78] Guidetti P et al. Changes in Mediterranean rocky-reef fish assemblages exposed to sewage pollution. Marine Ecology Progress Series. 2003;**253**:269-278

[79] Soto-Jiménez M, Páez-Osuna F, Morales-Hernández F. Selected trace metals in oysters (Crassostrea iridescens) and sediments from the discharge zone of the submarine sewage outfall in Mazatlán Bay (southeast Gulf of California): Chemical fractions and bioaccumulation factors. Environmental Pollution. 2001;**114**(3):357-370

[80] Bu-Olayan AH, Thomas BV. Dispersion model and bioaccumulation factor validating trace metals in sea bream inhabiting wastewater drain outfalls. International journal of Environmental Science and Technology. 2014;**11**:795-804

[81] Elser JJ et al. Global analysis of nitrogen and phosphorus limitation of primary producers in freshwater, marine and terrestrial ecosystems. Ecology Letters. 2007;**10**(12):1135-1142. DOI: 10.1111/j.1461-0248.2007.01113.x

[82] Powley HR et al. Direct discharges of domestic wastewater

are a major source of phosphorus and nitrogen to the Mediterranean Sea. Environmental Science & Technology. 2016;**50**(16):8722-8730

[83] MacIsaac JJ et al. The effect of sewage on uptake of inorganic nitrogen and carbon by natural populations of marine phytoplankton. Journal of Marine Research. 1979;**37**(1):51-61

[84] Savage C. Tracing the influence of sewage nitrogen in a coastal ecosystem using stable nitrogen isotopes. Ambio: A Journal of the Human Environment. 2005;**34**(2):145-150

[85] Ren JS et al. A dynamic growth model of macroalgae: Application in an estuary recovering from treated wastewater and earthquake-driven eutrophication. Estuarine, Coastal and Shelf Science. 2014;**148**:59-69

[86] Figueiredo da Silva J et al. Nearshore circulation revealed by wastewater discharge from a submarine outfall, Aveiro Coast, Portugal. Hydrology and Earth System Sciences. 2002;**6**(6):983-988

[87] Davies PA, Mofor LA, Jlao Y. Foam lines generated by a spreading buoyant plume. International Journal of Remote Sensing. 1993;**14**(14):2647-2663

[88] Murcia JJ et al. Study of the E. coli elimination from urban wastewater over photocatalysts based on metallized TiO2. Applied Catalysis B: Environmental. 2017;**200**:469-476

[89] Paulshus E et al. Diversity and antibiotic resistance among Escherichia coli populations in hospital and community wastewater compared to wastewater at the receiving urban treatment plant. Water Research. 2019;**161**:232-241

[90] Ganoulis J. Risk in coastal pollution. In: Water Resources Engineering Risk Assessment. Berlin Heidelberg: Springer; 1991

[91] Feitosa RC et al. Coupling bacterial decay and hydrodynamic models for sewage outfall simulation. Journal of Applied Water Engineering and Research. 2013;**1**(2):137-147

[92] Cisneros-Aguirre J, Manzanera M, Bursón E. Depuration of wastewater from table olive industries. Acta Scientific Agriculture. 2018;**2**(10):10-13

Chapter 4

Petroleum Wastewater Treatment

Ali Aghababai Beni, Mohammad Saleh Samie Adel, Mojgan Zaeimdar, Arezoo Ghadi, Vahid Hassani, Kiarash Jalalvandi and Sayed Asaad Abdollahi

Abstract

Petroleum hydrocarbons in refinery wastewater are considered the main cause of pollution. Wastewater from oil refineries contains large amounts of oil and fat in the form of suspended particles, light and heavy hydrocarbons, phenol, and other dissolved organic substances, which cause environmental pollution if they are discharged into the environment without treatment. Usually, conventional methods of treating petroleum wastes have a lot of costs; due to the existence of sufficient area for the construction of solar distillation ponds and suitable sunlight, as well as a large number of sunny days near the equator, the solar distillation method can be used. Membrane bioreactors based on biological decomposition and biological transformation of oils and waste oil materials have provided new solutions for the biological treatment of these wastewater. In addition to these methods, Fenton's advanced oxidation methods, electrochemical coagulation method, and membrane filtration method are mentioned in this chapter.

Keywords: petroleum wastewater, solar evaporation, membrane bioreactor, advanced oxidation Fenton, electrocoagulation, membrane filtration

1. Introduction

Water is one of the most important resources in the development of countries. During the twentieth century, the world's population tripled, and water use increased six-fold. The world's available water is only sufficient for the current population with minimal access to clean water [1]. Improper distribution in terms of space and time and an increase in population and per capita consumption of water have aggravated this problem. The world's population is increasing, and drinking water resources are decreasing, so the world may face the problem of water shortage in the future. Destructive activities and inefficient use of water resources, along with the increase in population and increase in water demand, have severely limited water resources in the last few decades. The United Nations states that the lack of water resources has caused the reduction of agricultural land, and the production of food in recent decades has been extremely risky. The lack of water resources is a serious threat to human life [2].

Climate change and the reduction of rain forests and the reduction of the thickness of the ozone layer all aggravate the water shortage. Lack of water also has side and indirect effects, such as increasing poverty and hunger, ecosystem destruction, desertification, climate change, and even world peace [3].

The per capita standard of drinking water consumption in different countries rarely exceeds 200 liters per day. But the results of researchers' studies show that per capita water consumption is much higher than the standards set in the countries [4]. That is because besides the direct consumption of water, humans consume water through the consumption of food, fruits, and even services and goods, and its amount is on average about 3400 L per day per person in the world, which is called virtual water [5]. Climatic conditions, place and time of production, management and planning, and culture and habits of people are some of the effective factors in the amount of virtual water [6].

The security of water resources has become another challenge with increasing demand. Harvesting and purifying water from surface and underground sources, as well as treating wastewater produced in underground aquifers, in addition to polluting aquifers, will also disrupt the natural water cycle [7]. On the other hand, due to the possibility of the spread of many diseases caused by the contamination of water and sewage, in order to preserve the health of human societies and prevent disruptions in the water cycle, sewage must be properly collected, treated, and returned to the natural water cycle [8]. The most important goals are to build urban and industrial wastewater treatment systems, maintain public health, protect the environment, prevent the pollution of water sources, and reuse treated wastewater in industry and agriculture [9].

Wastewater provides a valuable source of recoverable water. Although this source can contain dangerous compounds that endanger public health and the environment, at least 90% of the wastewater is water [10]. New technologies to treat wastewater and return it to water supply networks are an important factor in increasing limited water resources. Water treatment plants are an important part of the water recovery process. The main goal of water purification processes is to reduce the concentration of water pollutants by separation, destruction, and disinfection [11]. Many efforts have been made to maintain the quality of treated industrial wastewater, recover them, and prevent them from jeopardizing public health [12].

From another point of view, the disposal of untreated effluents from factories and industries creates many health risks for human societies [13]. In order to reduce these risks, wastewater treatment plans for factories must also be developed. Therefore, the use of water obtained from sewage treatment in agriculture can not only make an important contribution to the water supply of the society but is also considered a solution for environmental preservation and sustainable development [14].

The production fluid of oil wells is usually a combination of gas, oil, and water. Water with oil can be observed as free water or fine suspended droplets or both in the fluid. Since the production of excess water is an integral part of the production and preliminary processing of crude oil, in order to prevent environmental pollution, maintain reservoir pressure, and increase extraction from oil production wells, these waters, after preliminary treatment in the treatment systems of desalination units, are again sent to disposal wells that are intended to be injected for this purpose [15].

The production amount of salt water along with oil in desalination units from crude oil is significant. These effluents have created a big problem for the environment due to their specific quantitative and qualitative characteristics, which include soluble salts, the presence of petroleum substances, volatile and non-volatile organic substances, and other hazardous pollutants in large volumes [16]. Due to the presence of supersaturated soluble salts and suspended particles and corrosive agents, these wastewaters have a strong tendency to deposit, and if they are injected into the well without preliminary treatment, it may cause clogging of the flow path in the underground flow pipes and/or the opening of the well or cause the corrosion of the flow pipes as a result of the effluent leaks into the environment.

Oil refineries, as one of the complex process industries, consume significant amounts of water based on the size and configuration of the process for multiple operations (65 to 90 gallons of water per barrel of crude oil) and a large volume of wastewater with diverse natures from 1.6 to 0 [17]. They produce 4 times the amount of processed crude oil. The recycling and reuse of this significant amount of wastewater for various purposes, including the supply of water needed for cooling systems, process units, irrigation, and firefighting after the implementation of purification based on quality standards in oil refineries, is a significant matter [18]. Several post-refining approaches, depending on the nature of the type and size of process units in oil refineries, have evolved over the past decades with the aim of improving water and wastewater management [19]. These approaches include the investigation and implementation of traditional techniques such as distillation, evaporation, active carbon filtration, sand filtration, and chemical oxidation and advanced cases such as membrane separation under pressure, electrodialysis, ion exchange, and advanced oxidation processes [20].

The necessity of treatment includes engineering investigations and the use of appropriate technology, measuring the quantitative and qualitative parameters of wastewater in the outlet pool of the treatment plant, comparing with the declared standards, the fate of excess sludge, and ensuring the absence of unpleasant odors and floating objects in the outlet wastewater and no change in the color and turbidity of receiving water at the place of discharge [21].

2. The treatment of crude oil desalination unit wastewater with the solar evaporation method

Effluents from crude oil desalination units due to their specific qualitative and quantitative characteristics, which include highly soluble salts of 50 grams per liter, and the presence of petroleum substances, volatile and non-volatile organic substances, and other hazardous pollutants for the environment, as well as a large volume, create a big problem in the vicinity of oil units and the environment around them. At present, these wastewaters are usually re-injected into operating wells or abandoned wells without treatment or after initial treatment, including the removal of suspended particles and major associated oil substances, with the aim of increasing harvest or preserving the environment. Also, these wastewaters are sent to the evaporation ponds adjacent to these units after degreasing the crude oil desalination units to protect the environment, so that over time, with the solar evaporation method, their amount is reduced, and the volume of the pond to enter the production wastes is emptied again.

This method creates the risk of pollutant leakage into groundwater and release to air and harm to humans, birds, and other creatures around the pond. Considering the sufficient temperature and the high intensity of radiation most days of the year, the use of solar energy seems appropriate. Solar distillation is a relatively simple solution for brackish water sources. Distillation is one of the processes used to purify water, and for this purpose, any heat source can be used. In the solar distillation method, using the energy of the sun, evaporated water and pure water vapor after condensation are used as pure water. The use of solar distillation method is a solution for water supply in remote areas that face a shortage of drinking water and common resources such as heat and electricity grid.

The possibility of building in small capacities, no need or minimal need for fuel and electricity, and the absence of environmental pollution caused by fuel consumption are among the advantages that make the use of this system in areas with significant renewable energy potential and, at the same time, where electricity and fuel transmission is difficult justifiable. The first and simplest solar device built is a single pond solar still (solar still). The building of this device consists of a wooden pond whose floor is blackened by safe pigments.

In a general and apparent classification, pond solar desalination plants can be divided into two groups with a one-way slope and a two-way slope liquefaction surface (**Figure 1**). According to the investigations, the solar distillation pond with a liquefaction surface with a one-way slope has a higher efficiency, because the incoming radiation is more.

The main problem of this type of water desalination plant, like most solar water desalination plants, is the relatively low production of desalinated water. One of the other obstacles of desalters using the solar distillation method is the absorption of less solar energy in areas far from the equator, because in these areas, the liquefaction surface of the device is parallel to the horizon and the oblique radiation of solar rays, and hence, the absorption of solar energy is very low. After the desalination of sea water by the solar distillation system, researchers have performed chemical analyses to check the possibility of using the water produced by this system as drinking water and compared the results with drinking water. The results showed that the resulting distilled water could be mixed with well water to obtain drinkable water, and the quality of this water was acceptable. The results showed that impurities such as nitrate, chloride, iron, and solids soluble in water were removed by solar distillation method. Most of the research related to this topic is often reported for seawater desalination; the treatment of petroleum effluents with this method is a new topic. One of the parameters affecting the efficiency of pond solar desalination plants is the optimal depth of salt water. Research has been done on the effect of water depth inside the pond. In addition to the geometrical parameters of the pond, in recent years, most researchers have focused on the construction of solar energy absorbent beds and heat transfer to increase evaporation.

Zhang et al. [22] reported that highly polluted saline wastewater was treated by carbonized lotus seedpod with the solar evaporation method (**Figure 2**). In their research, COD was removed by more than 84%.

(a) (b)

Figure 1.
Pond solar desalination plants with a one-way slope (a) and a two-way slope, and (b) liquefaction surface.

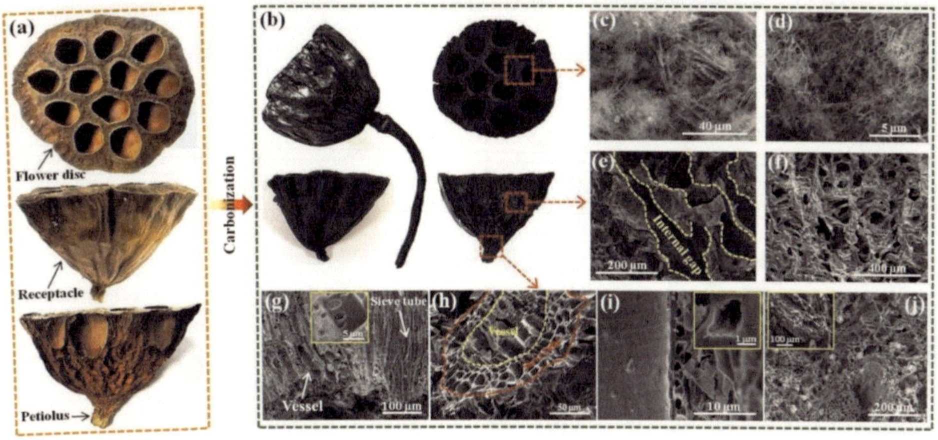

Figure 2.
Lotus seedpod-based solar steam generation. Digital photograph of a lotus seedpod (a) before and (b) after carbonization; (c–j) SEM images of the flower disc, receptacle, and petioles of the lotus seedpod after carbonization, respectively [22].

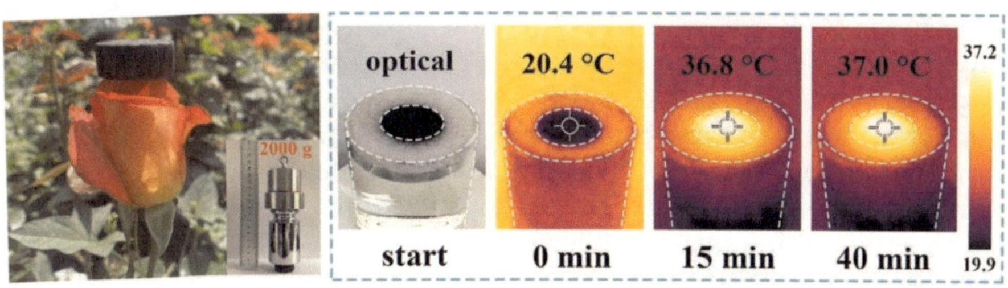

Figure 3.
Image of the ultra-light substrate based on a chitosan/bamboo fiber matrix and temperature rise in 40 min [23].

Sun et al. [23] synthesized a new photothermal substrate based on a chitosan/bamboo fiber matrix with high efficiency for use in water evaporators. According to **Figure 3**, this ultra-light substrate had the ability to increase the local temperature up to 37°C in less than 40 min. The evaporation rate in their research was 6.72 kg m^{-2} of purified water with a removal efficiency of 85%.

3. The treatment of petroleum wastewater with the membrane bioreactor

Membrane bioreactor technology or MBR refers to technologies in which wastewater is biologically treated, and then, the resulting biomass is physically separated from the mixed liquid using membrane processes. All these steps are performed in a single bioreactor (**Figure 4**). Therefore, in this method, the secondary sedimentation basin is removed from the system. Among the other advantages of this system we can mention the small amount of space they need, the lack of sludge production, and the high quality of the output effluent. These systems are used for purification; so far, urban and domestic wastewater as well as industrial wastewater such as food, pharmaceutical, oil and petrochemical industries have been used. A lot of research has been done on the treatment of petroleum industry wastewaters by MBR and methods of improving its performance. One of the most important wastewaters from oil

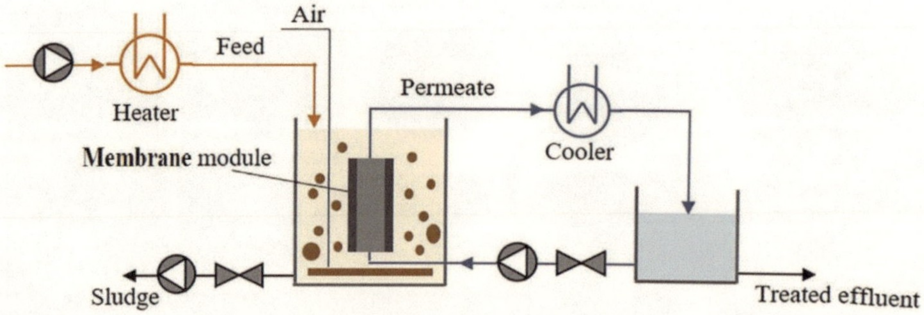

Figure 4.
Configurations of membrane bioreactor with the submerged membrane [9].

industries, which has many adverse effects on the environment, is the water produced in oil fields, which contains large amounts of salt and oil.

Soltani et al. [24] used a submerged MBR with a hollow fiber membrane to treat this wastewater. Due to the fact that this wastewater contains large amounts of salt, due to the increase in osmotic pressure, it destroys the cell wall of the normal microorganisms in the MBR system. In this study, the purified bacteria that were obtained from the areas of oil deposits in the sea, after exposure to the main sewage, were able to decompose 50% of phenanthrene, which is a complex and difficult-to-decompose aromatic compound with three benzene rings, after 45 days. Based on these results, these bacteria can break down other compounds in crude oil. By reducing the salt concentration in this experiment, contrary to expectation, the performance of bacteria purified from the environment with high salt concentration did not decrease. This confirms that these bacteria belong to the halotolerant group.

In a similar study, Xianling et al. [7] studied the purification of petroleum hydrocarbons in a membrane bioreactor by purifying different bacteria from oil-contaminated areas. In this system, it was found that COD removal efficiency was different in steady state, and despite the gradual increase of COD, the efficiency increased from 93 to 96.5%. The reason for this can be seen in the increase in MLSS concentration in these types of systems, which reached 16.2 g L^{-1} in this system.

The increase in the concentration of MLSS in such MBR systems is due to the use of a membrane that prevents the exit of the mixed liquid, and this concentration increases over time. After entering the petroleum wastewater into the reactor and physiological adaptation with it, the bacteria responsible for the decomposition begin to decompose the hydrocarbons, the final product of which is carbon dioxide and water. First, light hydrocarbons such as alkanes and aromatics with low molecular weight and then heavy hydrocarbons such as polycyclic aromatics are decomposed. This rate of decomposition of petroleum materials in an ultrafiltration bioreactor with transverse flow was observed up to 0.82 g of hydrocarbon per gram of MLVSS per day. In this study, all hydrocarbons with carbon atom numbers from C_{10} to C_{24} were removed with almost the same efficiency. Due to the complexity and diversity in the quality of petroleum wastes, in some researches, phenol has been used as a suitable indicator to investigate the removal of biodegradable compounds.

Zhou and Hong, by investigating the treatability of oil refinery wastewater by a fixed film bioreactor at a hydraulic retention time of 8 h and an initial phenol concentration of 30 mg/liter, reported a COD removal efficiency between 85 and 90% and a phenol removal efficiency of 100% [25].

The results of Viero et al. study confirm the high efficiency of phenol removal from petroleum wastewater in a submerged membrane bioreactor. In this research, which was carried out in three stages, during the operation period, a high organic loading rate was applied in the long term by mixing the flow of petroleum wastewater with phenol-rich wastewater to the bioreactor, and it was shown that the treatment of petroleum wastewater in a submerged membrane bioreactor with specific hydraulic retention time, in addition to removing organic matter, also caused high efficiency of phenol removal. The presence of a membrane in this bioreactor increases COD removal efficiency by 17% [9].

The operating conditions of a membrane bioreactor, such as the pressure applied to both sides of the membrane (Transmembrane Pressure: TMP), the amount of aeration, the speed of the transverse flow, and so on, affect the purification efficiency in this system. Increasing the transverse flow speed increases the turbulence of the liquid flow inside the reactor and increases the mass transfer coefficient in the concentration polarization layer and consequently the output flux. The relationship between the output flux (J) and the transverse flow velocity (V) in a bioreactor with transverse flow, in the treatment of petroleum wastewater in a refinery, is obtained as the following power relation [26]:

$$J = kV^n \tag{1}$$

n and k are affected by MLSS, and these two factors decrease with the increase of MLSS. The COD removal rate in this bioreactor is reported to be more than 93%. Of course, increasing the speed of the transverse flow and creating a turbulent flow removes the cake layer of organic materials on the surface of the membrane, which plays the role of an additional filter and prevents the passage of impurities, and as a result, the possibility of oil particles passing through the membrane increases. The pressure applied to the membrane also has a double effect. Although increasing TMP increases the amount of output flux, on the other hand, it causes more accumulation of oil droplets on the surface of the membrane as well as inside its pores and intensifies the clogging of the membrane. If this pressure increases, oil and oil droplets will pass through the pores of the membrane and reduce the quality of the output effluent. Aeration also causes the mixing of wastewater inside the bioreactor and provides better contact between particles and microorganisms. On the other hand, it creates a shear force that causes the biofilm flocs to break and separate from the membrane surface. Also, aeration increases dissolved oxygen in the bioreactor, which has a positive effect on COD removal.

4. The treatment of petroleum wastewater with the advanced oxidation Fenton method

Advanced oxidation basically refers to methods that destroy organic compounds by producing oxygen radicals. In these methods, they use strong oxidants, catalysts, radiation, and ozone to treat wastewater [4]. Fenton process, due to low operational cost compared to other advanced oxidation processes, low toxicity of iron ion and hydrogen peroxide, its simple technology, the possibility of its application in ambient temperature and pressure, high biocompatibility, short process duration, and low energy consumption, should be widely considered to reduce high pollution levels. Fenton's reaction is carried out in an acidic environment, and the optimal pH for this reaction is 2.8–3 [27]. Fenton process is defined on the basis of electron transfer between H_2O_2 and a metal ion (generally iron ion), which acts as a homogeneous catalyst [28]. According to the mechanism of research in an acidic environment due to the reaction of hydrogen peroxide with Fe(II) or Fe(III) ions, the oxidation-reduction mechanism of the Fenton process is as follows. Based on this mechanism, the OH radical produced by attacking organic materials (RH) causes their destruction [29]. The chemical relationships of the Fenton process are shown in relationships (2)–(5) [30]:

$$H_2O_2 + Fe^{2+} \rightarrow Fe^{3+} + HO^- + HO^\bullet \quad (2)$$

$$HO^\bullet + RH \rightarrow H_2O + R^\bullet \quad (3)$$

$$R^\bullet + Fe^{3+} \rightarrow R^+ + Fe^{2+} \quad (4)$$

$$R^\bullet + H_2O_2 \rightarrow ROH + OH^\bullet \quad (5)$$

Hydroxyl radical is able to decompose organic pollutants in a short period of time and non-selectively [31]. Among the methods of producing hydroxyl radicals, the use of ultrasonic waves in advanced oxidation methods is considered one of the new methods [32]. Usually, the destruction of organic pollutants using acoustic and thermal decomposition is attributed to the activity of radicals. The increase in the effect of thermal decomposition and the reaction with radicals cause an increase in sound decomposition. Water molecules are broken in this method, and hydroxyl and hydrogen radicals are released [33]. This phenomenon includes the formation and destruction of gas bubbles, which results in the generation of very high pressure and temperature, which includes the thermal decomposition of dissolved organic compounds and the production of free radicals such as $O^\cdot$ and $OH^\cdot$, $H^\cdot$ or some oxidants such as peroxide [34]. It is hydrogen that can react with organic compounds. The destruction of gas bubbles causes the formation of very high temperatures and pressure, which leads to the separation of water vapor in the reactive hydroxyl radical and hydrogen atoms with the presence of other species (H_2O, O_2) [35]. Although advanced oxidation processes alone are not effective, the sono-chemical oxidation process can be done by adding chemicals such as persulfate and catalytic particles and increasing the efficiency of the process [36]. In recent years, a compound called persulfate (${S_2O_8}^{2-}$) with an oxidation potential of 2.01 V has been known and introduced, which is capable of oxidizing toxic and resistant organic compounds, the advantages of which are cheapness and non-selective oxidation of organic compounds [37]. The high stability of the radical produced from it in different conditions, high solubility, a solid form, and, as a result, the ease of moving and storing, have shown the use of this substance in many research [38]. Regardless of these advantages, extensive studies on the use of persulfate show [13] that at room temperature, the ability of persulfate to decompose organic substances is low and slow; therefore, in order to accelerate the oxidation process with persulfate, it is necessary to perform an activation operation [12].

According to **Figure 5**, the activation of persulfate is carried out as an advanced oxidation process with heat, UV light, and transition metal (Me^{2+}). The final product of the activation process is the production of sulfate radical ($SO_4^{\cdot-}$) with an oxidation potential of 2.6 V. Relations (6) and (7) show thermal and chemical activation of ${S_2O_8}^{2-}$ [38]:

$$ {S_2O_8}^{2-} + \text{heat or UV} \rightarrow SO_4^{\cdot-} \quad (6)$$

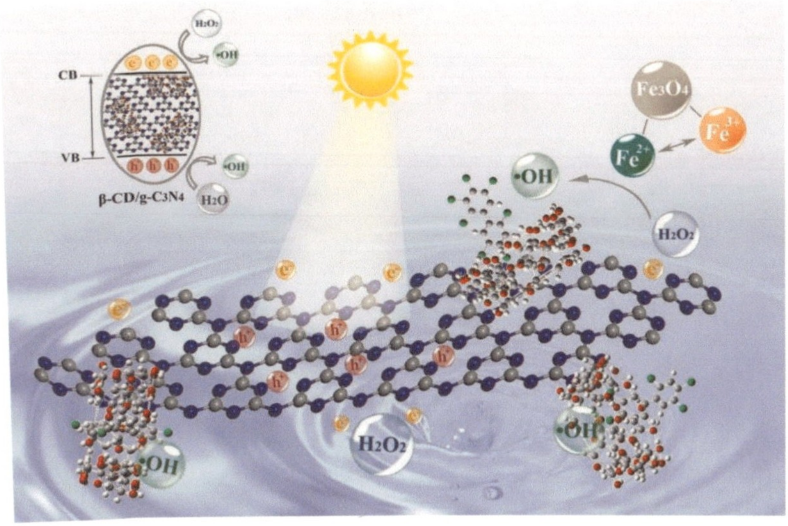

Figure 5.
Schematic of the Fenton process [32].

$$S_2O_8^{2-} + Me^{n+} \rightarrow Me^{(n+1)+} + SO_4^{2-} \tag{7}$$

Persulfate anion is considered a strong oxidant, and when activated, it can produce stronger oxidants such as sulfate radical. Since persulfate is produced slowly at normal room temperature, it is converted into radical sulfate by active photochemical or thermal decompositions, and it is used as a fast method in chemical decomposition processes [11]. During a study, Bing et al. investigated the effect of hydrogen peroxide, persulfate, and periodate in the oxidation of TiO_2 photocatalyst. The results showed that adding 2–10 moles of mineral oxidants, persulfate, periodate, and hydrogen peroxide had a faster decomposition rate compared to TiO_2/UV. In another study in which Hosseini et al. [39] investigated the decomposition of Blue 25 acid in aqueous media using Fe^{2+}/ultra-sonic and H_2O_2/ultra-sonic, the results showed that these two processes had a higher removal efficiency than the ultra-sonic process alone.

5. The treatment of petroleum wastewater with the electrocoagulation

Considering that spilling and leaking of oil into water is unavoidable in most cases and considering the adverse effects of water contaminated with petroleum derivatives on humans and the environment, so far, there have been various methods for purifying water, and separating these two important substances (water and oil) has been suggested by the researchers [40]; among them, methods of gravity separation, types of filters, reverse osmosis, biological processes, flotation with dissolved air, membrane bioreactors, adsorption with activated carbon, chemical coagulation, and electric flotation have been reported [16].

The electrocoagulation method is one of the effective separation methods of oil from the emulsion. It is the water that is optimal and affordable both technically and economically. In this method, as shown in **Figure 6**, the purification process is done in three stages: (I) The reaction of the electrolyte on the surface of the electrode and the

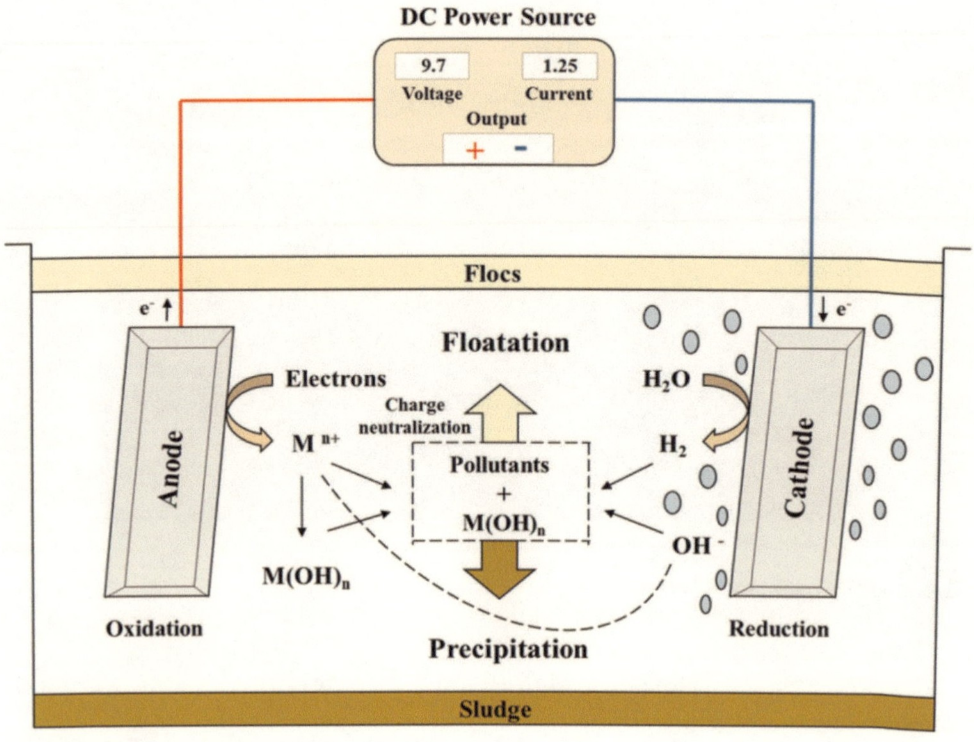

Figure 6.
Schematic diagram of the electrocoagulation process [41].

formation of coagulants by electrolytic oxidation in the aqueous phase. (II) Adsorption of colloidal particles on coagulants. (III) Removal of clots by sedimentation or flotation [42].

In this method, at the same time as the anode is corroded, electrolyte gases (generally H_2) are produced in the cathode, which leads to more flotation [43]. In electrocoagulation, metals such as iron and aluminum are usually used as anodes, which produce hydroxides, oxyhydroxides, and polymeric hydroxides when oxidized [44]. The metal hydroxides formed act as coagulants of liquid impurities, and the hydrogen bubbles formed on the cathode side provide foam formation. These products are usually much more effective than added chemicals and are able to destabilize colloidal suspensions and emulsions [45]. The electrocoagulation method has several advantages, including no need to add chemicals, simple equipment, convenient operation, low initial cost and low operating cost, short reaction time, rapid sedimentation of the created flocs, low sludge production [46], high safety, no need to transport and move chemicals, and coagulant production on site, and can act as an efficient method in separating petroleum compounds from water [21]. This method has been very efficient for purifying water contaminated with solids, dyes, heavy metals, and soluble organic and inorganic substances. As mentioned, iron and aluminum electrodes are used as anodes, and the reaction mechanism by iron is as follows [47]:

In the Andes:

$$4Fe \rightarrow 4Fe^{2+} + 8e^- \quad (8)$$

$$4Fe^{2+} + 10H_2O + O_2 \rightarrow 4Fe(OH)_3 + 8H^+ \quad (9)$$

In the Cathode:

$$8H^{+} + 8e^{-} \rightarrow 4H_2 \quad (10)$$

Total reactions:

$$4Fe + 10H_2O + O_2 \rightarrow 4Fe(OH)_3 + 8H^{+} \quad (11)$$

At the same time as the anode is corroded, electrolyte gases (generally H2) are produced in the cathode. Metals such as iron and aluminum are commonly used as anodes, which produce hydroxides, oxyhydroxides, and polymeric hydroxides upon oxidation [48]. The formed metal hydroxides act as coagulants of liquid impurities, and the formed hydrogen bubbles on the cathode side provide foam formation. These products are usually much more effective than added chemicals and are capable of destabilizing colloidal suspensions and emulsions. Nidheesh et al. [49] used electrocoagulation process with iron, aluminum, and steel electrodes to treat oil refinery wastewater, and the results showed that this process could be a suitable method to reduce sulfate and COD concentration from oil refinery wastewater. Also, in the research where *Asselin* et al. used electrocoagulation process with aluminum and steel electrodes to treat the oil effluents of ships, it was found that the optimal state was obtained with steel electrodes, and 93% removal for BOD, more than 95% for oil removal, and about 68% for COD removal are achieved in optimal conditions [41].

6. The treatment of petroleum wastewater with the membrane filtration method

Several separation processes, including ultrafiltration, nanofiltration, and reverse osmosis, have been employed for oil/water separation. Membrane ultrafiltration is one of the most important separation processes in the field of industrial petroleum wastewater treatment. When the solvent molecules are less than 0.5 microns, ultrafiltration is used [50]. Before oil emulsions enter the environment, it is necessary to remove the oil in them to an acceptable level, which is determined by the standards. Petroleum effluents and oil-water emulsions are two important environmental pollutants [21]. Unlike urban wastewater, industrial wastewater discharged into the environment does not have any fixed characteristics. The composition and characteristics of industrial effluents are significantly variable, and even in different parts of the industry, these flows are visibly different. Despite the physicality of the filtration process, chemical purification processes can also be used. A huge amount of oil refinery effluents are in the form of oil-in-water or water-in-oil emulsions, which are produced from different parts of the extraction, transportation, and refining processes [51].

Methods based on membrane separation include dehydration of oil emulsion by reverse osmosis, coagulation resulting from microfiltration, microfiltration, membrane distillation, and ultrafiltration. Among the benefits of membrane technology are lower cost, no need for any chemical additives, and the ability to create an acceptable quality flow. Ultrafiltration is used as an effective method to separate, purify, and saturate water-soluble solutes or water-dispersed substances. In any case, due to the deformation of oil droplets with operating pressure, oil droplets can pass through the holes with pressure and pollute the flow. Despite the reduction in the cost of energy consumption of the ultrafiltration process, the problems caused by washing in this process are very expensive [52].

By replacing membrane processes with traditional purification methods, product quality is improved, and process efficiency is increased. Microfiltration membranes purify colloidal particles and bacteria with a diameter of 0.1–10 μm. Ultrafiltration membranes can separate large soluble molecules such as proteins and petroleum substances from the solution. In reverse osmosis membranes, the solvents are dissolved in the membrane and penetrate through the membrane to a lower concentration and are mainly used in the field of desalination of underground water or sea water. The difference in cavity drops (or apparent cavity) creates significant differences in the field of membranes used. Reverse osmosis and ultrafiltration processes are often used in oil/water treatment. Tubular modules are used in the field of oily wastewater treatment due to their resistance to the clogging of emulsion particles, easy replacement of the membrane, and the ability to use the high linear speed of the oily emulsion on the membrane surface [53].

Wollborn et al. have reported that if the shear pressure is lower than the critical pressure, then the emulsion will reach the maximum possible volume [54]. Ma et al. showed that in porous hydrophobic membranes, due to the coagulation and sedimentation of oil on the membrane cavities, the separation of oil-in-water emulsions is reduced [55]. Using a microporous polytetrafluoroethylene flat membrane, Nittami et al. [56] have investigated the effect of emulsion droplet size, stirrer speed in penetration test, oil phase volume fraction, and surfactant concentration in the feed solution on oil flow flux. For industrial/petroleum effluents, the amount of oil in the flow passing through the membrane is higher than the acceptable amount of the standard discharge to the environment. Tong et al. [57] used commercial polyvinylidene fluoride to treat oil field effluents. At the beginning of membrane filtration, due to the lack of gelatin layer formation on the surface of the membrane, the quality of the water coming out of the membrane is not very favorable. As the process progresses and due to the concentration of pollutants and the polarization of the membrane surface, a gelatinous layer is formed on the membrane surface. The gelatin layer formed on the surface of the membrane prevents polluting particles from entering the membrane cavities and leads to a decrease in membrane flux. The flux recovery percentage of modified membranes reaches 100% after washing with 1% OP-10 surfactant solution. The relationship between flux and pore pressure is not completely linear due to resistance in addition to membrane resistance. When oil recovery has a downward trend with increasing pressure, the amount of flux reduction is greater. Porosity, pore size distribution, and membrane substrate structure play an important role in determining the flux through the membrane. Also, by increasing the concentration of titanium oxide nanoparticles in the polymer solution, the number of shell cavities increases [58]. Based on the observations of Sutrisna et al. [20], membrane fouling is a combination of pore-clogging by smaller oil droplets in the emulsion and sedimentation of the oil layer on the surface. To check the effect of membrane clogging, the permeability of pure water passing through each membrane is measured before and after washing the membrane. The results of Wang et al. [59] showed that liquid droplets passed through the pores of the membrane more easily with the increase in osmotic pressure. Of course, with the increase of the osmosis pressure, the operating cost and depreciation of the equipment increase. Also, membrane clogging occurs at high pressure due to the formation of a colloidal layer. As the colloidal layer increases, the resistance of the droplets passing through the membrane increases, so the membrane flux decreases. Organic-inorganic composite membranes such as Al2O3-PVDF are widely used in petroleum wastewater treatment. Flux recovery is better for membranes washed with alkaline solutions. Hashemi et al. [60]

investigated the ultrafiltration of oil effluent from engine houses using tubular modules (with a large diameter). Most of the constructed membranes reduced the oil content in the flow to less than 10 mg/DCM. The current passing through this membrane is suitable for discharge to the environment. Hollow fiber membranes are much more efficient than tubular and flat plate membranes due to their high surface area to volume [61]. Due to the high specific area and hydrophilicity of titanium oxide nanoparticles, the flux increases. Membrane wettability is one of the important factors that can affect the flux and anti-clogging ability of membranes. By increasing the titanium oxide particles, the contact angle of water with the membrane surface can be significantly reduced. As the repulsion decreases, the membrane flux increases. Therefore, a membrane with maximum porosity and hole size has maximum flux. Due to the inherent hydrophobicity of PVDF polymer, this type of membrane is used in petroleum wastewater treatment, organic/water separations, gas absorption and membrane distillation, and ultrafiltration [62]. Additives such as polyvinylpyrrolidone, polyethylene glycol, and lithium chloride are used to improve the morphology and performance of the membrane and its mechanical strength. Zhu et al. [19] used alumina to improve the hydrophilic property and antifouling ability of the polysulfone membrane. They added SZP particles to the porous polysulfone membrane, which ultimately led to the improvement of polysulfone membrane properties such as hydrophilicity, antifouling ability, and tensile strength [18]. Composite membranes are used to treat petroleum wastewater. Due to the increase in the hydrophilicity of the membrane with the increase of hydrophilic SZP particles, the hydrophilic layer formed on the surface of the membrane plays an important role in removing the gel-like layer [63].

One of the reasons for the reduction of the passing flux is concentration polarization, which is due to the increase in the concentration of oil particles on the surface of the membrane. As the membrane filtration continues, the concentration of the preservative on the membrane surface becomes higher than the feed concentration, which ultimately leads to concentration polarization (ultimately creating a gel layer on the membrane surface). Also, due to the presence of impermeable pores in the membrane for the passage of oil droplets, clogging occurs [64].

Due to the high specific area and hydrophilicity of titanium oxide nanoparticles, the flux of PVDF ultrafiltration hollow fiber membranes increases. By increasing the concentration of titanium oxide nanoparticles, the pores of the membrane are blocked due to the accumulation of particles, and the formation of a dense substrate decreases, and as a result, the average size of the cavity decreases [65].

As a hydrophilic surface modifier, Pluronic F127 can greatly reduce the water contact angle of the membrane. Due to the stability of oil droplets on the surface of the PES/Pluronic F127 membrane, the water contact angle for used membranes is higher than that for fresh membranes. During the ultrafiltration process, many oil droplets settle on the surface of the membrane or are adsorbed on the surface. After washing with water, the membrane surfaces are still hydrophobic, and the oil droplets are not removed from the membrane surface [17].

The effects of concentration polarization and membrane fouling at constant pressure are observed with a significant decrease in flux with time. In this case, the concentration polarization is omitted due to the large size of the emulsified oil particles. The decrease in the membrane flux is due to the clogging of the membrane through surface absorption or the settling of oil droplets on the surface of the membrane or inside the membrane cavities. Sodium dodecyl sulfate is used as detergent to wash the captured membranes. The membrane surface washed with SDS solution is very hydrophilic [66]. During the washing process, some SDS molecules distributed in the aqueous solution are absorbed on the membrane surface and lead to a decrease in surface tension [67]. Therefore, the stability of oil droplets is improved and prevents their sticking and coagulation. Polyethersulfone membranes offer very high thermal stability in addition to mechanical properties, but they also have disadvantages. The main problem of these membranes is their relative hydrophobicity [50].

7. Management of petroleum wastewater treatment

Wastewater or sewage refers to mainly liquid local, urban, or industrial wastes and discharges. The method of collecting and discarding it differs in each region, depending on the local awareness of the environment, and scientists believe that the future will belong to those who make the best use of water. One of the main axes of sustainable development in the petrochemical industry is the optimal use of resources, and the reuse of wastewater in terms of the increasing importance of water as a vital substance has been one of the goals of the management of petrochemical companies, so that even, if possible, the wastewater of the production units after performing purification can be used again in the green space irrigation sector or in the industrial sector.

The development that petrochemicals are trying to achieve is sustainable and all-round, and we deeply believe in the fact that without environmental preservation and optimal use of resources, the development of any industry is one-dimensional and unstable. Therefore, the reuse of wastewater is one of the interesting options in the petrochemical industry. The only concern of using wastewater is environmental pollution in the long term. Therefore, in order to solve existing environmental challenges and provide suitable solutions for the sustainable use of wastewater, determining the type of pollution caused by irrigation with wastewater and the resulting environmental effects should be fully investigated.

Wastewater management is planning, organization, care, and executive operations related to the production, collection, storage, transportation, recycling, processing, and disposal of wastewater, as well as education and information in the field of wastewater.

Environmental monitoring is a continuous process of care, examination, comparison, and accurate evaluation of environmental qualities, which is developed and carried out before, during, and after the implementation of projects. The most key things required for an environmental monitoring program to choose an effective method of oil waste treatment are: organizational structure, monitoring operations, timing, reporting, and financial status. By considering the conditions of the operating area and the facilities and characteristics of the petroleum wastewater and by choosing the appropriate treatment method, petroleum wastewater can be managed. For example, if there is enough space and suitable sunlight, the best option is to use the solar evaporation method, because petroleum wastewater can be managed by using a free energy with the lowest cost. Of course, if the oil effluent mainly contains volatile hydrocarbons, this method is not recommended, because pollution enters the air, or valuable hydrocarbons that can be recovered get lost, and this issue is not economically justified. The management of petroleum wastes in densely populated cities is faced with a lack of space to install and operate equipment. Therefore, in conditions where space is limited, a membrane bioreactor using microorganisms is a better choice. If the volume of wastewater is large, membrane filtration can be used. Of course, in filtration, membrane clogging is a major problem that limits the development of this method in petroleum wastewater management.

8. Conclusion

One of the problems of today's world is the pollution of underground water sources due to the pollutants imported from various industries, especially refineries. These harmful particles enter the water in different ways and pollute the water.

The volume of production effluents is increasing, and there are many petroleum substances in these effluents. Due to the fact that these hydrocarbons are difficult to decompose biochemically and cause damage and destruction to the environment, they must be purified before being discharged into the environment.

Therefore, due to the harmful risks that they cause to human health, the environment, plants, and aquatic organisms, the treatment of petroleum effluents has been given a lot of attention. In addition, due to the large volume of petroleum effluents, a method that can be easily performed, is economically viable, and is able to separate oil pollution with high efficiency is very important.

A specific separation method should be used for each type of industrial waste according to the physical nature of the oil waste. These petrochemical effluents are mainly in the form of oil-in-water emulsions; as a result, breaking emulsions and separating oil require a correct understanding of their physical properties and chemical composition. The choice of each of the methods depends on the economic conditions and the type and form of the oily pollutant in the water. Petroleum wastewater management is done based on the type of pollutant and its volume. Choosing the right treatment method depends on many variables (pollutant type, wastewater volume, pollutant concentration, ability to recycle and extract valuable materials, local energy sources). For example, if the wastewater volume is small and sunlight is available, solar evaporation is the best choice.

Author details

Ali Aghababai Beni[1*], Mohammad Saleh Samie Adel[2], Mojgan Zaeimdar[3], Arezoo Ghadi[4], Vahid Hassani[5], Kiarash Jalalvandi[6] and Sayed Asaad Abdollahi[2]

1 Department of Chemical Engineering, Shahrekord Branch, Islamic Azad University, Shahrekord, Iran

2 Department of Chemical Engineering, Shahreza Branch, Islamic Azad University, Shahreza, Isfahan, Iran

3 Department of Environment Engineering, North Tehran Branch, Islamic Azad University, Tehran, Iran

4 Department of Chemical Engineering, Islamic Azad University, Ayatollah Amoli Branch, Amol, Iran

5 Mechanical Engineering Department, Amirkabir University of Technology, Tehran, Iran

6 Department of Chemical Engineering, Faculty of Engineering, Razi University, Kermanshah, Iran

*Address all correspondence to: aliaghababai@yahoo.com

References

[1] Xie W, Tang P, Wu Q, Chen C, Song Z, Li T, et al. Solar-driven desalination and resource recovery of shale gas wastewater by on-site interfacial evaporation. Chemical Engineering Journal. 2022;**428**:132624. DOI: 10.1016/j.cej.2021.132624

[2] Pourehie O, Saien J. Solar driven homogeneous sodium hypochlorite/iron process in treatment of petroleum refinery wastewater for reusing. Separation and Purification Technology. 2021;**274**:119041. DOI: 10.1016/j.seppur.2021.119041

[3] Cao TND, Bui XT, Le LT, Dang BT, Tran DPH, Vo TKQ, et al. An overview of deploying membrane bioreactors in saline wastewater treatment from perspectives of microbial and treatment performance. Bioresource Technology. 2022;**363**:127831. DOI: 10.1016/j.biortech.2022.127831

[4] Das A, Adak MK. Photo-catalyst for wastewater treatment: A review of modified Fenton, and their reaction kinetics. Applied Surface Science Advances. 2022;**11**:100282. DOI: 10.1016/j.apsadv.2022.100282

[5] Srivastava A, Parida VK, Majumder A, Gupta B, Gupta AK. Treatment of saline wastewater using physicochemical, biological, and hybrid processes: Insights into inhibition mechanisms, treatment efficiencies and performance enhancement. Journal of Environmental Chemical Engineering. 2021;**9**:105775. DOI: 10.1016/j.jece.2021.105775

[6] Manetti M, Tomei MC. Extractive polymeric membrane bioreactors for industrial wastewater treatment: Theory and practice. Process Safety and Environment Protection. 2022;**162**:169-186. DOI: 10.1016/j.psep.2022.03.063

[7] Xianling L, Jianping W, Qing Y, Xueming Z. The pilot study for oil refinery wastewater treatment using a gas-liquid-solid three-phase flow airlift loop bioreactor. Biochemical Engineering Journal. 2005;**27**:40-44. DOI: 10.1016/j.bej.2005.06.013

[8] Paul T, Sinharoy A, Pakshirajan K, Pugazhenthi G. Lipid-rich bacterial biomass production using refinery wastewater in a bubble column bioreactor for bio-oil conversion by hydrothermal liquefaction. Journal of Water Process Engineering. 2020;**37**:101462. DOI: 10.1016/j.jwpe.2020.101462

[9] Kharraz JA, Khanzada NK, Farid MU, Kim J, Jeong S, An AK. Membrane distillation bioreactor (MDBR) for wastewater treatment, water reuse, and resource recovery: A review. Journal of Water Process Engineering. 2022;**47**:2-7. DOI: 10.1016/j.jwpe.2022.102687

[10] Domingues E, Fernandes E, Gomes J, Castro-Silva S, Martins RC. Advanced oxidation processes at ambient conditions for olive oil extraction industry wastewater degradation. Chemical Engineering Science. 2022;**263**:118076. DOI: 10.1016/j.ces.2022.118076

[11] Nidheesh PV, Behera B, Babu DS, Scaria J, Kumar MS. Mixed industrial wastewater treatment by the combination of heterogeneous electro-Fenton and electrocoagulation processes. Chemosphere. 2022;**290**:133348. DOI: 10.1016/j.chemosphere.2021.133348

[12] Ziembowicz S, Kida M. Limitations and future directions of application of the Fenton-like process in

micropollutants degradation in water and wastewater treatment: A critical review. Chemosphere. 2022;**296**:2-8. DOI: 10.1016/j.chemosphere.2022.134041

[13] Hu Y, Yu F, Bai Z, Wang Y, Zhang H, Gao X, et al. Preparation of Fe-loaded needle coke particle electrodes and utilisation in three-dimensional electro-Fenton oxidation of coking wastewater. Chemosphere. 2022;**308**:136544. DOI: 10.1016/j.chemosphere.2022.136544

[14] Samuel O, Othman MHD, Kamaludin R, Kurniawan TA, Li T, Dzinun H, et al. Treatment of oily wastewater using photocatalytic membrane reactors: A critical review. Journal of Environmental Chemical Engineering. 2022;**10**:108539. DOI: 10.1016/j.jece.2022.108539

[15] Pourehie O, Saien J. Homogeneous solar Fenton and alternative processes in a pilot-scale rotatable reactor for the treatment of petroleum refinery wastewater. Process Safety and Environment Protection. 2020;**135**:236-243. DOI: 10.1016/j.psep.2020.01.006

[16] Shokri A, Fard MS. A critical review in electrocoagulation technology applied for oil removal in industrial wastewater. Chemosphere. 2022;**288**:132355. DOI: 10.1016/j.chemosphere.2021.132355

[17] Chen W, Peng J, Su Y, Zheng L, Wang L, Jiang Z. Separation of oil/water emulsion using Pluronic F127 modified polyethersulfone ultrafiltration membranes. Separation and Purification Technology. 2009;**66**:591-597. DOI: 10.1016/j.seppur.2009.01.009

[18] Yin J, Tang H, Xu Z, Li N. Enhanced mechanical strength and performance of sulfonated polysulfone/Tröger's base polymer blend ultrafiltration membrane. Journal of Membrane Science. 2021;**625**:119138. DOI: 10.1016/j.memsci.2021.119138

[19] Zhu L, Song H, Zhang D, Wang G, Zeng Z, Xue Q. Negatively charged polysulfone membranes with hydrophilicity and antifouling properties based on in situ cross-linked polymerization. Journal of Colloid and Interface Science. 2017;**498**:136-143. DOI: 10.1016/j.jcis.2017.03.055

[20] Sutrisna PD, Kurnia KA, Siagian UWR, Ismadji S, Wenten IG. Membrane fouling and fouling mitigation in oil–water separation: A review. Journal of Environmental Chemical Engineering. 2022;**10**:107532. DOI: 10.1016/j.jece.2022.107532

[21] Akarsu C, Isik Z, M'barek I, Bouchareb R, Dizge N. Treatment of personal care product wastewater for reuse by integrated electrocoagulation and membrane filtration processes. Journal of Water Process Engineering. 2022;**48**:102879. DOI: 10.1016/j.jwpe.2022.102879

[22] Zhang Q, Fu Z, Chen S. Solar-driven purification of highly polluted saline wastewater into clean water by carbonized lotus seedpod. Separation and Purification Technology. 2022;**296**:1-8. DOI: 10.1016/j.seppur.2022.121401

[23] Sun X, Jia X, Weng H, Yang J, Wang S, Li Y, et al. Bioinspired photothermal sponge for simultaneous solar-driven evaporation and solar-assisted wastewater purification. Separation and Purification Technology. 2022;**301**:122010. DOI: 10.1016/j.seppur.2022.122010

[24] Soltani F, Toghraie D, Karimipour A. Experimental measurements of thermal conductivity of engine oil-based hybrid

and mono nano fl uids with tungsten oxide (WO 3) and MWCNTs inclusions. Powder Technology. 2020;**371**:37-44. DOI: 10.1016/j.powtec.2020.05.059

[25] Mutamim NSA, Noor ZZ, Hassan MAA, Yuniarto A, Olsson G. Membrane bioreactor: Applications and limitations in treating high strength industrial wastewater. Chemical Engineering Journal. 2013;**225**:109-119. DOI: 10.1016/j.cej.2013.02.131

[26] Caliani I, De Marco G, Cappello T, Giannetto A, Mancini G, Ancora S, et al. Assessment of the effectiveness of a novel BioFilm-membrane BioReactor oil-polluted wastewater treatment technology by applying biomarkers in the mussel Mytilus galloprovincialis. Aquatic Toxicology. 2022;**243**:106059. DOI: 10.1016/j.aquatox.2021.106059

[27] Salini PJ, Madhu G, Pawels R. Treatment of vehicle wash wastewater by Fenton process and its recycling. Materials Today: Proceedings. 2022;**57**: 2444-2451. DOI: 10.1016/j.matpr.2022.02.287

[28] Ciggin AS, Iravanian A, Doğruel S, Orhon D. Co-metabolism of olive mill wastewater in sequencing batch reactor under aerobic conditions after Fenton-based oxidation. Journal of Water Process Engineering. 2021;**43**:8. DOI: 10.1016/j.jwpe.2021.102277

[29] Domingues E, Fernandes E, Gomes J, Castro-Silva S, Martins RC. Olive oil extraction industry wastewater treatment by coagulation and Fenton's process. Journal of Water Process Engineering. 2021;**39**:8. DOI: 10.1016/j.jwpe.2020.101818

[30] Dogan EC, Kilicoglu O, Narci AO, Mert BK, Durna E, Akbacak UA, et al. Fenton and photo-Fenton processes integrated with submerged ultrafiltration for the treatment of pulp and paper industry wastewater. Journal of Environmental Chemical Engineering. 2021;**9**:105878. DOI: 10.1016/j.jece.2021.105878

[31] Srimoke W, Kanokkantapong V, Supakata N, Limmun W. Optimising zero-valent iron from industrial waste using a modified air-Fenton system to treat cutting oil wastewater using response surface methodology. Arabian Journal of Chemistry. 2022;**15**:104213. DOI: 10.1016/j.arabjc.2022.104213

[32] Wang H, Zhang C, Zhang X, Wang S, Xia Z, Zeng G, et al. Construction of Fe3O4@β-CD/g-C3N4 nanocomposite catalyst for degradation of PCBs in wastewater through photodegradation and heterogeneous Fenton oxidation. Chemical Engineering Journal. 2022;**429**:8-9. DOI: 10.1016/j.cej.2021.132445

[33] Ciggin AS, Sarica ES, Doğruel S, Orhon D. Impact of ultrasonic pretreatment on Fenton-based oxidation of olive mill wastewater—towards a sustainable treatment scheme. Journal of Cleaner Production. 2021;**313**:8. DOI: 10.1016/j.jclepro.2021.127948

[34] Lin R, Li Y, Yong T, Cao W, Wu J, Shen Y. Synergistic effects of oxidation, coagulation and adsorption in the integrated Fenton-based process for wastewater treatment : A review. Journal of Environmental Management. 2022; **306**:114460. DOI: 10.1016/j.jenvman.2022.114460

[35] Dinçer AR, Çifçi Dİ, Cinkaya DD, Dülger E, Karaca F. Treatment of organic peroxide containing wastewater and water recovery by Fenton-adsorption and Fenton-nanofiltration processes. Journal of Environmental Management. 2021;**299**:8. DOI: 10.1016/j.jenvman.2021.113557

[36] Bayrakdar M, Atalay S, Ersöz G. Efficient treatment for textile wastewater through sequential photo Fenton-like oxidation and adsorption processes for reuse in irrigation. Ceramics International. 2021;**47**: 9679-9690. DOI: 10.1016/j.ceramint.2020.12.107

[37] El Shahawy A, Mohamadien RH, El-Fawal EM, Moustafa YM, Dawood MMK. Hybrid photo-Fenton oxidation and biosorption for petroleum wastewater treatment and optimization using box–Behnken design. Environmental Technology and Innovation. 2021;**24**:101834. DOI: 10.1016/j.eti.2021.101834

[38] Domingues E, Silva MJ, Vaz T, Gomes J, Martins RC. Sulfate radical based advanced oxidation processes for agro-industrial effluents treatment: A comparative review with Fenton's peroxidation. Science of the Total Environment. 2022;**832**:8. DOI: 10.1016/j.scitotenv.2022.155029

[39] Hosseini A, Karimi H, Foroughi J, Sabzehmeidani MM, Ghaedi M. Heterogeneous photoelectro-Fenton using ZnO and TiO2 thin film as photocatalyst for photocatalytic degradation malachite green. Applied Surface Science Advances. 2021;**6**: 100126. DOI: 10.1016/j.apsadv.2021.100126

[40] Liu F, Zhang Z, Wang Z, Li X, Dai X, Wang L, et al. Experimental study on treatment of tertiary oil recovery wastewater by electrocoagulation. Chemical Engineering and Processing-Process Intensification. 2019;**144**: 107640. DOI: 10.1016/j.cep.2019.107640

[41] Das PP, Sharma M, Purkait MK. Recent progress on electrocoagulation process for wastewater treatment: A review. Separation and Purification Technology. 2022;**292**:121058. DOI: 10.1016/j.seppur.2022.121058

[42] An C, Huang G, Yao Y, Zhao S. Emerging usage of electrocoagulation technology for oil removal from wastewater: A review. Science of the Total Environment. 2017;**579**:537-556. DOI: 10.1016/j.scitotenv.2016.11.062

[43] AlJaberi FY, Alardhi SM, Ahmed SA, Salman AD, Juzsakova T, Cretescu I, et al. Can electrocoagulation technology be integrated with wastewater treatment systems to improve treatment efficiency? Environmental Research. 2022;**214**:10. DOI: 10.1016/j.envres.2022.113890

[44] Titchou FE, Zazou H, Afanga H, El Gaayda J, Akbour RA, Hamdani M. Removal of persistent organic pollutants (POPs) from water and wastewater by adsorption and electrocoagulation process. Groundwater for Sustainable Development. 2021;**13**:100575. DOI: 10.1016/j.gsd.2021.100575

[45] Asfaha YG, Tekile AK, Zewge F. Hybrid process of electrocoagulation and electrooxidation system for wastewater treatment: A review. Cleaner Engineering and Technology. 2021;**4**: 100261. DOI: 10.1016/j.clet.2021.100261

[46] Akter S, Suhan MBK, Islam MS. Recent advances and perspective of electrocoagulation in the treatment of wastewater: A review. Environmental Nanotechnology, Monitoring & Management. 2022;**17**:100643. DOI: 10.1016/j.enmm.2022.100643

[47] Emerick T, Vieira JL, Silveira MHL, Joaõ JJ. Ultrasound-assisted electrocoagulation process applied to the treatment and reuse of swine slaughterhouse wastewater. Journal of Environmental Chemical Engineering. 2020;**8**:104308. DOI: 10.1016/j.jece.2020.104308

[48] Pérez LS, Rodriguez OM, Reyna S, Sánchez-Salas JL, Lozada JD, Quiroz MA, et al. Oil refinery wastewater treatment using coupled electrocoagulation and fixed film biological processes. Physics and Chemistry of the Earth. 2016;**91**:53-60. DOI: 10.1016/j.pce.2015.10.018

[49] Nidheesh PV, Oladipo AA, Yasri NG, Laiju AR, Cheela VRS, Thiam A, et al. Emerging applications, reactor design and recent advances of electrocoagulation process. Process Safety and Environment Protection. 2022;**166**:600-616. DOI: 10.1016/j.psep.2022.08.051

[50] Tang J, Zhang C, Quan B, Tang Y, Zhang Y, Su C, et al. Electrocoagulation coupled with conductive ceramic membrane filtration for wastewater treatment: Toward membrane modification, characterization, and application. Water Research. 2022;**220**: 118612. DOI: 10.1016/j.watres.2022.118612

[51] Subtil EL, Almeria Ragio R, Lemos HG, Scaratti G, García J, Le-Clech P. Direct membrane filtration (DMF) of municipal wastewater by mixed matrix membranes (MMMs) filled with graphene oxide (GO): Towards a circular sanitation model. Chemical Engineering Journal. 2022;**441**:136004. DOI: 10.1016/j.cej.2022.136004

[52] Jiang D, Gao C, Liu L, Yu T, Li Y, Wang H. Application of nanoporous ceramic membrane derived from Fe/S/Si/Al/O-rich mining solid waste in oil–water separation and heavy metal removal of industrial high concentrated emulsifying wastewater. Separation and Purification Technology. 2022;**295**: 121317. DOI: 10.1016/j.seppur.2022.121317

[53] Nikbakht Fini M, Montesantos N, Maschietti M, Muff J. Performance evaluation of membrane filtration for treatment of H2S scavenging wastewater from offshore oil and gas production. Separation and Purification Technology. 2021;**277**:119641. DOI: 10.1016/j.seppur.2021.119641

[54] Wollborn T, Michaelis M, Ciacchi LC, Fritsching U. Protein conformational changes at the oil/water-interface induced by premix membrane emulsification. Journal of Colloid and Interface Science. 2022;**628**:72-81. DOI: 10.1016/j.jcis.2022.07.132

[55] Ma J, Wu G, Zhang R, Xia W, Nie Y, Kong Y, et al. Emulsified oil removal from steel rolling oily wastewater by using magnetic chitosan-based flocculants: Flocculation performance, mechanism, and the effect of hydrophobic monomer ratio. Separation and Purification Technology. 2023;**304**: 122329. DOI: 10.1016/j.seppur.2022.122329

[56] Nittami T, Tokunaga H, Satoh A, Takeda M, Matsumoto K. Influence of surface hydrophilicity on polytetrafluoroethylene flat sheet membrane fouling in a submerged membrane bioreactor using two activated sludges with different characteristics. Journal of Membrane Science. 2014;**463**:183-189. DOI: 10.1016/j.memsci.2014.03.064

[57] Tong Y, Zuo C, Ding W, Jiang S, Li W, Xing W. Sulfonic nanohydrogelled surface-modified microporous polyvinylidene fluoride membrane with excellent antifouling performance for treating water-oil separation of kitchen wastewater. Journal of Membrane Science. 2021;**628**:119113. DOI: 10.1016/j.memsci.2021.119113

[58] Davoodi S, Al-Shargabi M, Wood DA, Rukavishnikov VS, Minaev KM. Experimental and field

applications of nanotechnology for enhanced oil recovery purposes: A review. Fuel. 2022;**324**:124669. DOI: 10.1016/j.fuel.2022.124669

[59] Wang Y, He Y, Yu J, Li H, Li S, Tian S. A freestanding dual-cross-linked membrane with robust anti-crude oil-fouling performance for highly efficient crude oil-in-water emulsion separation. Colloids and Surfaces A: Physicochemical and Engineering Aspects. 2022;**654**:12. DOI: 10.1016/j.colsurfa.2022.130117

[60] Hashemi F, Hashemi H, Shahbazi M, Dehghani M, Hoseini M, Shafeie A. Reclamation of real oil refinery effluent as makeup water in cooling towers using ultrafiltration, ion exchange and multioxidant disinfectant. Water Resources and Industry. 2020;**23**:100123. DOI: 10.1016/j.wri.2019.100123

[61] Cifuentes-Cabezas M, Vincent-Vela MC, Mendoza-Roca JA, Álvarez-Blanco S. Use of ultrafiltration ceramic membranes as a first step treatment for olive oil washing wastewater. Food and Bioproducts Processing. 2022;**135**:60-73. DOI: 10.1016/j.fbp.2022.07.002

[62] Valizadeh K, Heydarinasab A, Hosseini SS, Bazgir S. Fabrication of modified PVDF membrane in the presence of PVI polymer and evaluation of its performance in the filtration process. Journal of Industrial and Engineering Chemistry. 2022;**106**:411-428. DOI: 10.1016/j.jiec.2021.11.016

[63] Aldana JC, Acero JL, Álvarez PM. Membrane filtration, activated sludge and solar photocatalytic technologies for the effective treatment of table olive processing wastewater. Journal of Environmental Chemical Engineering. 2021;**9**:13. DOI: 10.1016/j.jece.2021.105743

[64] Xiao X, Yu Z, Yang Z, Wang J, Xiang Q, Zhu X. Diamond & Related Materials Application of in-situ microbubble method on SEP @ MnO 2 / RGO composite membrane for efficient and long-acting treatment of oil field wastewater. Diamond and Related Materials. 2022;**130**:109499. DOI: 10.1016/j.diamond.2022.109499

[65] Chen Y, Liu H, Xia M, Cai M, Nie Z, Gao J. Green multifunctional PVA composite hydrogel-membrane for the efficient purification of emulsified oil wastewater containing Pb2+ ions. Science of the Total Environment. 2023;**856**:13. DOI: 10.1016/j.scitotenv.2022.159271

[66] Belibagli P, Isik Z, Özdemir S, Gonca S, Dizge N, Awasthi MK, et al. An integrated process for wet scrubber wastewater treatment using electrooxidation and pressure-driven membrane filtration. Chemosphere. 2022;**308**:14. DOI: 10.1016/j.chemosphere.2022.136216

[67] Li P, Yang C, Sun F, Li XY. Fabrication of conductive ceramic membranes for electrically assisted fouling control during membrane filtration for wastewater treatment. Chemosphere. 2021;**280**:1-10. DOI: 10.1016/j.chemosphere.2021.130794

Chapter 5

Phyco-Remediation of Sewage Wastewater by Microalgae

Radhakrishnan Vandana and Suchitra Rakesh

Abstract

Land and water resources are significant constraints in the present energy scenario. Phyco-remediation is crucial in attaining the UNDP's sixth sustainable development goal. The wastewater treatment by microalgae is highly economical, and the biomass generated can be further utilized for biofuel production. The successful coupling of microalgae with wastewater can overcome the expensive cultivation of microalgae and pollutants with wastewater and scale-up production of high-value products. A microalgae-based wastewater treatment process reduces BOD, inhibits coliforms, removes nutrients and contaminants, and removes heavy metals. In wastewater, nutrients are abundant, making it an ideal medium for growing microalgae. Microalgal biomass can produce a wide range of high-value products, such as biomethane, compost, biofuels, and animal feed.

Keywords: microalgae, biomass, sewage, wastewater, high-value products

1. Introduction

Water plays a vital role in all aspects of life, and the demand for wastewater treatment is currently a worldwide priority. Traditional sewage treatment consists of pre-treatment, physical, and chemical treatment of sewage water with high overhead expenses. The exploitation of natural resources worldwide to meet the energy demand has created a vast concern over environmental issues. Hence, wastewater treatment coupled with energy production and other valuable product generation. This method produces biomass that can be used further to produce valuable products and has a very low operational cost [1]. Recently, the use of microalgae for wastewater treatment has attracted much attention worldwide owing to its multifaceted benefits. Integrated biorefinery is a promising approach to microalgae cultivation for wastewater treatment with simultaneous production of high-value products and biofuel production [2, 3].

Furthermore, phycoremediation utilizes microalgae's ability to bio-sequester carbon dioxide, their high growth rates, high biomass production, high lipid productivities, and their ability to remove contaminants from wastewater and produce biofuels. Aside from bio-manure and biodiesel, algal biomass can also be used to produce bioethanol, hydrogen, and other valuable products [4]. Investigation into different microalgal species has established that they could bring down more than 98% of COD and BOD of sewage water. The impact of phycoremediation in treating sewage wastewater reduces greenhouse gas and sludge formation cost-effectively and energy-affluently.

Many studies have coded the usage of microalgae in various wastewater treatments like agricultural, municipal, dairy, piggery, and poultry wastewaters and industrial effluents [5]. Recently, some studies have reported on treating sewage wastewater and microalgae biomass generation for biofuel production [6]. *Chlorella vulgaris* is the most desired organism for simultaneous wastewater treatment and bioenergy production [7]. Pooja et al. [8] used *C. vulgaris* to simultaneously remove nutrients and pollutants from wastewater and recover biomass for biofertilizer application. The above study successfully proved the conversion of sewage to chemical fertilizer. Kumar et al. [9] used *C. vulgaris* for wastewater treatment and industrial flue gases for biomass production.

2. Conventional sewage treatment process

The main aim of sewage wastewater treatment is to remove the BOD, suspended solids, nutrients, pathogenic microbes, and toxicity. The typical sewage treatment process involves four to five steps, viz., preliminary, primary, secondary, tertiary, and disinfection. Preliminary sewage wastewater treatment removes large solid particles like rags, wood, heavy grit particles, and fecal matter. A well-designed sedimentation tank removes almost 70% of the settleable solids and 40% of BOD during the primary treatment of sewage [10]. A mixed population of heterotrophic bacteria further reduces the BOD in secondary sewage treatment. These bacteria facilitate the biological oxidation of BOD and can further remove almost 90% of pathogenic bacteria from sewage [11]. Tertiary sewage treatment removes all the organic ions, viz., ammonium, nitrate, and phosphate, either biologically or chemically [12].

In contrast, quaternary treatment aims to remove heavy metals, organic compounds, and soluble minerals [13]. Following tertiary treatment, disinfection kills all pathogenic microbes in the effluent. Disinfection can be achieved using a variety of physical and chemical methods. Ozone and UV light are the most preferred physical disinfection methods, while chlorine has been used extensively for disinfection [14].

3. Microalgae for sewage wastewater treatment

The use of microalgae for sewage wastewater treatment is cost-efficient, renewable source for biomass production and helps in carbon sequestration [15]. The microalgae can utilize the organic and inorganic carbon and inorganic nitrogen and phosphorus present in wastewater for their growth. Photosynthesis in microalgae helps heterotrophic bacteria degrade carbonaceous materials in wastewater treatment. Many studies have reported microalgae biomass production using nutrients removed from wastewater [16]. In addition, microalgae are efficient for carbon dioxide capture and nutrient removal from wastewater and are reported as a potential candidate for future energy production [17]. The microalgae can directly assimilate ammonia and phosphate from wastewater for their growth and metabolic functions [18]. Furthermore, the microalgae wastewater treatment process emits fewer greenhouse gases, as most of the nitrogen is being assimilated instead of converted to nitrogen oxide [19].

Sl. no.	Microalgae	Type of wastewater	Nutrient removal efficiency in %			References
			Phosphorous	COD	Nitrogen	
1	*Parachlorella kessleri*	Domestic sewage	65	95	70	[26]
2	*Chlorella fusca*	Urban sewage	—	45.48	24.6	[27]
3	*Chlorella pyrenoidosa*	Sewage treatment plant	94.2	87	99.5	[28]
4	*Chlamydomonas reinhardtii*	Swine farm sewage	13–14.5	—	42–83	[29]
5	*Scenedesmus obliquus*	Municipal sewage	47–98	—	79–100	[30]
6	*Tetraselmis indica*	Domestic sewage	60–93	72–94	78.46	[31]
7	*Nannochloropsis sp.*	Tannery effluent	99	84	82	[32]
8	*Chlorella vulgaris*	Industrial sewage	70	—	86	[33]
9	*C. pyrenoidosa*	Synthetic sewage	70.1	61	99.2	[28]
10	*Scenedesmus quadricauda*	Industrial sewage	75.33	—	77.50	[26]
11	*Chlorella sorokiniana* (WB1DG)	Biogas effluent	91.68	63.42	70.66	[34]
12	*C. sorokiniana* (P21)	Biogas effluent	92.11	73.78	67.33	[34]

Table 1.
Nutrient removal efficiency of microalgae from various wastewater.

Microalgae have recently been extensively studied for their ability to treat wastewater effluents. The performance of various microalgal species for wastewater treatment varies with the range of wastewater types [20]. Prandini et al. [21] successfully demonstrated nutrient removal from piggery wastewater by *Scenedesmus obliquus*. Kothari et al. [22] have used *Chlorella pyrenoidosa* for dairy effluent treatment. Many studies have reported *C. vulgaris* as an ideal candidate for municipal wastewater effluent treatment [23]. Other microalgae used for wastewater treatment are *Chlamydomonas sp., Nanochloropsis sp., Dunaliella sp., Botryococcus sp.*, etc. [24]. Microalgae at the secondary treatment phase or primary sewage wastewater to effluent standards are economical and eco-friendly approaches. The nutrient composition of primary sewage waste and secondary treatment effluent is almost the same but has different concentrations [25]. The concentration of nutrients, viz., nitrogen and phosphorus, in primary sewage waste is higher than in secondary treatment effluent (**Table 1**).

4. Mechanism of nutrient removal by microalgae

4.1 Carbon

As a primary carbon source, microalgae use CO_2, whereas in aqueous conditions, CO_2 splits into bicarbonate and carbonate ions depending on pH, temperature, and salinity [35]. Due to the low concentration of CO_2 in the aquatic environment, microalgae use a carbon concentration mechanism to minimize the loss of photosynthetic activity [36]. Microalgae convert inorganic carbon to organic carbon via the Calvin cycle, as it provides metabolic reactions to produce amino acids and lipids.

The carbon dioxide concentration in the wastewater is one of the essential factors that decide the growth of microalgae, i.e., low availability of inorganic carbon in wastewater limits microalgal growth. Hence, to improve microalgae growth, the wastewater is usually supplemented with carbon dioxide or bicarbonate salts [37].

Shen et al. [38] reported that *S. obliquus* at 5% CO_2 concentration removes total nitrogen from the wastewater within 2 days. In contrast, the total nitrogen recovery is less even on the third day under ambient and higher concentrations. Many studies have reported that at elevated CO_2 levels, biomass production and nutrient removal from wastewater via microalgae have improved [39, 40]. The microalgae tolerance to CO_2 is strain specific and has few species acclimatized with CO_2 concentrations up to 100% [41]. Microalgae metabolize organic carbon compounds from wastewater through photo-mixotrophy or strict heterotrophy [42]. Municipal wastewater is highly heterogeneous, with complex carbonaceous materials that limit its availability as an ideal carbon source for microalgae. Since municipal wastewater comprises majorly complex organic carbon compounds, their decomposition by heterotrophic microorganisms must be converted to viable carbon sources for microalgae [43]. It has been reported that supplementing inorganic carbon to wastewater enhances nutrient removal efficiency [42]. However, enriching the wastewater with organic carbon increases production costs.

4.2 Nitrogen

Microalgae can utilize nitrogen from various organic and inorganic sources. Ammonia is preferred among the various nitrogen sources as its assimilation and incorporation are more efficient. Ruiz-Marin et al. [30] reported that microalgae, viz., *S. obliquus* and *C. vulgaris* showed a preference for NH_3 in wastewater compared to other nitrogen sources. Membrane transporter proteins easily assimilate ammonium, and once translocated, the ammonium is directly incorporated into amino acids required for growth and other functions. Whereas transport of NO_3 and NO_2 is an energy-dependent process, they must first be reduced to ammonium via enzymatic reaction, requiring reductant NADH and ferredoxin [44]. In the microalgae wastewater treatment process, nitrification decreases the ammonium, and nitrate production is not desired as microalgae do not eliminate it if ammonium is present. Many studies have reported in steady-state wastewater treatment, almost 80% of the NH_3 is oxidized to NO_3, with a maximum of 40% assimilated by microalgae [45].

In photoautotrophic microalgae, inorganic carbon is fixed by the Calvin cycle and enters the glycolytic pathway as glucose-3-phosphate. Once converted to acetyl CoA, pyruvate is transported to mitochondria and enters the TCA cycle. Acetyl Co-A is further metabolized to CO_2, and ATP, reducing equivalents and carbon skeletons [46].

Organic carbon substrates are transported in the cytosol through the glycolytic or pentose phosphate pathways in heterotrophic mode. Glycerol can be used as an alternative carbon substrate, translocated across the membrane via passive diffusion into the cytosol of microalgae [42, 47].

4.3 Phosphorus

Phosphorus is an important element involved in many metabolic processes as well as structural component of microalgae [18]. In wastewater, inorganic P exists in many ionic forms and is mostly bioavailable than soluble organic P compounds for microalgae. Phosphorus is incorporated into organic compounds by phosphorylation of Adenosine diphosphate (ADP). It is an endergonic reaction that obtains energy either by oxidation of respiratory substrates or by photosynthetic electron transport chain [24]. If a wastewater in enriched with P, microalgae have the capacity to accumulate P beyond their metabolic needs and store as acid-insoluble polyphosphate granules via luxury uptake mechanism [48].

5. Factors affecting the microalgae wastewater treatment

5.1 Bacteria

The use of microalgae in wastewater treatment has been extensively studied. It is impossible to avoid other organisms like bacteria and fungi in wastewater, and wastewater sterilization is not feasible due to the huge volumes to be processed. The common bacterial species dominated in sewage wastewater are from the classes Bacteroidia, Flavobacteria, Betaproteobacteria, and Gamma proteobacteria. In primary sewage wastewater, variations in bacterial community composition are noticed at different inoculation ratios of microalgae to sludge [49].

Bacteria help the microalgae in wastewater treatment by providing CO_2 via heterotrophic metabolism of organic matter and later mineralizing it to inorganic compounds that can be consumed directly by microalgae [50]. In return, microalgae produce oxygen via photosynthesis, which is required for heterotrophic bacterial growth during organic matter degradation [44]. The activated sludge treatment microalgae facilitate nitrification by generating a sufficient quantity of dissolved O_2 [51]. The integration of a bacterial-microalgal approach for wastewater treatment is a promising approach as heterotrophic bacteria degrade the organic matter in the absence of aerated oxygen, as the microalgae provide O_2, and similarly, the need for CO_2 sparging is eliminated as bacterial respiration produces it [52].

5.2 Industrial contaminants

Microalgae can remove most industrial contaminants like heavy metals. Heavy metal contamination in wastewater is primarily due to industrial processing. The use of microalgae for wastewater treatment is termed phycoremediation, where algae uptake the nutrients, accumulate heavy metals, and degrade organic matter via symbiotic interaction with heterotrophic bacteria [53]. Microalgae has the potential to utilize waste as a nutritional source and reduce pollutants via enzymatic and metabolic processes. The microalgal metabolic pathways make them detoxify, transform and volatilize the heavy metal and xenobiotic pollution in wastewater [54]. Biosorption is the most commonly used mechanism by microalgae for either active or passive heavy metal uptake. Hence, biosorption is regarded as a cost-efficient way to eliminate heavy metals from industrial effluent [55]. The active algal biomass has a metal efflux metabolism-driven system for maintaining heavy metal concentration in intracellular space to avoid heavy metal toxicity. In microalgae, the heavy metal ions are distributed in cell vacuoles and organelles. In microalgae, the oxidation number of heavy metals is altered by various enzymatic reactions and makes them into less toxic forms. The microprecipitation of heavy metal removal in the form of phosphates and sulfates by active algal biomass is a practical approach to removing heavy metals from wastewater [56]. The microalgal cell wall has an overall negative charge due to the presence of various functional groups; this makes the algal cell an entire binding site for heavy metal cations and involved in metal exchange via an ion-exchange mechanism [57].

5.3 pH

pH is an important abiotic parameter that decides the efficiency of wastewater treatment. The increased pH of wastewater leads to an adverse effect on bacterial activity. If the assimilation of inorganic carbon by microalgae is increased, the medium leads to an alkaline environment. Under the alkaline situation, the beneficial activity of aerobic and facultative bacteria in wastewater is impaired. Many studies have reported the inactivation of bacterial activity at higher pH [58]. At the pH of 8.5–9.5, wastewater bacterial community like coliforms and other pathogenic microbes has been drastically reduced [59]. Many mechanisms lead to the elimination of bacterial community in wastewater, i.e., conformational changes in bacterial membrane structure, respiratory chain damage, and increased susceptibility to exogenous factors like light, temperature, etc. [60]. Sutherland et al. [61] reported a reduction in nutrient removal efficiency at higher pH from primary sewage wastewater treatment via a microalgae consortium. Martinez et al. [62] reported the disruption of the cell wall of *S. obliquus* while treated with municipal sewage effluent at pH > 11.

5.4 Temperature and light

The indigenous microbial community in the wastewater will also compete for nutrients and microalgae. Hence, to promote microalgal growth, the factors like temperature and light intensity has to be considered [63]. The rate of photosynthesis by microalgae is directly proportional to the optimum light intensity, as, beyond optimum, photoinhibition will take place [64]. The illumination saturation point for microalgae varies between 200 and 400 $\mu E\ m^{-2}\ s^{-1}$ [65]. The illumination period and amount of light intensity to microalgae-bacterial culture affect the nutrient removal efficiency from wastewater. The prolonged dark conditions during wastewater treatment via microalgae-bacteria consortium lower biomass recovery and chlorophyll. Gonzalez-Camejo et al. [66] reported in a bacterial-microalgal consortium for wastewater treatment, the lower light intensity of 40 $\mu E\ m^{-2}\ s^{-1}$ favors the activity of nitrifying bacteria. Whereas, higher light intensities of 85–125 $\mu E\ m^{-2}\ s^{-1}$ favor more microalgal growth over nitrifying bacteria.

The environmental temperature also plays a major role in nutrient removal efficiency from wastewater by microalgae. Ruiz-Martinez et al. [67] evaluated the ammonium removal efficiency of Scenedesmus sp. at different temperatures and found that the removal rate increases from 15 to 34°C. Similarly, Sforza et al. [68] assessed at a lower temperature of 15°C *Chlorella protothecoides* remove more $NH4^{+}$ N from effluent. The optimum temperature range of microalgae for wastewater treatment is between 10 and 30°C [65]. Usually, under normal conditions, higher temperature leads to a high growth rate and increased nutrient uptake by microalgae due to higher metabolic activity; these conditions are not always desired with wastewater treatment. Cultivating microalgae in wastewater at a lower temperature may also require less light intensity to minimize light saturation and photo-inhibition.

6. Future perspective and conclusion

In developing countries, almost 80% of the wastewater is discharged untreated to land or waterbodies, which may lead to serious health risks and environmental issues like eutrophication. Due to rapid industrialization and the global population, the requirement for fresh water and wastewater discharge is increasing daily. Conventional sewage treatment processes are not much desired due to their inability to reduce the nutrient concentration to acceptable levels in wastewater.

Microalgae-bacteria-based consortium to recover nutrients from wastewater is an alternative to conventional processes as both organisms establish a symbiotic relationship. Bacteria utilize the organic matter from sewage wastewater and produce CO_2 as a by-product. The microalgae, in turn, utilize the bacteria discharges CO_2 and produce carbohydrates and O_2, required for biomass production, and the latter serves as a terminal electron acceptor for bacterial respiration.

Integrating microalgae-based biorefinery is a promising approach with dual benefits, i.e., wastewater treatment and high-value biomass generation for biofuel production to make the wastewater treatment sustainable, eco-friendly, and economically viable. The microalgae in sewage wastewater treatment have a tri purpose of bioenergy production from generated microalgal biomass, phycoremediation, and organic farming with much less damage to soils and human health than chemical fertilizers.

Acknowledgements

The corresponding author would like to acknowledge the IntechOpen for providing the opportunity and accepting the chapter for the same.

Author details

Radhakrishnan Vandana and Suchitra Rakesh*
Biofuel Research Laboratory, Department of Microbiology, School of Life Sciences, Central University of Tamil Nadu, Thiruvarur, India

*Address all correspondence to: suchitrar@cutn.ac.in

References

[1] Ray B, Rakesh S. Phycoremediation of aquaculture wastewater and algal lipid extraction for fuel conversion. Highlights in BioScience. 2022;**5**:1

[2] Jothibasu K, Muniraj I, Jayakumar T, Ray B, Dhar DW, Karthikeyan S, et al. Impact of microalgal cell wall biology on downstream processing and nutrient removal for fuels and value-added products. Biochemical Engineering Journal. 2022;**187**:108642

[3] Ray B, Rakesh S. Advances on harvesting and extraction systems in microalgal biorefinery. In: Micro-algae: Next-generation Feedstock for Biorefineries. 2022. p. 239

[4] Khan SA, Hussain MZ, Prasad S, Banerjee UC. Prospects of biodiesel production from microalgae in India. Renewable and Sustainable Energy Reviews. 2009;**13**(9):2361-2372

[5] Plöhn M, Spain O, Sirin S, Silva M, Escudero-Oñate C, Ferrando-Climent L, et al. Wastewater treatment by microalgae. Physiologia Plantarum. 2021;**173**(2):568-578

[6] Ho L, Goethals PL. Municipal wastewater treatment with pond technology: Historical review and future outlook. Ecological Engineering. 2020;**148**:105791

[7] Mahdy A, Mendez L, Ballesteros M, González-Fernández C. Algaculture integration in conventional wastewater treatment plants: Anaerobic digestion comparison of primary and secondary sludge with microalgae biomass. Bioresource Technology. 2015;**184**:236-244

[8] Pooja K, Priyanka V, Rao BCS, Raghavender V. Cost-effective treatment of sewage wastewater using microalgae Chlorella vulgaris and its application as bio-fertilizer. Energy Nexus. 2022;**7**:100122

[9] Kumar PK, Krishna SV, Verma K, Pooja K, Bhagawan D, Himabindu V. Phycoremediation of sewage wastewater and industrial flue gases for biomass generation from microalgae. South African Journal of Chemical Engineering. 2018;**25**:133-146

[10] Horan NJ. Biological Wastewater Treatment Systems: Theory and Operation. Chichester: John Wiley & Sons; 1990

[11] Gray NF. Biology of Wastewater Treatment. Vol. 4. World Scientific. Covent Garden, London: Imperial College Press; 2004

[12] de la Noüe J, Laliberté G, Proulx D. Algae and waste water. Journal of Applied Phycology. 1992;**4**(3):247-254

[13] Oswald WJ. Microalgae and wastewater treatment. In: Borowitzka MA, Borowitzka LJ, editors. Microalgal Biotechnology. New York: Cambridge University Press; 1988b. pp. 357-394

[14] Abdel-Raouf N, Al-Homaidan AA, Ibraheem I. Microalgae and wastewater treatment. Saudi Journal of Biological Sciences. 2012;**19**(3):257-275

[15] Almomani F, Judd S, Bhosale RR, Shurair M, Aljaml K, Khraisheh M. Intergraded wastewater treatment and carbon bio-fixation from flue gases using Spirulina platensis and mixed algal culture. Process Safety and Environmental Protection. 2019;**124**:240-250

[16] Chawla P, Malik A, Sreekrishnan TR, Dalvi V, Gola D. Selection of optimum combination via comprehensive

comparison of multiple algal cultures for treatment of diverse wastewaters. Environmental Technology & Innovation. 2020;**18**:100758

[17] Arun S, Sinharoy A, Pakshirajan K, Lens PN. Algae based microbial fuel cells for wastewater treatment and recovery of value-added products. Renewable and Sustainable Energy Reviews. 2020;**132**:110041

[18] Borowitzka MA. Systematics, taxonomy and species names: Do they matter? In: The Physiology of Microalgae. Cham: Springer; 2016. pp. 655-681

[19] Alcántara C, Domínguez JM, García D, Blanco S, Pérez R, García-Encina PA, et al. Evaluation of wastewater treatment in a novel anoxic–aerobic algal–bacterial photobioreactor with biomass recycling through carbon and nitrogen mass balances. Bioresource Technology. 2015;**191**:173-186

[20] Chiu SY, Kao CY, Chen TY, Chang YB, Kuo CM, Lin CS. Cultivation of microalgal *Chlorella* for biomass and lipid production using wastewater as nutrient resource. Bioresource Technology. 2015;**184**:179-189

[21] Prandini JM, Da Silva MLB, Mezzari MP, Pirolli M, Michelon W, Soares HM. Enhancement of nutrient removal from swine wastewater digestate coupled to biogas purification by microalgae *Scenedesmus* spp. Bioresource Technology. 2016;**202**:67-75

[22] Kothari R, Pathak VV, Kumar V, Singh DP. Experimental study for growth potential of unicellular alga *Chlorella pyrenoidosa* on dairy waste water: An integrated approach for treatment and biofuel production. Bioresource Technology. 2012;**116**:466-470

[23] Gouveia L, Graça S, Sousa C, Ambrosano L, Ribeiro B, Botrel EP, et al. Microalgae biomass production using wastewater: Treatment and costs: Scale-up considerations. Algal Research. 2016;**16**:167-176

[24] Gonçalves AL, Pires JC, Simões M. A review on the use of microalgal consortia for wastewater treatment. Algal Research. 2017;**24**:403-415

[25] Mehrabadi A, Farid MM, Craggs R. Potential of five different isolated colonial algal species for wastewater treatment and biomass energy production. Algal Research. 2017;**21**:1-8

[26] Ji L et al. A comparative study of the growth and nutrient removal effects of five green microalgae in simulated domestic sewage. Water (Switzerland). 2021;**13**(24):3

[27] Arrojo MÁ et al. Potential of the microalgae *Chlorella fusca* (Trebouxiophyceae, Chlorophyta) for biomass production and urban wastewater phycoremediation. AMB Express. 2022;**12**(1):3

[28] Kumari P et al. Phycoremediation of wastewater by *Chlorella pyrenoidosa* and utilization of its biomass for biogas production. Journal of Environmental Chemical Engineering. Feb. 2021;**9**(1):104974

[29] Kong QX, Li L, Martinez B, Chen P, Ruan R. Culture of microalgae Chlamydomonas reinhardtii in wastewater for biomass feedstock production. Applied Biochemistry and Biotechnology. 2010;**160**(1):9-18

[30] Ruiz-Marin A, Mendoza-Espinosa LG, Stephenson T. Growth and nutrient removal in free and immobilized green algae in batch and semi-continuous cultures treating real wastewater. Bioresource Technology. 2010;**101**:58-64

[31] Amit CR, Ghosh UK, Nayak JK. Phycoremediation Potential of Marine Microalga *Tetraselmis indica* on Secondary Treated Domestic Sewage for Nutrient Removal and Biodiesel Production. Elsevier; 2019. pp. 405-436

[32] Saranya D, Shanthakumar S. An Integrated Approach for Tannery Effluent Treatment with Ozonation and Phycoremediation: A Feasibility Study. 2020

[33] Pandey BC, Sangita G. Effectiveness of Algae in Wastewater Treatment. No. August, 2022

[34] Padri M. Application of two indigenous strains of microalgal *Chlorella sorokiniana* in cassava biogas effluent focusing on growth rate, removal kinetics, and harvestability. Water (Switzerland). 2021;**13**(17):3

[35] Hill PS, Tripati AK, Schauble EA. Theoretical constraints on the effects of pH, salinity, and temperature on clumped isotope signatures of dissolved inorganic carbon species and precipitating carbonate minerals. Geochimica et Cosmochimica Acta. 2014;**125**:610-652

[36] Raven JA, Cockell CS, De La Rocha CL. The evolution of inorganic carbon concentrating mechanisms in photosynthesis. Philosophical Transactions of the Royal Society B: Biological Sciences. 2008;**363**(1504):2641-2650

[37] Kesaano M. Characterization and Performance of Algal Biofilms for Wastewater Treatment and Industrial Applications. Utah State University; 2015. Available from: https://digitalcommons.usu.edu/etd/4445

[38] Shen QH, Jiang JW, Chen LP, Cheng LH, Xu XH, Chen HL. Effect of carbon source on biomass growth and nutrients removal of Scenedesmus obliquus for wastewater advanced treatment and lipid production. Bioresource Technology. 2015;**190**:257-263

[39] Yao L, Shi J, Miao X. Mixed wastewater coupled with CO_2 for microalgae culturing and nutrient removal. PLoS One. 2015;**10**(9):e0139117

[40] Qi F, Xu Y, Yu Y, Liang X, Zhang L, Zhao H, et al. Enhancing growth of *Chlamydomonas reinhardtii* and nutrient removal in diluted primary piggery wastewater by elevated CO2 supply. Water Science and Technology. 2017;**75**(10):2281-2290

[41] Zhao B, Su Y. Process effect of microalgal-carbon dioxide fixation and biomass production: A review. Renewable and Sustainable Energy Reviews. 2014;**31**:121-132

[42] Perez-Garcia O, Escalante F, De-Bashan L, Bashan Y. Heterotrophic cultures of microalgae: Metabolism and potential products. Water Research. 2011b;**45**:11-36

[43] Lowrey J, Brooks MS, McGinn PJ. Heterotrophic and mixotrophic cultivation of microalgae for biodiesel production in agricultural wastewaters and associated challenges—A critical review. Journal of Applied Phycology. 2015;**27**(4):1485-1498

[44] Falkowski PG, Raven JA. Aquatic Photosynthesis. Princeton University Press; 2013

[45] Vargas G, Donoso-Bravo A, Vergara C, Ruiz-Filippi G. Assessment of microalgae and nitrifiers activity in a consortium in a continuous operation and the effect of oxygen depletion. Electronic Journal of Biotechnology. 2016;**23**:63-68

[46] Voet D, Voet J. Biochemistry. 4th ed. New Jersey, USA: John Wiley & Sons; 2011

[47] Yang C, Hua Q, Shimizu K. Energetics and carbon metabolism during growth of microalgal cells under photoautotrophic, mixotrophic and cyclic light-autotrophic/dark-heterotrophic conditions. Biochemical Engineering Journal. 2000;**6**(2):87-102

[48] Eixler S, Karsten U, Selig U. Phosphorus storage in *Chlorella vulgaris* (Trebouxiophyceae, Chlorophyta) cells and its dependence on phosphate supply. Phycologia. 2006;**45**(1):53-60

[49] Su Y, Mennerich A, Urban B. Synergistic cooperation between wastewater-born algae and activated sludge for wastewater treatment: Influence of algae and sludge inoculation ratios. Bioresource Technology. 2012;**105**:67-73

[50] de Godos I, Vargas VA, Blanco S, González MCG, Soto R, García-Encina PA, et al. A comparative evaluation of microalgae for the degradation of piggery wastewater under photosynthetic oxygenation. Bioresource Technology. 2010;**101**(14):5150-5158

[51] Wang M, Yang H, Ergas SJ, van der Steen P. A novel shortcut nitrogen removal process using an algal-bacterial consortium in a photo-sequencing batch reactor (PSBR). Water Research. 2015;**87**:38-48

[52] Mathew MM, Khatana K, Vats V, Dhanker R, Kumar R, Dahms HU, et al. Biological approaches integrating algae and bacteria for the degradation of wastewater contaminants—A review. Frontiers in Microbiology. 2021;**12**:5

[53] Majumder S, Gupta S, Raghuvanshi S, Gupta S, Majumder S. Removal of dissolved metals by bioremediation. In: Heavy Metals in Water: Presence, Removal and Safety. 2014. pp. 44-56

[54] Gautam RK, Sharma SK, Mahiya S, Chattopadhyaya MC. Contamination of Heavy Metals in Aquatic Media: Transport, Toxicity and Technologies for Remediation. 2014

[55] Ahmad S, Pandey A, Pathak VV, Tyagi VV, Kothari R. Phycoremediation: Algae as eco-friendly tools for the removal of heavy metals from wastewaters. In: Bioremediation of Industrial Waste for Environmental Safety. 2020. pp. 53-76

[56] Ungureanu G, Santos S, Boaventura R, Botelho C. Biosorption of antimony by brown algae *S. muticum* and *A. nodosum*. Environmental Engineering & Management Journal (EEMJ). 2015;**14**(2):6

[57] Monteiro CM, Castro PM, Malcata FX. Capacity of simultaneous removal of zinc and cadmium from contaminated media, by two microalgae isolated from a polluted site. Environmental Chemistry Letters. 2011;**9**(4):511-517

[58] Ansa EDO, Lubberding HJ, Gijzen HJ. The effect of algal biomass on the removal of faecal coliform from domestic wastewater. Applied Water Science. 2012;**2**(2):87-94

[59] Awuah E. Pathogen Removal Mechanisms in Macrophyte and Algal Waste Stabilization Ponds. Wageningen University and Research. Leiden, The Netherlands: Taylor & Francis/Balkema; 2006

[60] Bosshard F, Bucheli M, Meur Y, Egli T. The respiratory chain is the cell's Achilles' heel during UVA inactivation in *Escherichia coli*. Microbiology. 2010;**156**(7):2006-2015

[61] Sutherland DL, Howard-Williams C, Turnbull MH, Broady PA, Craggs RJ.

The effects of CO_2 addition along a pH gradient on wastewater microalgal photo-physiology, biomass production and nutrient removal. Water Research. 2015;**70**:9-26

[62] Martınez ME, Sánchez S, Jimenez JM, El Yousfi F, Munoz L. Nitrogen and phosphorus removal from urban wastewater by the microalga Scenedesmus obliquus. Bioresource Technology. 2000;**73**(3):263-272

[63] Lee J, Cho DH, Ramanan R, Kim BH, Oh HM, Kim HS. Microalgae-associated bacteria play a key role in the flocculation of *Chlorella vulgaris*. Bioresource Technology. 2013;**131**:195-201

[64] Williams PJLB, Laurens LM. Microalgae as biodiesel & biomass feedstocks: Review & analysis of the biochemistry, energetics & economics. Energy & Environmental Science. 2010;**3**(5):554-590

[65] Singh SP, Singh P. Effect of temperature and light on the growth of algae species: A review. Renewable and Sustainable Energy Reviews. 2015;**50**:431-444

[66] Gonzalez-Camejo J, Barat R, Pachés M, Murgui M, Seco A, Ferrer J. Wastewater nutrient removal in a mixed microalga–bacteria culture: Effect of light and temperature on the microalgae–bacteria competition. Environmental Technology. 2017;**39**(4):503-515

[67] Ruiz-Martínez A, Serralta J, Seco A, Ferrer J. Effect of temperature on ammonium removal in Scenedesmus sp. Bioresource Technology. 2015;**191**:346-349

[68] Sforza E, Ramos-Tercero EA, Gris B, Bettin F, Milani A, Bertucco A. Integration of *Chlorella protothecoides* production in wastewater treatment plant: From lab measurements to process design. Algal Research. 2014;**6**:223-233

Chapter 6

Predictive Control, a Strategy for Dissolved Oxygen Control in a Wastewater Treatment Plant

Jose A. Muñoz Hernandez, Luis Eduardo Leguizamon and Helmer Muñoz Hernandez

Abstract

This chapter presents a strategy for controlling the concentration of dissolved oxygen (DO) in the bioreactor of a pilot wastewater plant (WWTP). The control strategy being developed is model-based predictive control (MPC). To apply the control algorithm, the estimation of the oxygen transfer function (KLa) is first performed, then the model and linearization technique are determined and finally the MBPC controller is designed. The results are simulated in MATLAB® and executed in the plant control and supervision system Supervisory Control and Data Acquisition (SCADA) in LabVIEWTM. This chapter is organized as follows: Section 1 presents a brief introduction, and Section 2 determines and describes the model to be used and its respective linearization, as well as the results obtained for the KLa parameter. Finally, Section 3 describes the design methodology of the generalized predictive control (GPC) controller proposed by Clarke, using the Model Predictive Control Toolbox and the EPSAC strategy developed by De Keyser. It should be noted that the simulations in each of the sections were performed in MATLAB® and executed in the control and supervision system with the MATLAB® script interface in LabVIEWTM.

Keywords: predictive control, oxygen control, wastewater treatment plant, modeling, linearization

1. Introduction

The WWTP treats water from a pipeline of a sector in a community, and the activated sludge used comes from the wastewater plant (WWTP) of a nearby residential complex. Its main objective is to serve as a didactic and experimental means for learning, simulation, and research in the area of control of this biological process. Oxygen transfer is an important factor for aerobic biological water purification processes; hence, the need to obtain a good estimation of this parameter. Consequently, this chapter discusses a strategy for the control of dissolved oxygen (DO). The proposed control strategy is the Model-Based Predictive Control (MPC). In this strategy, a model of the process is used to predict how the system will behave under a proposed sequence of control actions u(t). Traditionally, a linear model is used where an optimal control action can be calculated; however, when nonlinear models are used in the algorithm, a numerical search algorithm is used to calculate an optimal sequence uopt(t).

2. Process and modeling of wastewater systems

In this part, the generalities of the process are described, and some techniques for obtaining important parameters in the modeling process of a wastewater treatment plant are proposed and a linear version of the ASM1 model is proposed.

2.1 General concepts

In the application of predictive control techniques based on models, it is necessary to have quality models; otherwise, you will have erroneous predictions and the control system will not be good. The predictions are carried out using a model of the process, with which the specification of a reference trajectory is obtained. De Keyser proposed some process models that can be used in the implementation of predictive control strategies (EPSAC Model) [1]. The output of the model can be seen in **Figure 1**.

Figure 1 shows schematically the representation of the system taking into account the effect of noise n (colored noise), where the response to an input u can be obtained with the Eq. (1):

$$y(t) = x(t) + n(t) \tag{1}$$

The disturbance is represented by n(t) and corresponds to colored noise, a random signal that removes the offset. x is the result of the process model, while y corresponds to the measured value. The colored noise signal n(t) can be represented by Eq. (2):

$$n(t) = \frac{C(q^{-1})}{D(q^{-1})} e(t) \tag{2}$$

With, e(t): White noise (uncorrelated noise with zero mean value).

The transfer function (TF) between e(t) and n(t) describes the disturbance class and corresponds to the noise filter and is given by Eq. (3):

$$\frac{C(q^{-1})}{D(q^{-1})} = \frac{(1 + cq^{-1})}{(1 + dq^{-1})(1 - q^{-1})} \tag{3}$$

The output of model x can be written as Eq. (4):

$$x(t) = f[x(t-1), x(t-2), \ldots, u(t-1), u(t-2), \ldots] \tag{4}$$

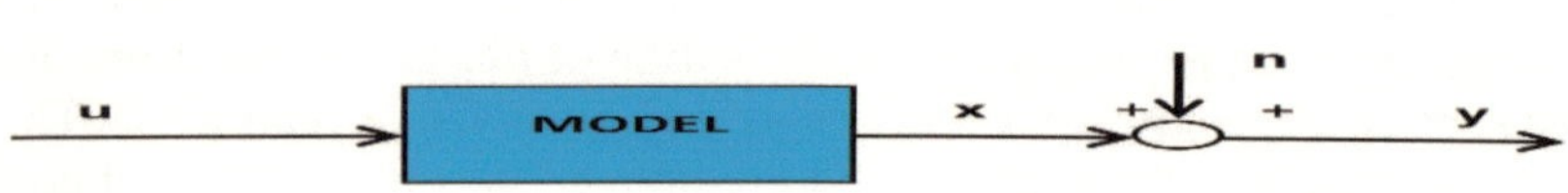

Figure 1.
Process model representation.

A special kind of EPSAC Model is the CARIMA Model [2]. This model links the process model and the disturbance model, as shown in Eq. (5):

$$\begin{aligned} A\left(q^{-1}\right)y(t) &= B\left(q^{-1}\right)u(t) + \frac{C(q^{-1})}{1-q^{-1}} \\ y(t) &= \frac{B(q^{-1})}{A(q^{-1})}u(t) + \frac{C(q^{-1})}{A(q^{-1})(1-q^{-1})} \\ y(t) &= x(t)u(t) + n(t)e(t) \end{aligned} \tag{5}$$

x(t) represents a pulse transfer function and B(q-1) and A(q-1) are given by Eq. (6):

$$\begin{aligned} B\left(q^{-1}\right) &= b_1 q^{-1} + \ \dots \ + b_{nb} q^{-nb} \\ A\left(q^{-1}\right) &= 1 + a_1 q^{-1} + \ \dots \ + b_{na} q^{-na} \end{aligned} \tag{6}$$

The model can also be obtained using neural networks or Fuzzy modeling techniques. The model obtained for the wastewater treatment plant is presented below.

2.2 Oxygen transfer function estimation, KLa

In a wastewater treatment system, oxygen must be available at a rate equivalent to the oxygen demand load exerted by the wastewater entering the plant. The process consists of bringing the wastewater in contact with oxygen, transferring it across the gas-to-liquid interface to dissolve it in the liquid, and then transferring the dissolved oxygen through the liquid to the microorganisms. The dynamics of DO concentration change can be represented by Eq. (7):

$$\frac{dSo}{dt} = K_L a\left(So_{sat} - So\right) \tag{7}$$

where So is the dissolved oxygen concentration DO, So_{sat} is the oxygen saturation concentration, K_L is the oxygen transfer coefficient, and a is the total interfacial contact area per unit volume of liquid. Since it is admittedly impossible to measure the interfacial area a, the total term KLa is estimated [3]. To obtain the KLa parameter, the integration, differentiation, and with oxygen consumption methods were initially used in transient regime [4], assuming negligible biomass concentrations, proceeding first to deoxygenate the wastewater, bringing the DO to a value close to zero. Then aeration is restarted, measuring the increase in DO concentration over time using the DO sensor. In **Figure 2**, it can be seen how the DO concentration increases at different airflow rates until reaching a steady state.

The determination of KLa with the integration method proposes to separate variables and integrate Eq. (7). Assuming that KLa does not depend on the sampling time, it is obtained Eq. (8):

$$\ln\left(So_{sat} - So\right) = -K_L at \tag{8}$$

With this equation, a straight line can be obtained from a semi-logarithmic graph (Sosat - So) as a function of time, where KLa is its slope, as shown in Eq. (9):

$$K_L a = -\frac{\ln\left(C_f/C_i\right)^*}{t_f - t_i} 60 \tag{9}$$

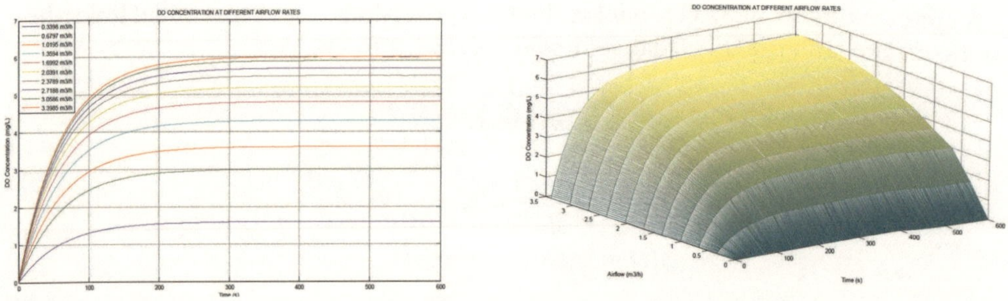

Figure 2.
DO concentration. Left: Function on time. Right: Function on time and airflow.

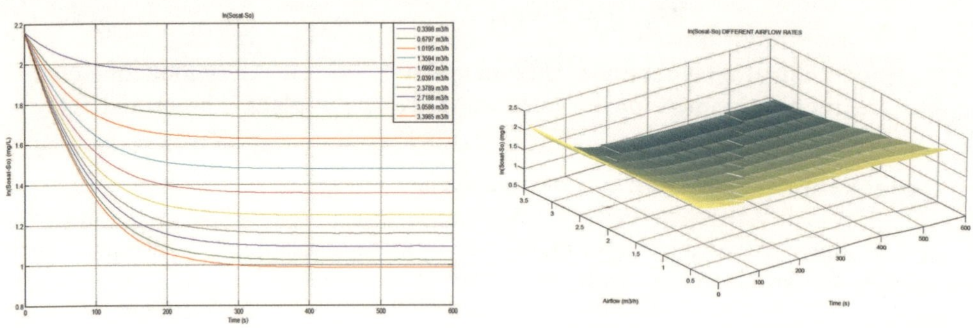

Figure 3.
Slope (-KLa). Left: Function on time. Right: Function on time and airflow.

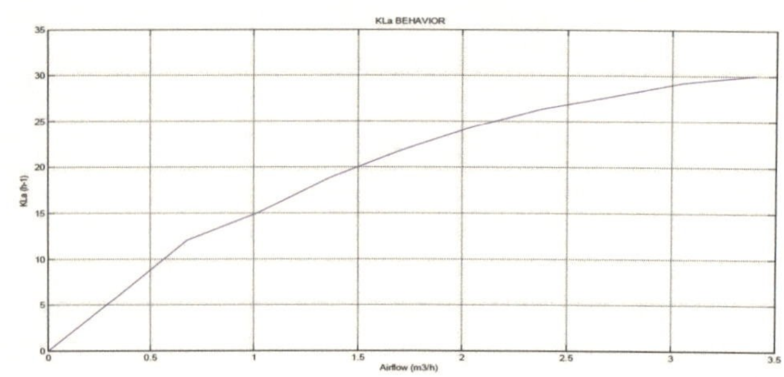

Figure 4.
KLa integration method.

where ti and tf are the initial and final time selected for the slope, Ci = (Sosat - So) in ti and Cf = (Sosat - So) in tf. It is multiplied by 60 to convert from minutes to hours, when the samples have been taken in minutes. The results obtained can be seen in **Figure 3**. Here, it is observed that the greater the airflow, the greater the slope (KLa).

With these results, **Figure 4** shows the curve obtained from KLa, whose behavior resembles a first-order system.

The differentiation method is based on the difference of (Sof - So), where Sof is the last sampled OD data. KLa is obtained from Eq. 10:

$$K_L a = \frac{d\left(\ln\left(So_f - So\right)\right)^*}{dt} 60 \tag{10}$$

The results are shown in **Figure 5**, where it is observed that the slope for the different levels of airflow can be obtained in the first 300 seconds, during which period the slope remains constant.

In a similar way, the KLa curve is obtained, as a function of the airflow, observing a behavior similar to that of a first-order system, as shown in **Figure 6**.

In activated sludge processes, when the wastewater under treatment has a significant concentration of microorganisms, it is necessary to take into account the oxygen consumption (respiration) [4], since the mass balance presented in Eq. (7) is affected as show in Eq. (11):

$$\frac{dSo}{dt} = K_L a(So_{sat} - So) - R \tag{11}$$

where R is the rate of oxygen utilization (respiration rate). Then Eq. (11) can be written as Eq. (12):

$$\frac{dSo}{dt} = (K_L a So_{sat} - R) - K_L a So \tag{12}$$

Eq. (12) indicates that the derivative term dSo/dt as a function of So provides a straight line whose slope is equal to KLa and its cutoff point with the ordinate is (KLa. Sosat - R), value from which R can also be calculated. **Figure 7** shows the derivative

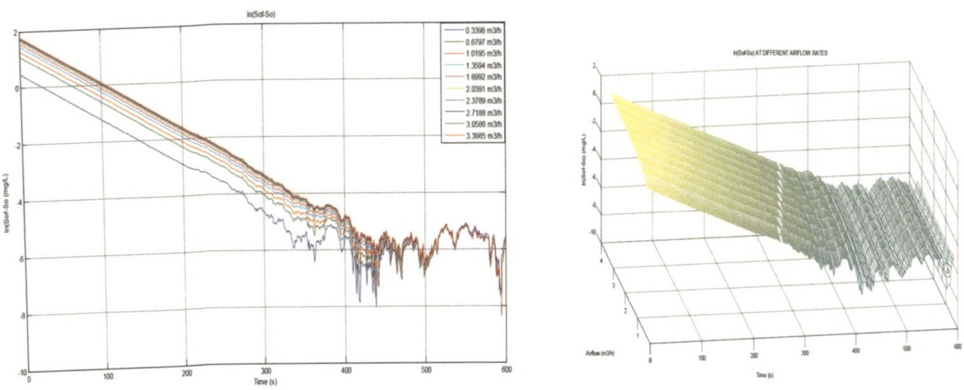

Figure 5.
Slope ln(Sof - So). Left: Function on time. Right: Function on time and airflow.

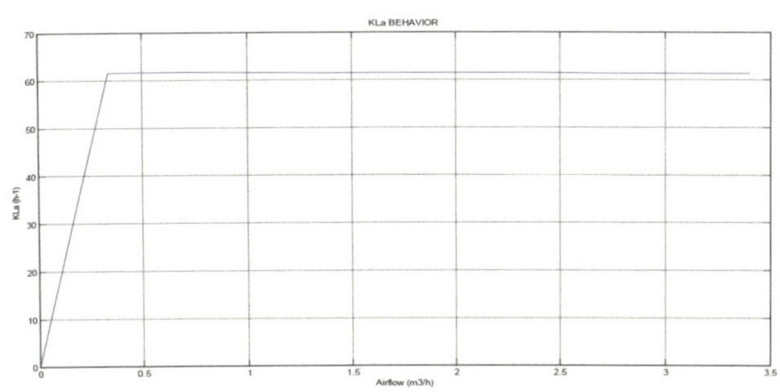

Figure 6.
KLa differentiation method.

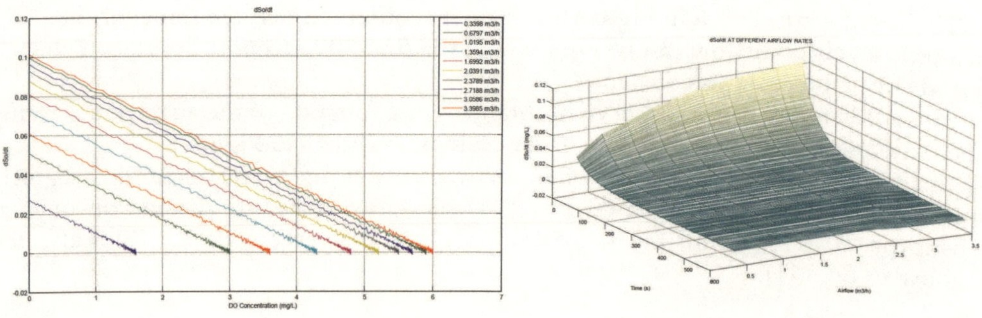

Figure 7.
Slope (dSo/dt). Left: Function on time. Right: Function on time and airflow.

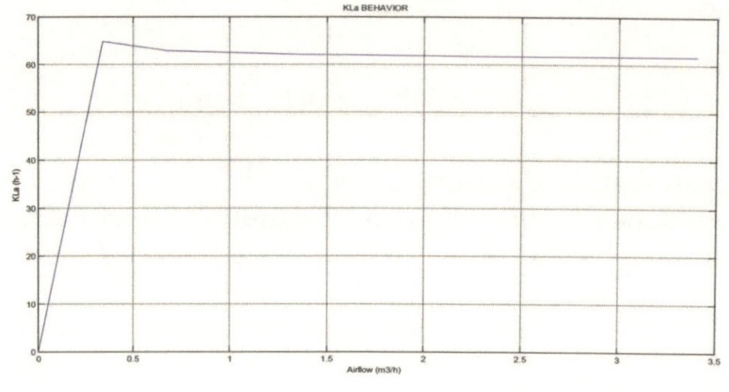

Figure 8.
KLa oxygen consumption method.

dSo/dt with respect to So, the value of KLa (slope), and the value of R (cutoff point on the ordinate).

Figure 8 shows that the behavior of KLa as a function of the airflow obtained by this method also follows a dynamic similar to that of a first-order system.

The oxygen transfer function KLa describes the rate at which oxygen is transferred to the activated sludge by the aeration system. This function is nonlinear and depends on several factors, the main one being the airflow rate. Eq. (7) can be written as Eq. (13):

$$\frac{dSo}{dt} = K_L a\big(q_A(t)\big)\left(So_{sat} - So\right) \tag{13}$$

where qA(t) is the airflow rate entering the bioreactor. Here, it is assumed that KLa depends on the nonlinearity of the airflow rate. A typical function of KLa is shown in: **Figure 4**, **Figure 6**, and **Figure 8**, where it is observed that the slope of KLa changes when the airflow rate changes [3]. Based on the behavior of KLa observed in these figures, it can be assumed that it corresponds to a first-order system that can be represented by the differential equation, Eq. (14):

$$b\frac{dK_L a\big(q_A(t)\big)}{d\big(q_A(t)\big)} + K_L a\big(q_A(t)\big) = au\big(q_A(t)\big) \tag{14}$$

where b is the time constant of the system, a is the gain of the system, and u(qA (t)) is the input of the system. Applying the Laplace transform to Eq. (14), the transfer function of the system is modeled with Eq. (15):

$$\frac{K_L a(s)}{u(s)} = \frac{a}{bs + 1} \tag{15}$$

The step response of the system is obtained with of the inverse Laplace transform, as shown in Eq. (16):

$$K_L a(q_A(t)) = a\left(1 - e^{\frac{-q_A(t)}{b}}\right) \tag{16}$$

The values of the parameters a and b are determined by identification using the Smith model [5]. In this model, it is proposed that the steady state gain a can be easily obtained from the graph of the step response of the system, while the system constant b corresponds approximately to the time in which 63.2% of the final steady state response is reached [6].

The gains (a) at steady state are determined to be 30.05, 61.38, and 61.66 for the integration, differentiation, and oxygen consumption methods, respectively. On the other hand, the constants of qA(t) are 1.386, 0.214, and 0.2042. These constants are calculated by linear interpolation or by a cubic spline. The identified models for integration, differentiation, and with oxygen consumption correspond to the following transfer functions, as shown in Eq. (17), Eq. (18), and Eq. (19):

$$\frac{K_L a(s)}{u(s)} = \frac{30.05}{1.386s + 1} \tag{17}$$

$$\frac{K_L a(s)}{u(s)} = \frac{61.38}{0.214s + 1} \tag{18}$$

$$\frac{K_L a(s)}{u(s)} = \frac{61.66}{0.2042s + 1} \tag{19}$$

The verification of the models is performed by comparing the process signal (red) with the model signal (blue), as shown in **Figure 9**, in which it can be seen that there is a good tracking and consequently the error is low. The value of KLa is found by applying Eq. (16), for a given value of airflow qA(t).

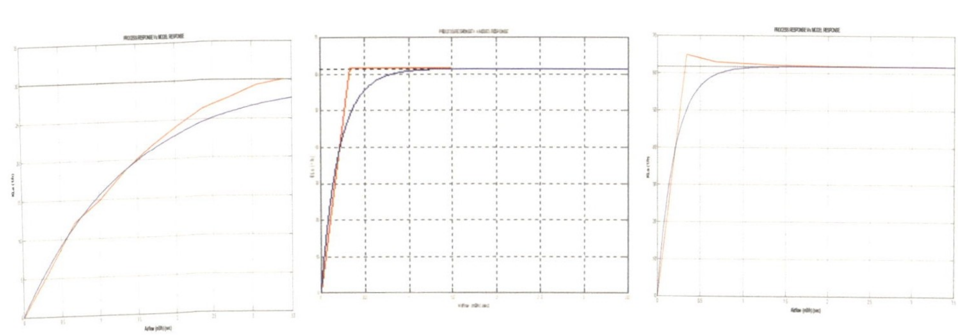

Figure 9.
Transfer function KLa. Left: integration. Center: differentiation. Right: with oxygen consumption.

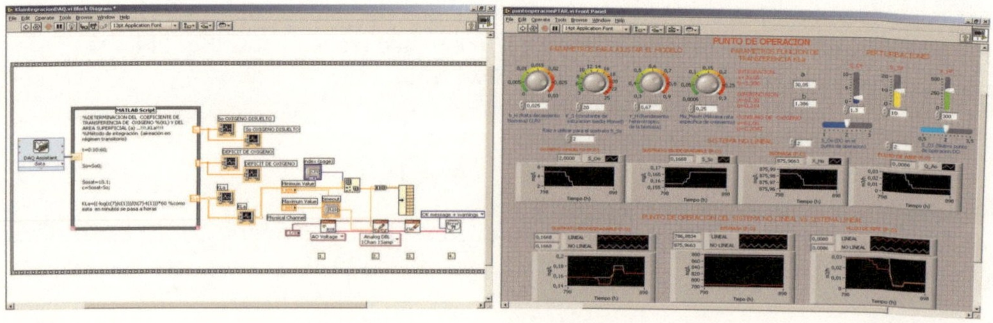

Figure 10.
Left: block diagram integration. Right: front panel of the virtual instrument (VI).

For data acquisition, supervision, and control of the WWTP, a Supervisory Control and Data Acquisition (SCADA) system is designed in LabVIEWTM [7]. The block diagram and the front panel of the virtual instrument (VI) in the control and monitoring system are shown in **Figure 10**.

2.3 Modeling and linearization of the WWTP

The use of the ASM1 model [8] is proposed to obtain the mathematical models (mainly for the biological model and the dissolved oxygen model). Among the main variables to be controlled in a WWTP are the control of the organic content in the effluent (BOD) and the control of nutrients (nitrogen and phosphorus). However, given the low concentrations found of these last two compounds in the wastewater to be treated, this chapter will focus on DO control in the bioreactor. This leads us to conclude that the proposed model can be simplified, eliminating for example the equations that have to do with the rate of change of nitrogen concentration and some other differential equations and variables that would not be necessary to take into account in this study.

Figure 11 shows the configuration and ratio of the activated sludge WWTP inlet and outlet flows, where part of the effluent biomass (XR) is recycled to the bioreactor. The effluent from the bioreactor (QBS) feeds the clarifier (settler), used to separate substrate and biomass. On the other hand, part of the biomass in the clarifier is fed back to the bioreactor (QR), while the excess biomass (QW) is removed from the process.

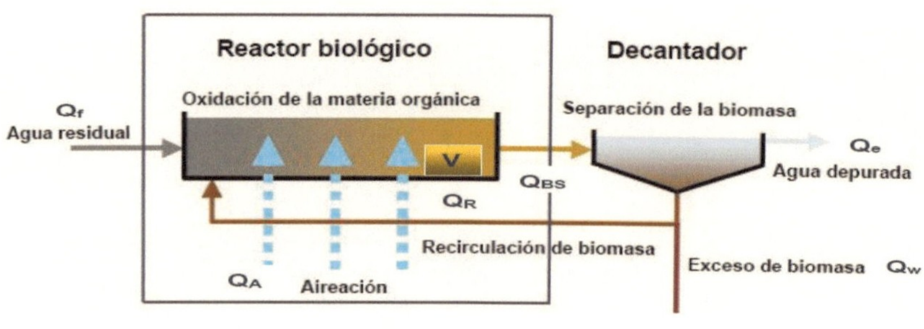

Figure 11.
Configuration and flow ratio at the WWTP.

where,
Qf = Inflow (300 l/min = 18 m3/h).
QBS = Flow from bioreactor to settler (QBS = 1.5 Qf).
QR = Biomass recirculation flow (QBS – Qf), approximately 50% del Qf.
Qe = Output flow (Qe = 0.6QBS).
QW = Residual biomass flow (QW = QBS – QR – Qe), approximately 20% of QR.
QA = Airflow.
V = Bioreactor volume (1.2 m3).

The mass balances for substrate, dissolved oxygen, and heterotrophic biomass concentrations in the bioreactor are represented by the differential equations Eqs. (20)-(22).

- Substrate balance:

$$\frac{dSs}{dt} = \frac{Q_f}{V}(Ss_f - Ss) - \frac{\mu_H}{Y_H}\left(\frac{Ss}{K_S + Ss}\right)\left(\frac{So}{K_{OH} + So}\right)X_H \tag{20}$$

- Oxygen balance:

$$\frac{dSo}{dt} = \frac{Q_f}{V}So_f - \frac{Q_f + Q_R}{V}So + \frac{Y_H - 1}{Y_H}\mu_H\left(\frac{Ss}{K_S + Ss}\right)\left(\frac{So}{K_{OH} + So}\right)X_H + a\left(1 - e^{\frac{-Q_A}{b}}\right)(So_{sat} - So) \tag{21}$$

- Heterotrophic biomass balance:

$$\frac{dX_H}{dt} = \frac{Q_f}{V}X_{Hf} - \frac{Q_w}{V}\left(\frac{Q_f + Q_R}{Q_w + Q_R}\right)X_H + \mu_H\left(\frac{Ss}{K_S + Ss}\right)\left(\frac{So}{K_{OH} + So}\right)X_H - b_H X_H \tag{22a}$$

These equations represent a nonlinear system, where $K_L a = a\left(1 - e^{\frac{-Q_A}{b}}\right)$ in accordance with the foregoing, being QR, Qw y QA the manipulated variables, where,
Ss = Substrate concentration in the bioreactor.
Ssf = Substrate concentration in inflow.
So = Oxygen concentration.
Sof = DO concentration in the inflow.
XH = Heterotrophic biomass concentration.
XHf = Heterotrophic inflow biomass.
YH = Yield Heterotrophic biomass.
μ_H= Maximum specific biomass growth rate.
KS = Average Monod saturation constant for the substrate.
bH = Biomass decay rate.
KOH = Mean Monod saturation constant for oxygen for heterotrophs.

To simplify the system of equations, the following constants are formed:

$$D = \frac{Q_f}{V}, D_1 = \frac{\mu_H}{Y_H}, D_3 = \frac{Q_f + Q_R}{V}, D_4 = \frac{Y_H - 1}{Y_H}\mu_H, D_5 = \frac{Q_w}{V}\left(\frac{Q_f + Q_R}{Q_w + Q_R}\right)$$
$$K_1 = \frac{Ss}{K_S + Ss}, K_2 = \frac{So}{K_{OH} + So} \tag{22b}$$

When replacing, you have Eqs. (23)–(25):

$$\frac{dSs}{dt} = D(Ss_f - Ss) - D_1K_1K_2X_H \tag{23}$$

$$\frac{dSo}{dt} = DSo_f - D_3So + D_4K_1K_2X_H + a\left(1 - e^{\frac{-Q_A}{b}}\right)(So_{sat} - So) \tag{24}$$

$$\frac{dX_H}{dt} = DX_{Hf} - D_5X_H + \mu_HK_1K_2X_H - b_HX_H \tag{25}$$

The results obtained with the nonlinear model are shown in **Figure 12**, which shows the accumulation of biomass until the adaptation of the microorganisms in the bioreactor, once they begin to grow the substrate decreases. On the left without airflow and on the right supplying 3.3985 m3/h, for integration (blue), differentiation (red), and with oxygen consumption (green).

Eqs. (23)–(25), linearized by Taylor series, result in Eqs. (26)–(28):

$$\begin{aligned} \Delta\dot{Ss} &= -(D + D_1K_{2o}K_3X_{H_o})\Delta Ss - (D_1K_{1o}K_4X_{H_o})\Delta So \\ &-(D_1K_{1o}K_{2o})\Delta X_H \end{aligned} \tag{26}$$

$$\begin{aligned} \Delta\dot{So} &= (D_4K_{2o}K_3X_{H_o})\Delta Ss + \left[D_4K_{1o}K_4X_{H_o} - D_3 - a\left(1 - e^{\frac{-Q_{Ao}}{b}}\right)\right]\Delta So + \ldots \\ &(D_4K_{1o}K_{2o})\Delta X_H + \left[\frac{a}{b}(So_{sat} - So_o)e^{\frac{-Q_{Ao}}{b}}\right]\Delta Q_A \end{aligned} \tag{27}$$

$$\begin{aligned} \Delta\dot{X}_H &= (\mu_HK_{2o}K_3X_{H_o})\Delta Ss + (\mu_HK_{1o}K_4X_{H_o})\Delta So \\ &+(\mu_HK_{1o}K_{2o} - D_5 - b_H)\Delta X_H \end{aligned} \tag{28}$$

Here Ss_o, So_o, X_{Ho}, Q_{Ao}, K_{1o}, and K_{2o} are the parameters and constants evaluated at the point of operation. The constants K3 y K4 correspond to the partial derivatives of K1 and K2, respectively. The incremental variables are: $\Delta Ss = Ss - Ss_o$, $\Delta So = So - So_o$ y $\Delta X_H = X_H - X_{Ho}$.

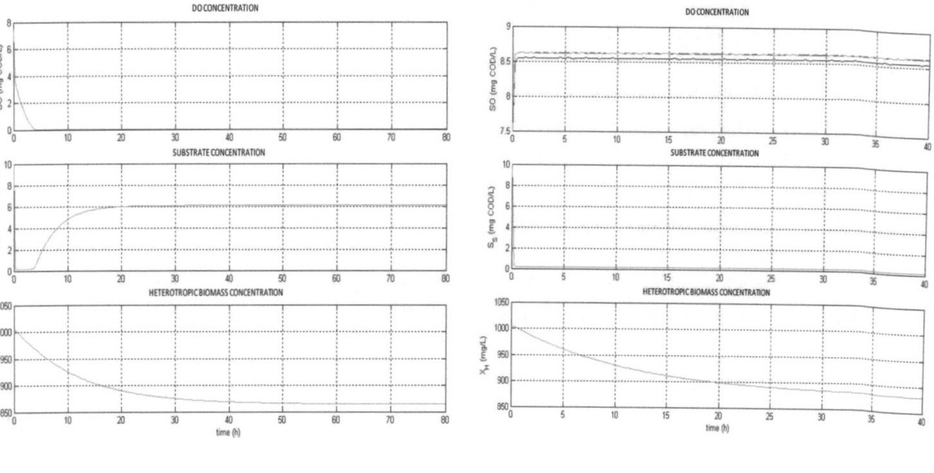

Figure 12.
Concentrations nonlinear system Top: OD. Center: substrate. Bottom: biomass.

As a result, the linear model is obtained with Eq. (29):

$$\begin{bmatrix} \Delta\dot{Ss} \\ \Delta\dot{So} \\ \Delta\dot{X}_H \end{bmatrix} = \begin{bmatrix} Ss_1 & Ss_2 & Ss_3 \\ So_1 & So_2 & So_3 \\ X_{H1} & X_{H2} & X_{H3} \end{bmatrix} \begin{bmatrix} \Delta Ss \\ \Delta So \\ \Delta X_H \end{bmatrix} + \begin{bmatrix} D_1 & 0 & 0 & 0 \\ 0 & D_1 & 0 & So_4 \\ 0 & 0 & D_1 & 0 \end{bmatrix} \begin{bmatrix} \Delta Ss_f \\ \Delta So_f \\ \Delta X_{Hf} \\ \Delta Q_A \end{bmatrix} \tag{29}$$

where,

$$\begin{aligned} Ss_1 &= -(D + D_1 K_{2o} K_3 X_{H_o}) \\ Ss_2 &= -(D_1 K_{1o} K_4 X_{H_o}) \\ Ss_3 &= -(D_1 K_{1o} K_{2o}) \\ So_1 &= (D_4 K_{2o} K_3 X_{H_o}) \\ So_2 &= \left[D_4 K_{1o} K_4 X_{H_o} - D_3 - a\left(1 - e^{\frac{-Q_{Ao}}{b}}\right)\right] \\ So_3 &= (D_4 K_{1o} K_{2_O}) \\ So_4 &= \left[\frac{a}{b}(So_{sat} - So_o) e^{\frac{-Q_{Ao}}{b}}\right] \\ X_{H1} &= (\mu_H K_{2o} K_3 X_{H_o}) \\ X_{H2} &= (\mu_H K_{1o} K_4 X_{H_o}) \\ X_{H3} &= (\mu_H K_{1o} K_{2o} - D_5 - b_H) \end{aligned} \tag{30a}$$

The linear model obtained in incremental variables is Eq. (30).

$$\begin{bmatrix} \Delta\dot{Ss} \\ \Delta\dot{So} \\ \Delta\dot{X}_H \end{bmatrix} = \begin{bmatrix} -14.8622 & -0.1117 & -0.0028 \\ -4.8220 & -0.5977 & -0.0009 \\ 9.7902 & 0.0748 & -0.0856 \end{bmatrix} \begin{bmatrix} \Delta Ss \\ \Delta So \\ \Delta X_H \end{bmatrix} + \begin{bmatrix} 0.25 & 0 & 0 & 0 \\ 0 & 0.25 & 0 & 143.9339 \\ 0 & 0 & 0.25 & 0 \end{bmatrix} \begin{bmatrix} \Delta Ss_f \\ \Delta So_f \\ \Delta X_{Hf} \\ \Delta Q_A \end{bmatrix} \tag{30b}$$

3. Predictive control strategy

3.1 Concepts

Model-based predictive control, MBPC, corresponds to a control strategy that uses a model of the system dynamics to predict the future behavior of the system over a finite time window called horizon. The predictive control strategy corresponds to an optimal control algorithm, which asks how best to control the system and calculates the future control action "u(t)" as a function of a penalty or cost function. The optimization of predictive control is limited to a moving time interval and is carried out continuously online.

Based on the model predictions and the actual measurement or estimation of the system state, the optimal control inputs are calculated, based on the defined control objective, and subject to the imposed constraints. After a time interval, the measurement, estimation, and calculation processes are repeated using a shifted horizon.

Among the advantages is that the controller can anticipate future disturbances and can be applied to systems with high nonlinearities without the need to perform system liberalization and finally can explicitly consider operational, physical, or safety constraints of the system.

The following are the three ingredients of an optimal control problem.

Model: the model will be used to predict the evolution of the state for a given sequence of inputs, as shown in Eq. (31):

$$x_{t+1} = f(x_t, u_t) \tag{31}$$

Given a sequence of inputs, if you have a model, you can predict the evolution of the trajectory, that is, the evolution of the state in the future.

Objetive is used to assign a cost to a given trajectory and qualifies how good a given trajectory is for the task to be performed. The objective is written as a cost function, which depends on the sequence of states x and inputs u, mapping it to a scalar, as shown in Eq. (32):

$$J(x_{1:T}, u_{1:T}) = \sum_{t \in [T]} g_t(x_t, u_t) \tag{32a}$$

where,

$$x_{1:T} := (x_1, \ldots, x_T); \quad u_{1:T} := (u_1, \ldots, u_T)$$

Restrictions codify the allowed domains for states and entries:

$$\begin{aligned} &x_{t+1} = f(x_t, u_t), \forall t \in [T-1] \\ &x_t \in \mathcal{X}_t, u_t \in \mathcal{U}_t; \forall t \in [T] \\ &x_1 = x_{init} \end{aligned} \tag{32b}$$

3.2 Solution of the optimization problem

In general, there is no closed solution to this type of problem, and in general numerical methods must be used to solve MATLAB and Python.

Using an optimal control strategy does not guarantee success due to several reasons. Firstly, the prediction model will always have errors, since it is an abstraction of the real system. Errors accumulate over time resulting in divergent predictions. If the control sequence u is applied to the open-loop prediction model, it will follow a different trajectory than the actual one and the task will not be carried out accurately. Secondly, even if you have a perfect model, knowing how the system behaves to the inputs that are applied, but you have a very long task horizon, if you want to apply an optimal control strategy to the entire sequence that is too long, you have the problem of having many time steps to consider in the optimization. Thus, you have a problem that is too long and too difficult to solve in a limited amount of time.

3.3 Model-based predictive control

Model-based predictive control adds an idea to the optimization problem explained in the previous section. Instead of optimizing for the complete trajectory in the future, the following receding horizon control is done:

You start at the current time step, considering only the current state.

From there, you only look ahead and only consider a limited preview, e.g. the next 50 steps.

Only the first entry is applied, then plan again.

Figure 13 shows a general scheme of the predictive control strategy, where it is observed that first the state is measured, then the task is assigned to the optimizer, to optimize for example the next 50 steps in time in the future, and from these 50 steps in time, the optimal solution is taken, taking only the first input to be applied to the system, to then measure the new state, that is, re-planning from there. The advantage is that errors can be taken into account because replanning introduces feedback and reduces the size of the problem, because you only need to plan for the next 50 time steps, for example.

Figure 14 shows a general scheme corresponding to the predictive control strategy, MBPC. The method consists of a computerized algorithmic control software based on optimization, where the best control policy is chosen so that it meets a defined criterion, for example, after 10 samples is desired that the output y is on the set point.

Figure 14 shows that past information is available, past control policy data (u(t-1), u(t-2) ...) and past process output information (y(t-1), y(t-2) ...). Two cases are shown on how the process output will evolve in the future for two future inputs u to the process and the reference output r is shown. N2 corresponds to the prediction

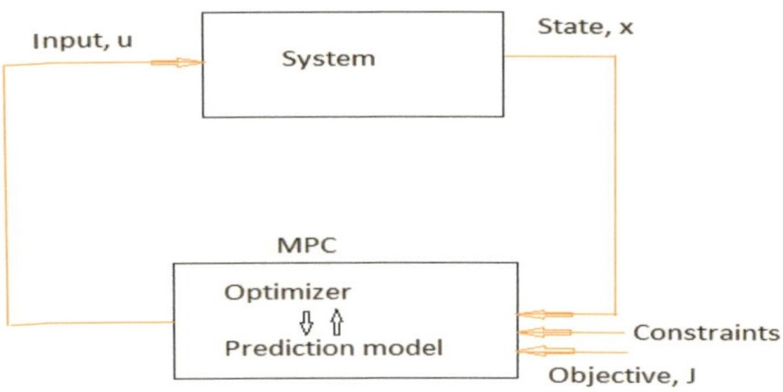

Figure 13.
General block diagram of a MBPC strategy.

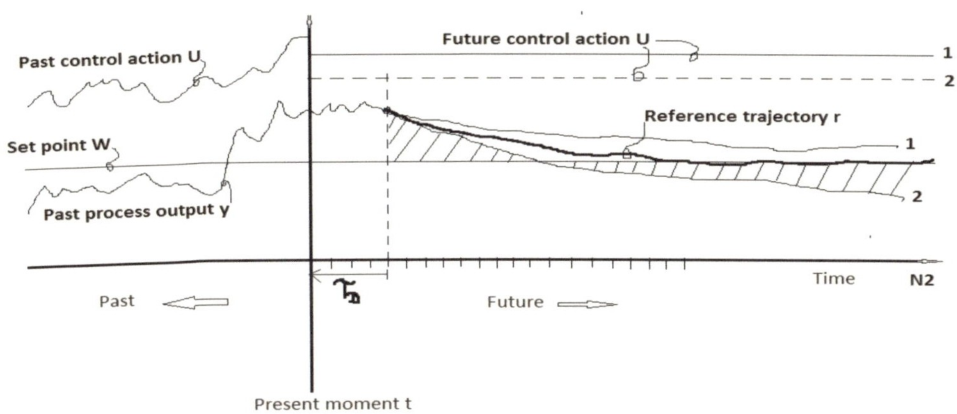

Figure 14.
General scheme of a MBPC strategy.

horizon and corresponds to one of the MBPC design parameters. In this strategy, u is selected such that the shaded area corresponding to the evolution of the process output in the future is minimal.

The MBPC algorithm can be described as follows:

- **Input:** Objective (cost function) J, Dynamics model f, horizon T, initial guess $\hat{u}_{1:T}$;

  1. $u_{1:T} \leftarrow \hat{u}_{1:T}$;

  2. **While** task not completed do

     $x_{init} \leftarrow$ Get current state ();

     $u_{1:T\leftarrow}$ Solve optimization problem

     $u \leftarrow first\ (u_{1:T})$

     $ApplyInput(u)$ to the system

  3. *Start again*

The Diophantine equations propose the generalized predictive control (GPC) strategy, while the EPSAC strategy [1] proposes filtering techniques, where the prediction of the process output is given by:

$$\{y(t+k/t), k = 1, 2, \ldots, N2\} \tag{33a}$$

and is based on available measurements of the control input u and the output y at instant t:

$$\{y(t), y(t-1)\ldots, u(t-1), u(t-2), \ldots\} \tag{33b}$$

What the control algorithm must calculate is the last value of the input u(t). Similarly, future values of the control input u must be postulated:

$$\{u(t/t), u(t+1)/t, u(t+2/t)\ldots\} \tag{33c}$$

- Output prediction: The process output prediction is given by two terms, namely the model output prediction and the disturbance prediction, as shown in Eq. (33):

$$y(t+k/t) = x(t+k/t) + n(t+k/t)$$

Model prediction can be obtained using the parallel method as it is simpler and is used in stable processes, as shown in **Figure 15**.

- **Disturbance prediction:** In Eq. (34), the disturbance prediction can be obtained by defining a hypothetical filter nf (the signal does not exist), and this value must be stored in the computer memory, as it will be needed at time instant t + 1:

$$\begin{aligned} n_f &= \frac{D(q^{-1})}{C(q^{-1})} n(t) \\ &= -c_1 n_f(t-1) - c_2 n_f(t-2) - \ldots \\ &\ldots + n(t) + d_1 n(t-1) + d_2 n(t-2) + \ldots \end{aligned} \tag{34}$$

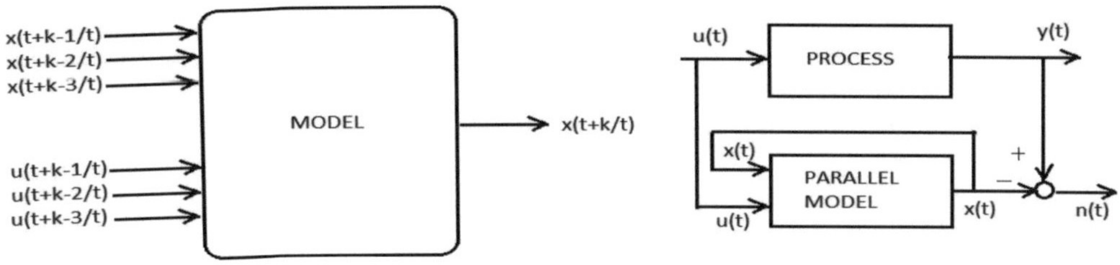

Figure 15.
Parallel model prediction.

All the information in Eq. (34) is known, it is from the past and is in the computer's memory, and the value of n(t) can be found from (1) for k = 0, as shown in Eq. (35):

$$n(t) = y(t) - x(t) \tag{35}$$

The future prediction of nf is given by Eq. (36):

$$n_f(t+k/t) \equiv 0, k = 1 \dots N_2 \tag{36}$$

In Eq. (37), the average of this signal is zero (0), because,

$$n_f(t) = \frac{D(q^{-1})}{C(q^{-1})}\frac{C(q^{-1})}{D(q^{-1})}e(t) = e(t) \tag{37}$$

White noise is an uncorrelated signal, so it cannot be predicted (μe = 0), while n(t) corresponds to a correlated colored noise signal, where there is trend, then its value can be predicted (the prediction corresponds to the mean, μe $\neq$ 0). The forward prediction of n(t) is given by Eq. (38):

$$\begin{aligned} &for\ k = 1 : N_2 \\ &n(t+k/t) = \frac{C(q^{-1})}{D(q^{-1})} n_f(t+k/t) \end{aligned} \tag{38}$$

The best prediction of the future random n signal is obtained using Eq. (39):

$$\begin{aligned} n(t+k/t) = d_1 n(t+k-1/t) - d_2 n(t+k-2/t) - \dots \\ +n_f(t+k/t) + c_1 n_f(t+k-1/t) \end{aligned} \tag{39}$$

- **Base/Optimizing response:** according to the superposition principle, the forward response of the system can be written as Eq. (40):

$$y(t+k/t) = y_{base}(t+k/t) + y_{optimize}(t+k/t) \tag{40}$$

- $y_{base}(\boldsymbol{t}+\boldsymbol{k}/\boldsymbol{t})$: it takes into account the effects of past control actions, {u(t-1), u (t-2) ... }, the effects of a basic future control scenario (u_base (t + k/t)), which is chosen for linear systems, and in nonlinear systems it is different, and finally the effect of future perturbations n(t + k/t).

- $y_{optimize}(\boldsymbol{t}+\boldsymbol{k}/\boldsymbol{t})$: it takes into account the effect of optimization of future control actions, $\{\delta u(t/t), \delta u(t+1/t), \dots \delta u(t+N_u-1/t)\}$.

Figure 16 shows the past, base, and sought control actions u(t + k/t). In order to reduce the number of computations and the size of the matrices, another design parameter is introduced, the control horizon Nu.

The control algorithm finds by optimization (minimizing the prediction errors), the future values of δu(t + k/t) (vector with Nu elements), adds them to the control signal ubase, to obtain the values of u(t + k/t).

yoptimize(t + k/t) can be calculated using the impulse response coefficients up to the value of Nu, with the last coefficient given by the value of the step response coefficient given by Eq. (41):

$$\begin{aligned} y_{optimize}(t+k/t) &= h_k\delta u(t/t) + h_{k-1}\delta u(t+1/t) + \dots \\ &\dots + g_{k-N_u+1}\delta u(t+N_u-1/t) \end{aligned} \tag{41}$$

According to [1], the basic equation for the EPSAC algorithm can be written in Eq. (42):

$$\boldsymbol{Y} = \overline{\boldsymbol{Y}} + \boldsymbol{G}.\boldsymbol{U} \tag{42}$$

$$\left\{ \begin{aligned} Y &= [y(t+N_1/t) \dots y(t+N_2/t)\,]^T \\ \overline{Y} &= [y_{base}(t+N_1/t), \ \dots y_{base}(t+N_2/t)\,]^T \\ U &= \left[\delta u\left(\tfrac{t}{t}\right) \dots \delta u(t+N_u-1/t) \dots\right]^T \end{aligned} \right\}$$

The objective of the predictive controller consists of finding the control vector, {u (t + k/t), k = 0 … N2–1}, that minimizes the following cost function, as shown in Eq. (43):

$$\sum_{k=N_1}^{N_2} [r(t+k/t) - y(t+k/t)]^2 \tag{43}$$

Figure 17 shows the error between the forward trajectory of the process and the postulated reference.

EPSAC proposes the following solution for an unconstrained linear system, as shown in Eqs. (44) and (45):

$$\sum_{k=N_1}^{N_2} \left[r\left(t+\frac{k}{t}\right) - y\left(t+\frac{k}{t}\right)\right]^2 = \left[\boldsymbol{R} - \overline{\boldsymbol{Y}} - \boldsymbol{G}.\boldsymbol{U}\right]^T \left[\boldsymbol{R} - \overline{\boldsymbol{Y}} - \boldsymbol{G}.\boldsymbol{U}\right] \tag{44}$$

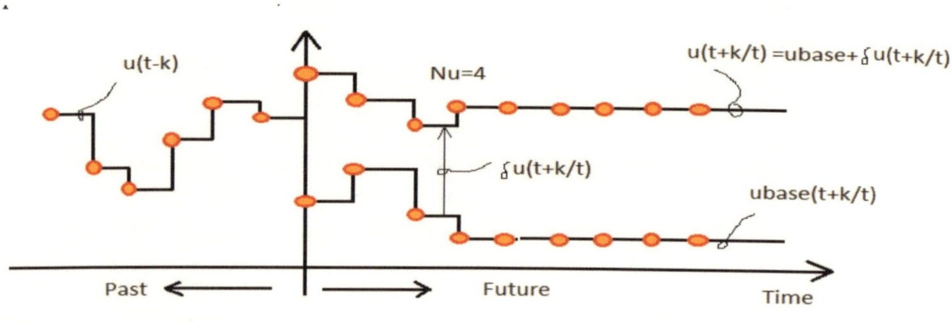

Figure 16.
u, past, base, and optimum control actions.

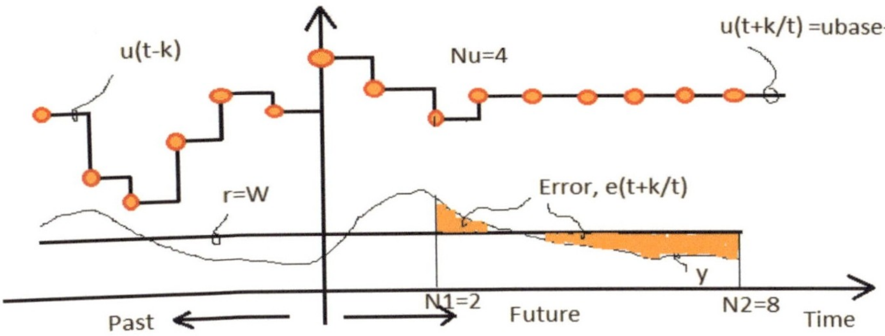

Figure 17.
Objective of the predictive control strategy.

$$U^* = \left[G^T G\right]^{-1}\left[G^T(\boldsymbol{R} - \overline{\boldsymbol{Y}})\right] \tag{45}$$

The actual control action applied to the process is shown in Eq. (46):

$$u(t) = u_{base}(t/t) + \delta u(t/t) = u_{base}(t/t) + U^*(1) \tag{46}$$

3.4 Predictive control of DO concentration in the WWTP

The basic MPC strategy consists of estimating the error of the future to determine the value of the control signal, predicting the future outputs, by means of the past outputs and the control signals. GPC control will be used, starting from the CARIMA prediction model based on a transfer function (TF) model [9], Eq. (44):

$$y(t) = x(t) + n(t) = \frac{B(z^{-1})}{A(z^{-1})}u(t-1) + \frac{C(z^{-1})}{\Delta D(z^{-1})}e(t) \tag{47}$$

where,

x(t) = process transfer function, n(t) = disturbance transfer function, Δ = integrator in the disturbance, $e(t)$ = white noise, and $u(t\text{-}1)$ = control action (intrinsic delay in the process). With $A(z^{-1})$, $B(z^{-1})$, $C(z^{-1})$, and $D(z^{-1})$, polynomials are obtained by identification. Following the procedure described in (Aguado A), Eq. (47) is transformed into Eq. (48):

$$\hat{y}(t+i/t) = G_i\Delta u(t+i-1) + \Gamma_i\Delta u^f(t-1) + F_i y^f(t) \tag{48}$$

This expression allows us to know at instant t, the value of the predicted output at instant $t + i$. Here, G_i yΓ_i y F_i are polynomials FIR in z^{-1}. Δu^f and y^f are filtered inputs and outputs, respectively. The vector expression of the prediction model is shown in Eq. (49):

$$Y = Gu + \Gamma\Delta U^f + FY^f \tag{49a}$$

where,

$$\Delta U^f = \left[\Delta u^f(t-1) \ldots \Delta u^f(t-2) \ldots \Delta u^f(t-nt)\right]^T \tag{49b}$$
$$Y^f = \left[y^f(t) \ldots y^f(t-1) \ldots y^f(t-na)\right]^T$$

The N present and future control actions are calculated from the minimization of the quadratic cost index, as shown in Eq. (50):

$$J(u) = \sum_{i=1}^{N} \alpha_i [\hat{y}(t+i/t) - w(t+i)]^2 + \sum_{j=1}^{N} \lambda_j [\Delta u(t+j-1)]^2 \tag{50}$$

The index in vector form is shown in Eq. (51):

$$J(u) = (Y - W)^T \alpha (Y - W) + u^T \lambda u \tag{51}$$

α and λ are diagonal matrices of NxN, with:

$$Y = [\hat{y}(t+1/t) \dots \hat{y}(t+2/t) \dots \hat{y}(t+N/t)]^T \tag{52}$$

$$u = [\Delta u(t) \dots \Delta u(t+1) \dots \Delta u(t+N-1)]^T \tag{53}$$

$$W = [W(t+1) \dots W(t+2) \dots W(t+N)]^T \tag{54}$$

Substituting Eq. (49) into equation Eq. (51), we obtain the expression that calculates the N future changes of the control action that minimizes the quadratic cost index, as shown in Eq. (55):

$$u = \left(G^T \alpha G + \lambda\right)^{-1} G^T \alpha \left(W - \Gamma \Delta U^f - F Y^f\right) \tag{55}$$

Although the N control actions are computed, the linear controller only implements the first $\Delta u(t)$, converting u into:

$$\Delta u(t) = hW - h\Gamma \Delta U^f - hFY^f \tag{56}$$

where h is the first row of $\left(G^T \alpha G + \lambda\right)^{-1} G^T \alpha$.

The Z-transform of $\Delta u(t)$ allows us to obtain the configuration of the GPC controller, as shown in **Figure 18**, represented by Eq. (57):

$$u(z) = \frac{T(z^{-1})}{(T(z^{-1}) + R(z^{-1})z^{-1})\Delta} \left[(H(z)W(z) - \frac{S(z^{-1})}{T(z^{-1})} y(z) \right] \tag{57}$$

From the linear system found, the gain matrix of the KMPC controller was found using the Model Predictive Control Toolbox:

$KMPC = [$ 0.0058156 1.9218e-006 -6.0503e-006 -1.209e-007 1.8134e-006 2.4445e-006 $]$

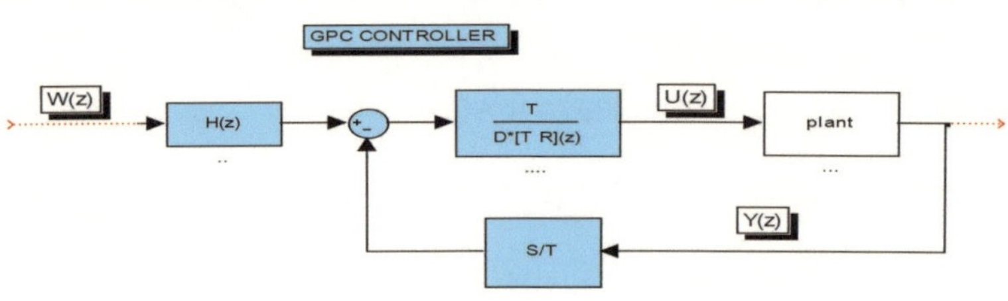

Figure 18.
GPC controller (green).

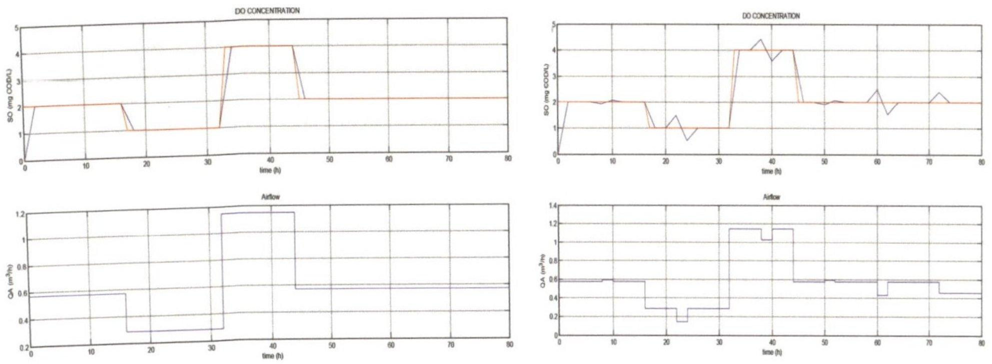

Figure 19.
Left: no valve restrictions. Right: with valve restrictions.

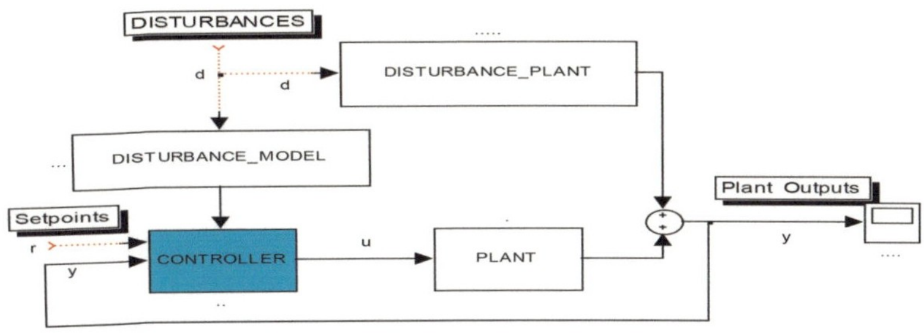

Figure 20.
GPC controller, with disturbances.

To determine the robustness of the GPC, simulations were performed with and without perturbation for different reference values of the DO. In all of them, the performance achieved was adequate. By way of explanation, the case is reproduced when there is no disturbance in **Figure 19**, where the set point (red) of 2, 1, 4, and 1 mg/l of DO is reached in the upper part, while the lower part shows the control actions sent to the valve that regulate the airflow.

If disturbances to the controller and to the process output are involved, the block diagram shown in **Figure 20** [10] is obtained.

When the plant is subjected to disturbances, as shown in **Figure 21**, a faster control action is observed, causing the response to have a slight overshoot. However, the DO set point is reached.

The MPC controller design methodology has a great acceptance in the control of the different variables that are present in the wastewater treatment processes, thanks to its great robustness and the availability of software that allows the development and execution of the algorithms necessary for its implementation.

This chapter shows the possibility of controlling the concentration of DO in a WWTP from a GPC, whose algorithm has been programmed in MATLAB and executed in a LabVIEW supervisory system.

4. Conclusions

The MPC controller design methodology is widely accepted in the control of the different variables that occur in wastewater treatment processes, thanks to its great

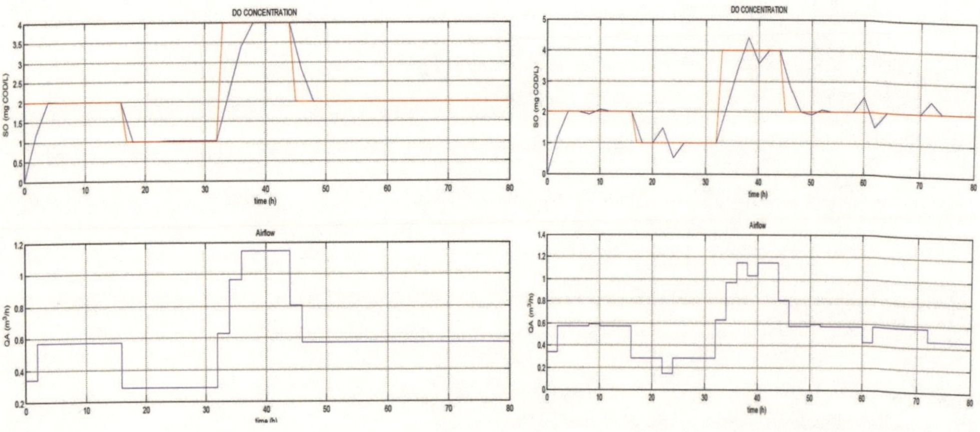

Figure 21.
Left: no valve restrictions. Right: with valve restrictions.

robustness and the availability of software that allows the development and execution of algorithms necessary for its implementation. One of its advantages is to guarantee lower energy consumption due to the fact that the steps of the control action are smaller and of an anticipated type, resulting in lower opening values of the airflow valve.

This work shows how to control the concentration of DO in a WWTP from a GPC, whose algorithm has been programmed in MATLAB and executed in a LabVIEW SCADA system. It was based on the step response of the system and the description in state space, with or without restrictions in the airflow (manipulated variable) and/or in the concentration of DO (controlled variable), either in the presence or not of disturbances in the system (concentrations of dissolved oxygen, substrate, and biomass in the inlet flow to the bioreactor). In the different simulations and tests in the implemented controller, it was found that when there are no restrictions neither in the airflow valve nor in the DO sensor, even if disturbances are present, the set point is adequately reached in all cases. On the contrary, when physical restrictions are present in the airflow and/or the oxygen concentration, there are moments in which the desired value of the controlled variable is not adequately reached.

Conflict of interest

The authors declare no conflict of interest.

Author details

Jose A. Muñoz Hernandez[1*], Luis Eduardo Leguizamon[2]
and Helmer Muñoz Hernandez[3]

1 Faculty of Agronomic Engineering, University of Tolima, Ibagué, Colombia

2 Industrial Management Center, SENA, Bogotá, Colombia

3 Faculty of Administrative and Accounting Economic Sciences, Universidad del Sinú, Montería, Colombia

*Address all correspondence to: amunoz@ut.edu.co

References

[1] De Keyser R. Adaptative and predictive control. In: EeSA-Department: Electrical energy, Systems & Automation. Ghent: Ghent University; 2008

[2] Egno Ngongi W, Jialu D, Wang R. Design of generalised predictive controller for dynamic positioning system of surface ships. International Journal of Intelligent Systems Technologies and Applications. 2020; **19**(1):17-35

[3] Lindberg CF. Control and Estimation Strategies Applied to the Activated Sludge Process. Stockholm: Uppsala University; 1997. pp. 81-121

[4] Ramalho RS. Tratamiento de aguas residuales. Barcelona: Reverte; 1996. pp. 199-294

[5] Gutierrez A, Briers J. Sistemas de identificación. Ibague: Coruniversitaria-Katholieke-Gent, Universiteit; 1999

[6] Aguado A, Martinez M. Identificación y control adaptativo. Madrid: Pearson Prentice Hall; 2003. pp. 235-280

[7] Leguizamón Castellanos LE. Control predictivo de la concentración de DO en el biorreactor de una planta de tratamiento de aguas residuales con lodos activados. Ibague: Universidad de Ibague-Leuven-Gent, Universiteit; 2007

[8] Diaz Delgado C, Fall C, Quentin E. Agua potable para comunidades rurales, reuso y tratamientos avanzados de aguas residuales domestica, Virtualpro, Ed., RIPDA-CYTED. 2003

[9] Clarke DW, Scattolini R. Constrained receding-horizon predictive control. Proceedings. 1991;**138**(4):347-354

[10] Morari M, Ricker N. Model Predictive Control Toolbox, For Use with MATLAB. The Mathworks., Natick: The Mathworks; 2005

Chapter 7

Simulation of Permanent Movement in Collectors Non-Standardized Sewerage

Mihail Luca

Abstract

Sewage collectors for domestic waste water and rainwater have flow sections that are standardized in shape and size. The collector with vertical dimension H > 1.50 m is admitted to be visitable and for H ≤ 0.80 m it is not visitable. Sewage collectors can be made in demanding situations with non standardized flow sections. A series of natural and anthropogenic action can change the geometric and hydraulic parameters of the flow section over time. Thus, the flow section no longer respects the initial geometric and hydraulic parameters and becomes a non-standardized section. Knowing the hydraulic parameters at such a flow section becomes difficult for monitoring the mining process. The circular, ovoid, circular bell, parabolic bell type flow section transforms into sections consisting of straight lines and circular arcs or only circular arcs. The factors that produce are erosion, siltation and cementation of transported material, repair and rehabilitation works and others. The erosion phenomenon also causes a change in the roughness on the watered perimeter of the section. In order to obtain the hydraulic operating parameters of the visitable sever collectors with non-standard flow section, several calculation programs have been elaborated on the forms of permanent water movement.

Keywords: hydraulic calculation, visitable collector, polycentric sections, variable roughness, calculation program

1. Introduction

The transport of waste water required the definition of flow cross-sections with simple geometric shapes, which would achieve high transport velocities towards the lower part of the section. The most used geometric shapes for sewer collectors are circle, ovoid and bell with various variants (circular, elliptical and parabolic). Hydraulic calculation methods were developed for these geometric shapes, which were standardized and included in design regulations [1].

The hydraulic calculation for dimensioning and verification of sewer collectors has evolved over time according to the contribution of scientific research. From the grapho-analytical calculation used about 50 years ago [2–4], we have reached the use of special calculation programs based on the classic geometric shapes of sewer collectors [5, 6].

Current hydraulic research on sewer manifolds is focused on several areas of interest. One area of research analyzes the velocity profile in the flow section and its optimization to increase the sediment transport velocity by using the maximum entropy principle, CFD or numerical methods for special situations and so on [7, 8]. Another field of research analyzes ways to increase the flow rate in the flow section taking into account the geometric and hydraulic parameters [9, 10]. Research is mainly carried out on closed and free-flowing channel sections (mainly circular and ovoid) [10, 11].

An important field of research lately is represented by the monitoring of the functional parameters of the sewerage networks and especially the hydraulic parameters (minimum allowed speeds, flow rates, water depths, degree of filling and others). Modern investigation methods such as CFD [11], fuzzy logic [12], with improved reliability of hydraulic data [13] and concrete applications on urban sewage networks are used in the research. The processes of washing the sediments deposited in the sewage collectors is a field of research targeted in the last period of time [5, 14].

The creation of new sewage systems and the rehabilitation and modernization of the existing ones led to the appearance of pipes with different sections than the classic ones. Also, the degradation over time of the geometry of the flow section, through the transformation from a shape made of arcs of a circle to a shape with arcs of a circle + straight line segments imposed a new field of research, namely that of collectors with sections of non-standard flow [15].

For now, this field is less addressed, but it has become necessary considering the monitoring of the operation of these collectors, which occupy the main positions in an urban sewage network. A series of researches have been carried out for galleries where the flow is free-level, and the circular section has been transformed into a section consisting of arcs of a circle and line segments [16]. Accurate knowledge of geometric and hydraulic operating parameters in various flow regimes requires a more detailed approach to hydraulic parameters.

2. Sewer collectors with non-standard sections

The sewage system consists of a sequence of closed channels (sewage collectors) in which the flow of water generally takes place with a free level. Sewer collectors can be classified according to their position and role in the urban sewer scheme as follows: service, secondary, main and discharge. The main and exhaust collectors have large dimensions and are executed on site or by using prefabricated sections. The geometric form of the main collectors is standardized for the typification of the hydraulic calculation and the technological process of execution. The most commonly used standardized geometric shapes are, ovoid, bell and circular (**Figure 1**). The construction material of the main collectors of the old type was stone and brick, while the modern ones use reinforced concrete and composite structures. On the route of the collectors there are a series of constructions (fireplaces, overflow chambers, basins, pumping stations, underpasses and others) that ensure a correct exploitation process. The correct operation of the collector is achieved by corroborating all its structural components to the temporal variation of the transported flow [1].

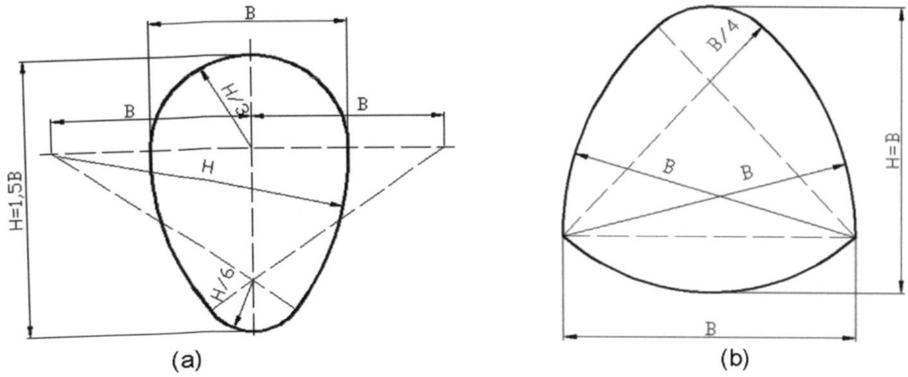

Figure 1.
The profile and the geometrical parameters of drawing the standardized sewerage collectors: (a) Ovoid type; (b) semi-eliptical bel type.

Depending on the height H of the collector section, the following are allowed in Romania: non-visitable collectors with $H \leq 0.80$ m; semi-visitable collectors with $0.80 \text{ m} \leq H < 1.50$ m; visitable collectors with $H > 1.50$ m [2, 15].

The natural and anthropogenic factors in the site intervene on the collectors with a complex of actions, which influence their behavior over time. Among the natural factors can be listed the geotechnical and hydrogeological characteristics of the site, seismic action, land subsidence and so on. Anthropogenic factors acting on the collector show an evolution over time as influence and impact. They influence the behavior of the sewer collector through the design concept, the execution technology, the degree of application of the maintenance works and annual repairs and so on. The most influential anthropogenic factors that degrade the flow section are wastewater concentration, chemical parameters, transport speed, protection material of the flow section and absence of maintenance works and so on. Some of the natural factors in conjunction with the anthropogenic ones cause the appearance of continuous degradation processes of the collector structure and change its functional parameters over time [16, 17].

Non-standard flow sections are determined by the following situations:

- the production by some production companies of pipes that do not respect the standardized dimensions;
- degradation of the flow section of the collector through phenomena of hydrodynamic erosion and clogging: in certain situations cementation of the clogging materials occurs; these phenomena substantially change the geometric shape of the flow section;
- the repair works, but especially those for the rehabilitation of the flow section, determine the modification of the geometric shape; the execution technologies and the machines used may require a certain shape of the flow section in the lower area;
- the transformation of the main collectors with unitary operation (domestic waste water + rainwater) into collectors with separate operation; this mode of

operation, imposed by the EU regulations of recent years, determines the geometric modification of the flow section and implicitly the appearance of a non-standard shape.

Hydrodynamic erosion and clogging of the flow section change the initial circular, ovoid or bell-type geometric shape of the collector into a new shape with non-standard parameters. The new shape presents curves differentiated from the original ones, but also the presence of straight lines in the flow section. The research carried out in situ, in a gallery with a circular section, where the water flow was at a free level, highlighted the modification of the geometric shape by the flattening of the lower area and the curvature of the watered perimeter. At the same time, the phenomenon of hydrodynamic erosion and the execution works changed the roughness of the watered perimeter, causing the appearance of three to four areas of roughness variation starting from the radius and continuing on the wall (**Figure 2a**) [16].

Thus, the flow section transforms geometrically over time, where the shape made of circular arcs changes into a mixed shape (arcs + straight line segments) (**Figure 2b**) [16].

The research carried out [17] in the sewage networks in Romania showed that the flow section of the visitable collectors was substantially modified compared to the designed one. Thus, an ovoid section can become a polycentric section through the erosion processes of the flood bed and the walls (**Figure 3a**). The same situation also occurs at the bell-type sections (**Figure 3b**), where the erosion phenomena and the consolidation of the deposited sediments have substantially transformed the flow section.

The application of some rehabilitation works to the collectors can geometrically transform the flow section by creating a transport or support area for the execution machinery. Monitoring of flow parameters (flow rate, minimum speed and degree of filling) in non-standardized sections can no longer be performed correctly. In some cases, the geometric modification of the section determined the reduction of the speed and the transported flow by about 15–28% [17].

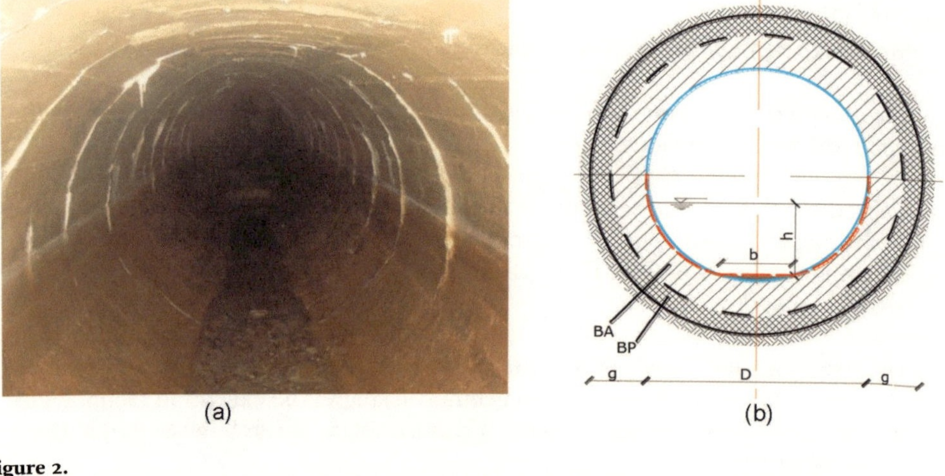

Figure 2.
Standardized circular section for a free-level flow modified by hydrodynamic erosion and technological repair processes: (a) General view of the registered changes; (b) Modification scheme of the lower area (blue line – standard section, red line – modification areas of the section), BA – Reinforced concrete, BP prestressed concrete [16].

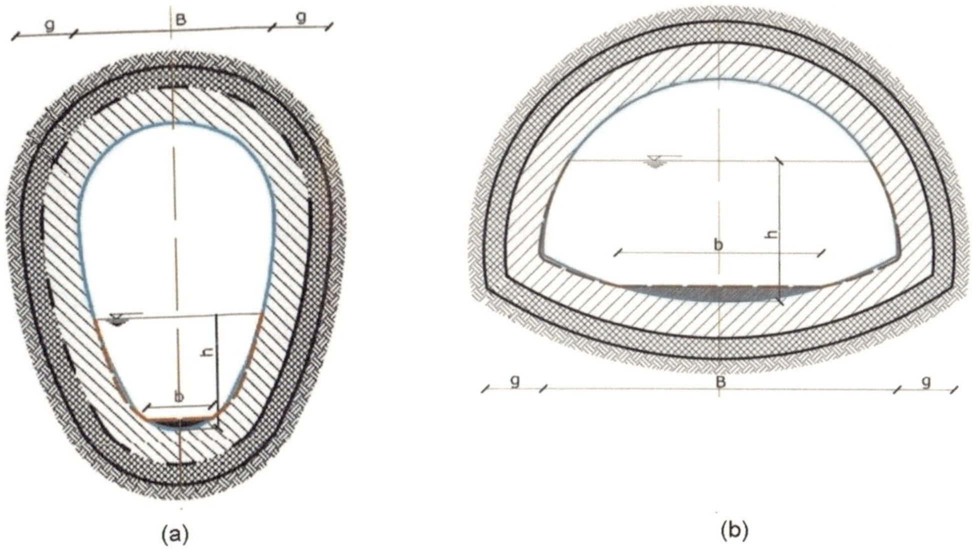

Figure 3.
Scheme of modification sewer of standardized section at the visitable sewage collectors: (a) ovoid section (blue line – standard section, red line – Modification areas of the section); (b) circular bell section [17].

3. Hydraulic calculation of non-standard sewer collectors

3.1 Types of wastewater movement in the sewage network

The movement of waste water with a free level in a sewer network with a closed section can be of non-permanent and permanent type. The operating safety of a sewage system is most affected by the non-permanent (transient) flow regime. This, depending on the speed of variation of the hydraulic parameters - quantitatively, the flow Q, and qualitatively, the depth, h, or the free water level share, z - can be considered as being slowly variable, or rapidly variable.

The non-permanent regime can be simulated by using complex hydraulic-mathematical models formed, in general, from: 1 - motion equations; 2 - continuity equations; 3 - initial conditions; 4 - contour conditions. The initial conditions are represented by a permanent flow regime, which can be considered of varied gradual non-uniform type. To be able to approach any hydraulic simulation, the basic data of the sewer network must be obtained, namely geometric and hydraulic parameters of the uniform permanent movement for any form of collector (standardized and non-standard).

To generalize the analysis of the main sewer collectors in the flow regimes existing in practice and for all types of collector sections (standardized and non-standardized) used, their geometric and hydraulic characteristics

$$A = f_{\mathrm{A}}(h), P = f_{\mathrm{P}}(h), R_{\mathrm{h}} = f_{\mathrm{Rh}}(h), B = f_{\mathrm{B}}(h), z_{\mathrm{G}} = \mathrm{f}_{\mathrm{zG}}(h), W = f_{\mathrm{W}}(h), K = f_{\mathrm{k}}(h) \tag{1}$$

were considered by the matrix M, of the form:

$$\mathrm{M} = [\mathrm{h} \quad \mathrm{A} \quad \mathrm{P} \quad \mathrm{B} \quad \mathrm{R_h} \quad \mathrm{Z_G} \quad \mathrm{W} \quad \mathrm{K}]. \tag{2}$$

The M matrix was determined based on a series of mathematical models and computer programs developed for the characteristics of sewer components in various flow regimes. The elements of the matrix M are NC-dimensional column vectors, where NC represents the number of distinct values assigned to the value pairs

$$\{h, B, A, R_h, z_G, W, K\}_i, \ i = 1, 2, \dots, N_C \tag{3}$$

In carrying out the simulations along the length of a collector, it is more convenient to perform the substitution

$$h = z - z_a \tag{4}$$

where z_a is the apron elevation of the channel and z - the elevation of the free water level. Thus, in the characteristics (Eq. (1)) similar characteristics will be generated, but which presents the z rate as an independent variable,

$$A = f_A(z), P = f_P(z), R_h = f_{Rh}(z), B = f_B(z), z_G = f_{zG}(z), W = f_W(z), K = f_k(z) \tag{5}$$

The main sewer collectors were considered on the analysis sections as prismatic/cylindrical beds, with a constant slope. The connection of collector sections with different shapes and/or sizes, as well as the confluence with lower-order collectors, is achieved by using connecting or intersection manholes. In this case, adequate contour conditions must be allowed in the inspection manhole section.

3.2 Hydraulic calculation elements for sewer collectors

Geometric parameters of the flow section at a sewer collector are as follows (**Figures 1** and **4**): section shape, constructive height (*D, H*), constructive width (*B*), radii of curvature (*r*), angles, geodesic slope of apron collector (*I*) section and others. Hydraulic parameters considered in the calculation are the following: total flow (*Q*) and specific (*q*), flow modulus (*K*), average speed (*v*), calculation degree of filling (*a*) and maximum imposed (a_{max}), minimum (v_{min}) and maximum imposed speed (v_{max}), hydraulic slope (*J*), the wetted area of the flow section (*A*), wetted perimeter (*P*),

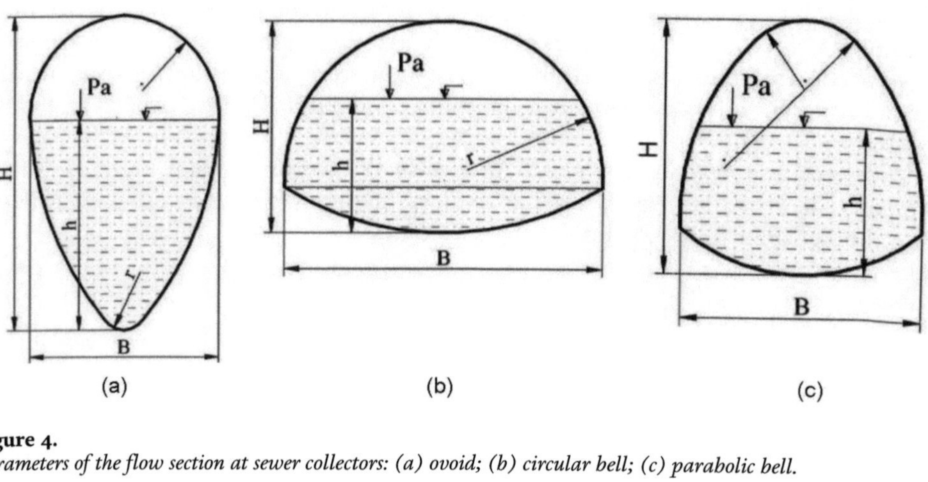

Figure 4.
Parameters of the flow section at sewer collectors: (a) ovoid; (b) circular bell; (c) parabolic bell.

hydraulic radius (R or R_h), roughness (k) or roughness coefficient (n - Manning), depth of the centre of gravity of the watered area (z_G) and others.

The calculation of dimensioning and verification of sewerage collectors goes through the two forms of movement of water specific to permanent movement: a – uniform movement; b – non-uniform movement with gradually varied non-uniform and rapidly varied non-uniform types [1, 3].

The calculation relations used for uniform permanent motion are as follows:

$$Q = AC\sqrt{RI} \tag{6}$$

$$v = C\sqrt{RI} \tag{7}$$

In the case of sewer pipes with partially filled sections, the analytical expressions for each of the functions presented in Eq. (1) are used in the calculation and are customized for each geometric shape of the collector section.

The solution methods are analytical, with the use of numerical calculation programs and graph analytic, with the use of graphs and tables specialized on section shapes in the case of calculations with relatively low precision.

The calculation relation of parameters for non-uniform gradually varied permanent motion is as follows [4]:

$$dl = \frac{1 - \frac{\propto Q^2 B(h)}{gA^3(h)}}{I - \frac{Q^2}{K^2(h)}} dh \tag{8}$$

$$l_2 - l_1 = \frac{1}{I}\int_{h_1}^{h_2} f(h)dh \tag{9}$$

In the case of closed channels and with special shapes of the section (circular, ovoid and bell), the determination of the geometric and hydraulic parameters (A, P, R, C, K, Q, v, and so on) is carried out depending on the degree of filling defined by the ratio $a = h/H$. By introducing a correlation of the type $Q/Q_p = K/K_p = f(a)$, where Q_p and K_p are the flow, respectively the flow modulus, at the full section, the equation of the gradually varied non-uniform movement can be written in the form [4]:

$$\frac{|I|L}{H} = \frac{f^2(a)\left(1 - \frac{\propto Q^2 B}{gA^3}\right)}{f^2(a) \pm \frac{Q^2}{Q_p^2}} AC\sqrt{Ri} \tag{10}$$

and by integration over a section of length L and in two sections 1–1 and 2–2 we obtain:

$$\frac{|I|L}{H} = \varphi\left(a_1, \frac{Q}{Q_p}\right) - \varphi\left(a_2, \frac{Q}{Q_p}\right) - \frac{\alpha Q^2}{gH^5}\left[\Psi\left(a_1, \frac{Q}{Q_p}\right) - \Psi\left(a_{21}, \frac{Q}{Q_p}\right)\right] \tag{11}$$

The functions φ and ψ present different integration domains, according to the sign of the slope and the ratio between the slope of the bottom of the bed i and the hydraulic slope J. To solve practical problems, the functions φ and ψ are integrated by different methods. For calculations with a relatively good precision, the functions φ and ψ were integrated and presented in tabular or graphical form [4].

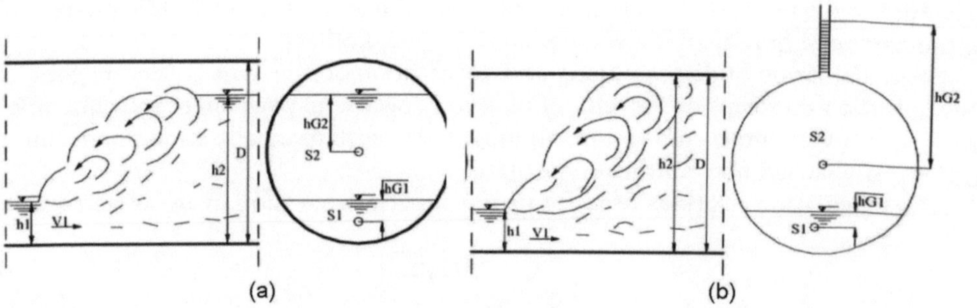

Figure 5.
Analysis scheme of the hydraulic jump in closed channels and flow with free level: (a) case h2/D < 1; (b) case h2/D > 1 [4].

Rapidly varying non-uniform permanent motion in free-flowing sewer collectors presents special calculation features determined by the magnitude of the second conjugate depth of the hydraulic jump. The hydraulic jump calculation relation has the form:

$$\frac{\alpha_0 Q^2}{gA_1} + h_{G1}A_1 = \frac{\alpha_0 Q^2}{gA_2} + h_{G2}A_2\,, \tag{12}$$

where h_{G1}, h_{G2}, are the depths of the centre of gravity corresponding to the watered areas A_1 and A_2.

On a collector section with diameter D, or height H, the second conjugate depth of the hydraulic jump, depending on its size, can impose a free-level or pressurized flow downstream on a certain bed length. This aspect determines the differentiation of hydraulic calculation problems.

In the first case, for the ratio $h_2/D < 1$ (h_2 – second conjugate depth), the flow in the collector will be aerated, and the movement of water downstream of the jump will be free level (**Figure 5a**). The hydraulic jump develops as in an open bed, and the conjugate depths are determined by solving the equation (Eq. (12)).

In the second case, when $h_2/D > 1$, respectively $h_2 > D$, the collector enters under pressure downstream (after the jump) and at the same time introduces a vacuum upstream of the jump (**Figure 5b**). If the air flow is not compensated by a process of aerating the collector, instability of the movement occurs, manifested by the oscillation of the parameters of the jump and of the liquid current.

3.3 Hydraulic calculation elements for non-standard sewer collectors from uniform movement

The monitoring of the operation of the sewer collectors is carried out by taking the characteristic hydraulic parameters from the imposed control sections. Simulating the operation of a collector, and especially the main ones, requires the calculation of hydraulic parameters for permanent and non-permanent movement. For a collector with a non-standard section and with a variable roughness on the wetted perimeter, it is necessary to go through several stages of analysis and calculation, respectively:

Stage I. Field analysis and retrieval of geometric and hydraulic data from the flow section of the collector. Carrying out a topographical study along the length of the simulation (longitudinal profile and transverse profiles) [16, 17]. Carrying out an in situ research on the way of roughness variation on the perimeter of the flow section. Data processing and creation of a database specific to each collector.

Stage II. Determination of geometric and hydraulic parameters for uniform permanent movement by using calculation programs specific to sections formed by curves and straight lines. The results obtained will be included in a database specific to each analyzed collector.

Stage III. Determination of geometric and hydraulic parameters for non-uniform gradual varied and rapidly varied permanent motion by using calculation programs. The results obtained will be included in a database specific to each analyzed collector.

The flow sections in sewer collectors geometrically made of curved lines were treated as "*single bed*" [18], where they are composed only of circular arcs (**Figure 6a**). Also, some sections were considered "*multi-bed*" being composed of straight line segments only. In a particular study case [16], a "*single bed*" composed of both circular arcs and straight segments was analyzed (the case of gallery-type collectors formed by a rectangle and a semicircular area on the apron (**Figure 6b**).

Later, a generalized analysis model was developed for „*single bed*" sections composed of both curved lines (generally circular arcs) and straight line segments). In the analysis, it was considered that the tube that forms the collector has symmetry along a longitudinal vertical plane (**Figure 6a,b**). It follows that the contour of the cross-sectional half of the tube is completely determined by an odd number of points, N_M, conveniently established by a yOz coordinate system. The number of points is defined by the relationship:

$$(y_C, z_C)_k \ k = 1, 2, \ldots, N_M \tag{13}$$

The coordinates (Eq. (13)) are related to a Cartesian system of coordinate axes, yOz, defined as follows: the Oz axis, vertical included in the longitudinal plane of symmetry; the origin O at the point on the minimum height tube eraser, and the horizontal axis Oy, normal to the axis Oz, included in the transverse half-section of the tube.

The hydraulic calculation model of sewer collectors for the permanent and non-permanent flow regime uses the following hydraulic characteristics depending on the water depth, h or the elevation of the free water level, z: $B(h)$ or $B(z)$, $A(h)$ or $A(z)$, $P(h)$ or $P(z)$, $R_h(h)$ or $R_h(z)$, $z_G(h)$ or $z_G(z)$, $W(h)$ or $W(z)$ and $K(h)$ or $K(z)$.

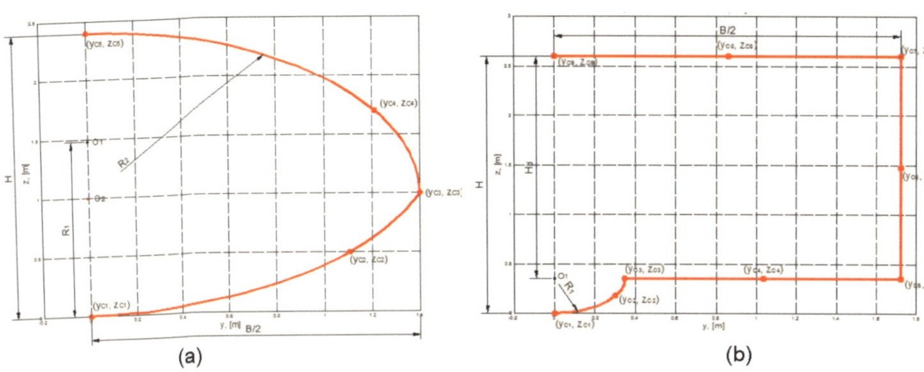

Figure 6.
Calculation schemes for non-standard collector sections: (a) semicircular bell; (b) rectangular collector with semicircular groove.

In sewer collectors with sections composed of $N_a \geq 2$ arcs of a circle and/or line segments, the analytical expressions for each of the functions $A(h)$, $P(h)$, $R_h(h)$, $z_G(h)$, $W(h)$ and $K(h)$ must be rendered by N_a functions defined on portions of the perimeter, respectively:

$$A_i(h), P_i(h), R_{hi}(h), B_i(h), z_{Gi}(h), W_i(h), K_i(h), \text{with } z_i - z_{C0} \leq h \leq (z_{i+1} - z_{C0}), \quad i = 1, \dots, N_a \tag{14}$$

where z_{C0} (usually $z_{C0} = 0$) is the height of the tube eraser; z_i and z_{i+1} ordinates of the points that delimit the lower, respectively upper, circle arc or line segment i.

To determine the functions (Eq. (14)) the following steps must be solved:

A. Deduction of analytical expressions for circles and/or supporting lines of N_A circle arcs and/or line segments.

B. Determination of the $N_A + 1$ points that border the lower and upper N_A circle arcs and/or straight line segments, (y_i, z_i), $(i = 1, 2, \dots, N_A + 1)$.

C. Deduction of analytical expressions, by portions, for the hydraulic characteristics of sewer collectors.

For the analytical expressions describing circles and support lines of circle arcs and line segments, the N_M coordinate points (Eq. (13)) were considered. From these were formed N_A groups of three successive points - M_{i1}, M_{i2}, M_{i3} - of coordinates:

$$(y_i, z_i)_j, \text{with } (j = 1, 2, 3), (i = 1, 2, \dots, N_A) \text{ and } z_{i1} \leq z_{i2} \leq z_{i3}. \tag{15}$$

Each group of three such points determines:

- the circle arc i, centre C_i (b_i, a_i) and radius R_i – when the points M_{i1}, M_{i2}, M_{i3} are non-collinear;
- the line segment i, with the coefficients of the line (a_i, b_i, c_i) - when the points M_{i1}, M_{i2}, M_{i3} are collinear (the line segment i can be considered an arc of a circle with radius R_i tending to infinity).

Based on the mentioned, the equations describing the circle arcs and the straight line segments that form the geometric figure of the studied collector were deduced.

To determine the coordinates of the points bordering the N_A circle arcs and straight line segments, proceed as follows:

The N_A distinct geometric shapes (circle arcs and/or straight line segments) that structure the cross-section of the tube are bordered by $N_A + 1$ points, coordinates

$$(y_i, z_i) \; (i = 1, \dots, N_A + 1), \tag{16}$$

For extreme points, coordinates of the type (16) are selected directly from the set of coordinates (13)

$$(y_1, z_1) = (y_C, z_C)_1 = (y_{C1}, z_{C1}), \; (y_{N_A+1}, z_{N_A+1}) = (y_C, z_C)_{N_M} = (y_{CN_M}, z_{CN_M}) \tag{17}$$

while the coordinates

$$(y_{i+1}, z_{i+1})\ (i = 1, \dots, N_A - 1), \tag{18}$$

correspond to the intersection points between each grouping of circles or lines related to adjacent geometric elements i and $i + 1$.

Depending on the geometric shape (arc of a circle or line segment) of elements i and $i + 1$, four distinct analysis situations are distinguished:

1. Both elements, i and $i + 1$, are arcs of a circle.

2. Element i is an arc of a circle and element $i + 1$ is a line segment.

3. Element i is a line segment and element $i + 1$ is an arc of a circle.

4. Both elements, i and $i + 1$, are line segments.

For the four analysis situations, the equations were deduced that allow obtaining the coordinates of the points bordering the N_A circle arcs and straight line segments for the geometric shape of the studied collector.

The hydraulic-mathematical model is customized according to the shape and structure of the non-standard section. In the case of a bell-type collector with a section modified by a new manufacturing technology, or by rehabilitation works, with the simulation scheme shown in **Figure 6a**, the following calculation steps were carried out:

Stage 1. Definition and calculation of plot coordinates of geometric parameters (y, z) for the modified cross-section. It is assumed that the bell-shaped section (**Figure 6a**) is formed at the bottom by a circle segment with radius R1, with centre O_1 $(0, R_1)$, a chord of length B and height ζ. Also, the bell-type section is described at the top (dome) by a semicircle with radius $R_2 = B/2$ from the centre O_2 $(0, \zeta)$ (**Figure 6b**). The way of working for this type of section is as follows:

I.1. Determining the plotting coordinates of the flow area.

I.2. Determination of the drawing coordinates of the hydraulic radius.

I.3. Determining the drawing coordinates of the centre of gravity of the flow area and so on.

The constructive parameters ζ and R_1 are evaluated according to the representative sizes B and H of the flow section with the relations:

$$\zeta = H - B/2, R_1 = \frac{B\left[\beta^2 + (2-\beta)^2\right]}{4\beta(2-\beta)}, \text{with } \beta = \frac{B}{H} \tag{19}$$

The semicircular geometric profile of the collector is completely determined by five coordinate points (y_C, z_C).

$$(y_C, z_C) \in \left[(0,0), \left(\sqrt{\zeta(4R_1 - \zeta)}/2, \zeta/2\right), (B/2, \zeta), \left(B\sqrt{3}/4, B/4 + \zeta\right), (O, H)\right]. \tag{20}$$

Stage 2. Mathematical modeling of hydraulic parameters specific to the non-standard collector (modified geometric shape). The analytical expressions for the hydraulic characteristics of the sewer pipes were derived on Na portions, on which the water depth h satisfies the condition $z_i < h \leq z_{i+1}$, ($i = 1, \ldots, N_A$).

The non-uniform distribution of the roughness on the wetted perimeter of the tube was taken into consideration by assigning a specific value to the roughness coefficient (the Manning coefficient was accepted), n_i, on each circle arc element or right segment i, $i = 1, \ldots, N_A$. A series of roughness coefficient values have been determined by research on main sewer collectors in operation.

The expressions of the geometric parameters proposed to be evaluated for the circular bell type collector are the following ($z_{c0} = 0$ was considered):

2.1. Hydraulic parameters of the flow section: $A(h)$ or $A(z)$, $P(h)$ or $P(z)$, $R_h(h)$ or $R_h(z)$, $B(h)$ or $B(z)$ and $z_G(h)$ or $z_G(z)$.

2.2. Mathematical modeling of the hydraulic characteristics for the collector with modified geometric shape: the velocity module, $W(h)$ or $W(z)$ and the flow modulus, $K(h)$ or $K(z)$.

To solve the mathematical model, a numerical calculation program Collector.Non-Standard_Uniform Motion_Unique Bed (Col.NS_UM_UB), was developed in the MATLAB programming environment.

Each type of non-standard collector geometry has an identification code attached to it in the uniform motion simulation program. Through this code, the computer selects the menu and work subroutines.

The program entry data are as follows: E.1 - Work menu (Specific code); E.2 – General data (Dt_H, Dt_B, respectively the specific calculation distance in the Cartesian system [calculation step, (m)]: E.3 – Geometric characteristic data of the section (D, B, H); E.4 - Roughness coefficient values on characteristic perimeter lengths, n_i.

The output data from the program is as follows:

O.1. The geometry of the section represented by y, z - the coordinates of the connection points on the contour of the non-standard collector.

O.2. Geometric and functional characteristics of the collector section: data for partially filled section (h, B, A, R, z_G, W and K) and full section (h_p, B_p, A_p, R_{hp}, z_{Gp}, W_p and K_p). The calculated data are represented in a table depending on the length gap chosen.

O.3. Graphic representations for contour geometry and hydraulic characteristics:
the characteristic curves of the geometry of the collector section $B = f_B(h)$, $z_G = f_{zG}(h)$;
the hydraulic characteristics of the flow section of the collector $A = f_A(h)$, $P = f_P(h)$, $R_h = f_{Rh}(h)$.
the hydraulic-functional characteristics of the collector $W = f_W(h)$ and $K = f_K(h)$.
the hydraulic characteristics of partial filling of the collector $Q/Q_p = K/K_p = f_Q(h/H)$ and $v/v_p = W/Wp = f_v(h/H)$.

The simulation results are presented in the form of customized tables on the operating domains and on the calculation sections of the collector, as well as by graphical representations of the calculated functions.

3.4 Hydraulic calculation in non-uniform permanent motion

The hydraulic parameters (flow Q and depth h, or elevation z) are a function only of the one-dimensional spatial coordinate l in the gradually varied permanent motion. The hydraulic calculation model in the varied gradual non-uniform permanent movement consists of: 1 - the differential equation of the varied gradual movement; 2 - contour conditions. A synthesis of the developed hydraulic model is presented next.

A calculation section of the main sewerage collector, which is delimited by two successive homes with a feeding role, is considered as a prismatic/cylindrical bed and with a constant slope. On this section, a permanent flow regime (Q = const) can be admitted and the fundamental equation of the varied gradual non-uniform movement has the form (Eq. (8)).

Eq. (8) is a first-order differential equation with partial derivatives, in which case it can be transposed in the form:

$$\Psi(h)dh = dl \tag{21}$$

where:

$$\Psi(h) = \left(1 - \frac{\alpha Q^2}{g} \bullet \frac{B(h)}{A^3(h)}\right) / \left(1 - \frac{Q^2}{I - \frac{Q^2}{K^2(h)}}\right). \tag{22}$$

Equations of the type (Eqs. (8) and (22)) are generally solved with the following simple boundary condition:

$$l = l^0; h = h^0 , \tag{23}$$

where l^0 and h^0 represent known values for the coordinate l and the depth h, respectively.

In this case, the Cauchy problem defined by the relations [Eqs. (21)and (23)] presents the following particular solution explained in relation to the variable l by the relation:

$$l = l^0 + \Phi(h) - \Phi(h^0) \tag{24}$$

where the function $\mathbf{F}(h)$ has the expression,

$$\Phi(h) = \int \Psi(h)dh \tag{25}$$

and represents a primitive of the function $\mathbf{Y}(h)$.

The solution of the equation (Eq. (24)) is of the form

$$l = f_s(h, h^0, l^0) \tag{26}$$

and can be explained later in the h variable in the form

$$h = f_l^{-1}(l, l^0, h^0) \tag{27}$$

The main sewer collectors have complex geometric shapes, where the characteristics $A(h)$, $B(h)$ and $K(h)$ are expressed by complicated functions, which are defined by portions, or graphically or tabular. The primitive function (Eq. (25)) can be determined by numerical methods, with the use of quadrature formulas or Taylor series expansions.

The concrete analytical expressions for the contour conditions in the permanent regime are established on the basis of the hydraulic-functional characteristics of the accessory constructions existing along the length of the collector (manholes with various functions). For some types of manholes, the hydraulic characteristic was determined taking into account their structural form and the way of access and evacuation of waste water.

To solve the mathematical model, a numerical calculation program named Collector_Movement_Permanent_Gradual_Variate (Col_MPGV) was developed in the MATLAB programming environment.

The results obtained are represented by hydraulic parameters of the movement along the collector and in imposed calculation sections, as well as the graphic representation of the curves of the free surface of the water. Thus, the water depth h or free level elevation z, degree of filling $a = h/H$, water speed v and others can be checked along the collector.

4. Results and discussion

The Col.NS_UM_UB program was tested and validated for the „standard semicircular bell" profile with parameters: B/H = 600/380 mm, uniform roughness on the perimeter $n = 0.0135$ and geodesic slope $I = 0.0004$. The calculation steps for height H and width B were Dt_H = 0.02 m and Dt_B = 0.02 m, respectively. The number of points considered was $N_M = 5$.

The data obtained were compared with those obtained by using the classical calculation method for the standard semicircular bell section. The results obtained (**Table 1**) and analyzed comparatively show a good agreement between the values calculated with the Col.NS_UM_UB program and the values calculated with the classic model used in the design of standardized sewer collectors.

The calculation program allows obtaining a high accuracy of the obtained data and a significant reduction of the calculation time. The depths corresponding to the maximum values for the hydraulic radius R_h and the velocity module W are identical, because the roughness is uniform throughout the perimeter of the sewer collector.

In the next stage, the uniform movement was simulated for a sewer collector with non-standardized semicircular bell-type profile (**Figure 6a**), for which the geometric and hydraulic parameters were calculated. The section of the collector is bell type with dimensions B/H = 2.800/2.400 m and is made of reinforced concrete. The roughness on the perimeter is uneven in the range of values $n = 0.015–0.019$, specific to collectors made of reinforced concrete and in operation for a long period of time.

The input data to the program (I.1, I.2, I.3, I.4) in a synthetic presentation were: calculation code for non-standardized semicircular bell; calculation steps Dt_H = 0.02 m, Dt_B = 0.02 m, $B = 2.800$ m, $H = 2.400$ m, $n = 0.017–0.019$ (Manning) and $N_M = 5$.

Type	A_p, (m^{-2})	R_{hp}, (m)	Q_p, (m^{3}s^{-1})	v_p, (ms^{-1})
CC	0.1738	0.1104	0.591	0.3416
CP	0.1735	0.1107	0.594	0.3406

CC, classical calculation; CP, calculation program.

Table 1.
Comparative analysis of calculation program verification data.

The exit data from the program, in a synthetic presentation, were as follows:

O.1. Contour geometry: NA = 2, where the coordinates of the connection points y, z, on the collector contour are shown in bold characters in **Table 2**; code with analysis specification, Kod = 2_circle arcs; a = [1.4800, 1.0000]: ordinates of the centres of the support circles of the arcs, (m); b = [0.0000, 0.0000]: the abscissas of the centres of the support circles of the circle arcs, (m); R = [1.4800, 1.4000]: radii of the support circles of the circular arcs, (m).

Five representative points N_M of coordinate's yC, z (**Table 2**) were considered on the contour of the collector to define the shape. The spikes were evaluated with the equation (Eq. (20)).

O.2. The geometric and hydraulic-functional characteristics: N_C = 120. The N_C elements of each of the vectors **hC, BC, AC, RhC, zGC, WC** and **KC**, which constitute the columns of the matrix M, are tabulated (**Table 3**). Also, the values of the parameters for the full section h_p, B_p, A_p, R_{hp}, z_{Gp}, W_p and K_p are presented in **Table 3**, last line in bold. The results presented in **Table 3** are used as a database for the simulation of the gradually varied movement in the collector.

The absolute and relative maximum values of the hydraulic parameters W, K and for the absolute water depth h and relative h/H at which these parameters are recorded in the circular bell collector are centralized and presented in **Table 4**. Due to the roughness coefficient n, which is uneven on the perimeter of the collector, the

Coordinates	1	2	3	4	5
yC, z (m)	**0**	1.109	**1.400**	1.212	**0**
yC, z (m)	**0**	0.500	**1.000**	1.700	**2.400**
n = [0.0190–0.0150]					

Table 2.
The coordinates of the representative points on the contour non-standardized semicircular bell type collector.

hC (m)	**BC (m)**	**AC (m^2)**	**RhC (m)**	**zGC (m)**	**WC (ms^{-1})**	**KC (m^3s^{-1})**
0.010	0.343	0.002	0.007	0.006	1.863	0.004
0.050	0.764	0.025	0.033	0.030	5.435	0.139
0.291	1.763	0.349	0.185	0.174	17.098	5.977
0.512	2.239	0.795	0.313	0.304	24.282	19.318
0793	2.262	1.483	0.460	0.467	31.392	46.570
1074	2.796	2.253	0.590	0.627	37.172	83.768
1.496	2.618	3.405	0.726	0.849	43.515	148.167
1.918	2.114	4.417	0.778	1.044	46.234	204.224
2.359	0,666	5.106	0.690	1.187	43.382	218.791
2.380	0.473	5.117	0.674	1.190	42.752	218.791
2.400	**0.000**	**5.124**	**0.635**	**1.192**	**41.223**	**211.225**

Table 3.
Matrix elements M = [hC BC AC RhC zGC WC KC] obtained for the non-standard semicircular bell collector (extracted from 120 values).

Parameters		Geometric parameters				Hydraulic parameters		
		A (m^2)	B (m)	z_G (m)	P (m)	R_h (m)	W (m/s)	K (m^3/s)
1		2	3	4	5	6	7	8
Maximum values	Abs.	2.800	5.124	1.193	8.070	0.778	46.304	225.68
	Rel.	—	—	—	—	1.226	1.123	1.068
h, (m)		1.014	2.400	2.400		1.938	1.998	2.259
h/H, (−)		0.422	1.000	1.000	1.000	0.807	0.832	0.941

Table 4.
Maximum values of the geometric and hydraulic parameters obtained for the non-standard ovoid type section B/ H = 2.800/2.400 m.

depths corresponding to the maximum values for the hydraulic radius R_h and the velocity modulus W are different.

Q.3. Graphic representations for the contour geometry and the hydraulic and functional characteristics of the analyzed collector: graphs for the geometric characteristic curves, the hydraulic characteristic curves and the hydraulic-functional characteristic curves.

In **Figure 7** the geometric characteristic curves $A = f_A(\mathrm{h})$, $B = f_B(h)$, $z_G = f_{zG}(h)$ and the hydraulic ones $P = f_P(h)$, $R_h = f_{Rh}(h)$ are presented.

Figure 8a shows the hydraulic-functional characteristics $W = f_W(h)$ and $K = f_K(h)$. **Figure 8b** shows the hydraulic-functional characteristics when the collector is partially filled, respectively $f_Q(h/H) = Q/Q_p = K/K_p$ and $f_v(h/H) = v/v_p = W/W_p$.

Performing a simulation of the operation of a collector with standard or non-standard section in permanent movement requires the serial use of the three simulation programs made for uniform movement, for the hydraulic characteristics of the inspection chambers and for non-uniform movement.

The results obtained by running the program and which define the parameters of the permanent uniform movement at this sewer collector can be used in the analysis of the non-uniform permanent movement, respectively the non-permanent movement. The Col.NS_UM_UB program was also run for a collector with a section

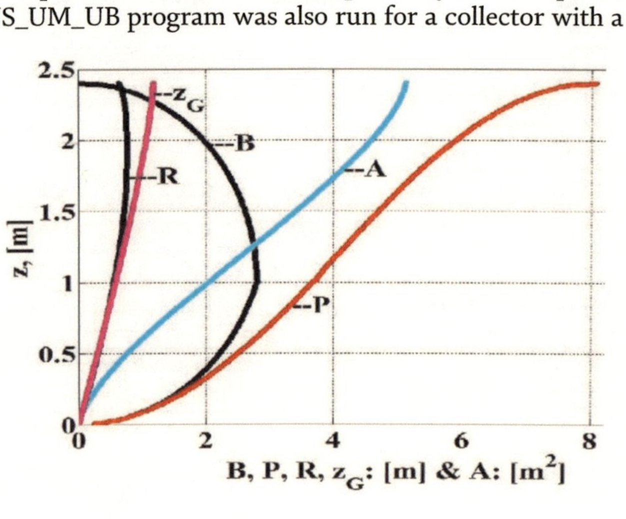

Figure 7.
Geometric and hydraulic characteristics $A = f_A(h)$, $B = f_B(h)$, $z_G = f_{zG}(h)$ $P = f_P(h)$, $R_h = f_{Rh}(h)$ *at the non-standardized semicircular bell collector* B/H = 2.80/2.40 *m.*

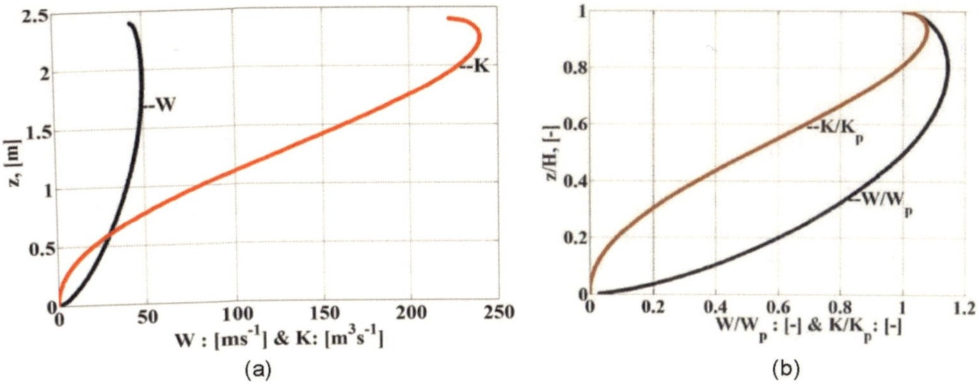

Figure 8.
Hydraulic and functional characteristics at the non-standardized semicircular bell collector B/H = 2.80/2.40 m: a – For W = f_W (h) *and* K = f_K (h)*; hydraulic flow characteristics* f_Q (h/H) = Q/Q_p = K/K_p *and* f_v (h/H) = v/v_p = W/W_p *at partial filling.*

consisting of circle arcs + straight line segments (**Figure 6b**). The obtained results were used to simulate the non-permanent movement in the case of the transport of rainwater on a collector with dimensions 3.450/2.250 m and length 2416 m [15, 19].

The separation of domestic wastewater from rainwater in some urban sewerage networks in Romania has determined the modification of the geometric shape of the main collectors. The modernization principle consists in creating a pipe inside the collector, which will separately transport household waste water. The modified remainder of the collector section will carry storm water only. The monitoring of the hydraulic and functional parameters of the collector with the modified geometric shape, especially during the evacuation of torrential rains, requires the use of special simulation programs. Thus, the research carried out and the creation of a set of programs for simulating the permanent and non-permanent movement of waste water in non-standard sewerage collectors contribute to the technical basis of the monitoring system.

5. Conclusions

The main sewerage collectors of domestic and rainwater with a “visitable” type of construction have a long operating life, and the action of natural and anthropogenic factors in the site determines the evolution of complex processes of degradation of the flow section. Physico-chemical degradation processes cause the flow section to change to a shape different from the designed one.

The flow section of the visible sewer collectors can be modified as a geometric shape over time by hydrodynamic erosion and clogging phenomena, but also by technological repair and rehabilitation works. Degradation processes determine a new geometric shape consisting of curves and lines that change the value of the hydraulic parameters in the flow section.

The monitoring of the hydraulic parameters of the collector during the exploitation process requires a correct measurement and interpretation of them with high precision, a situation that cannot be achieved with collectors that present geometric shapes modified over time or non-standardized sections.

The calculation programs developed for uniform and non-uniform permanent movement have the advantage that they allow obtaining the geometric and hydraulic parameters of the sewer collector with non-standardized section with high accuracy and in a relatively short working time. The developed calculation programs have the advantage of taking into account the roughness variation on the wetted perimeter of the flow section.

The application of the developed calculation programs requires the presence of a database, which must be achieved through measurements and data retrieval from the inside of the visitable sewage collector. The database must contain a cross-sectional topographic survey and a detailed longitudinal profile, as well as an analysis of the values and variation of the roughness on the perimeter of the collector.

The hydraulic-mathematical model created can be generalized to obtain the geometric and hydraulic characteristics [$A = f_A(h)$, $P = f_P(h)$, $R_h = f_{Rh}(h)$, $B = f_B(h)$, $z_G = f_{zG}(h)$, $W = f_W(h)$ and $K = f_K(h)$] for single-bed closed channel sections consisting of curves, circular arcs and straight line segments. The simulations carried out and on concrete cases existing in practice confirm the viability of the calculation program.

Nomenclature

h (m)	water depth
z (m)	free water level share
A (m^2)	cross sectional area of flow
P (m)	wetted perimeter
R_h (m)	hydraulic radius
B (m)	width of water surface
b (m)	width at the bottom of the channel
C ($m^{0,5}s^{-1}$)	Chézy coefficient
g ($m\ s^{-2}$)	gravitational acceleration
n ($s/m^{1/3}$)	roughness coefficient (Manning)
k (m)	roughness
v ($m\ s^{-1}$)	average speed
Q (m^3s^{-1})	flow
a (–)	degree of filling
W ($m\ s^{-1}$)	speed module
K (m^3s^{-1})	flow module
J (–)	hydraulic slope
z_a (m)	apron share of the collector
D (m)	collector diameter
H (m)	collector height
r (m)	radii of curvature
l (m)	calculation length
L (m)	total length
v_p ($m\ s^{-1}$)	speed at full section
Q_p (m^3s^{-1})	flow at full section
z_G, h_G (m)	ordinate to the centre of gravity of the watered area

Author details

Mihail Luca
"Gheorghe Asachi" Technical University of Iasi, Iasi, Romania

*Address all correspondence to: mluca2015@yahoo.com

References

[1] NE 133-2013. Regulations for the Design, Execution and Operation of Water and Sewage Systems. Vol. 2. Sewage Systems. Bucharest, Romania: Matrix Publishing House; 2013. pp. 266-305

[2] Kiselev PG. Straightened for Hydraulic Calculations. Bucharest: Technical Publishing House; 1977. pp. 122-179. (in Romanian)

[3] Blitz E, Pislarasu I, Vladimirescu G. Sewer. Hydrotechnical engineer's Manual. Bucharest, Romania: Technical Publishing House; 1970. pp. 984-992. (in Romanian)

[4] Luca M. Technical Hydraulics – Permanent Movement in Channels. Vol. 1. Iasi, Romania: Tehnopress Publishing House; 1998. p. 262. (in Romanian)

[5] Schaffner J, Obverlack M, Kirchheim N. The application of numerical modelling (3-D) for the calculation of flush waves in sewer channels. In: 6th International Conference on Urban Drainage Modelling (UDM04), Dresden, Germany. 2004. pp. 1-8

[6] Safari MJS, Aksoy H, Unal NE. Self-cleansing Design of Circular Urban Drainage Channels. In: Proceedings of the 3rd International Conference "Water across Time in Engineering Research" Constantza. Romania: "Ovidius" University Press Constanza; 2016. pp. 245-252

[7] Jiang Y, Chen J. Analysis of the velocity distribution in partially-filled circular pipe employing the principle of maximum entropy. PLoS One. 2016;**11**: 1-7. DOI: 10.1016/S0273-1223 (97) 00575-1

[8] Bardiaux JB, Bonakdari H, Larrarte F, Mose R, Vazaquez J. Velocity profiles through a sewer channels: Using CFD to obtain velocity fields. Dresdner Wasserbauliche Mitteilungen. 2006;**32**: 32-335

[9] Reguiero-Picalo M, Naves J, Anta J, Puertas J, Suarez J. Experimental and numerical analysis of egg shaped sewer pipes flow performance. MDPI – Water. 2016;**8**:1-9. DOI: 10.3390/w8120587

[10] Alihosseini M, Thamsen P, U. Numerical and experimental investigation of flow in partially filled sewer pipes. Technische Mechanik. 2019;**39**(1):113-124. DOI: 10.24352/UB.OVGU-2019-011

[11] Jarman DS, Faram MG, Butler D, Tabor G, Stovin VR, Burt D, et al. Computational fluid dynamics as a tool for urban drainage system analysis. A review of applications and best practice. In: 11th International Conference on Urban Drainage, Edinburg, Scotland, UK. 2008

[12] Fuchs L, Beeneken T, Sponemman P, Scheffer C. Model based real-time control of sewer system using fuzzy logic. Water Science & Technology. 1997;**36**(8–9):343-347. DOI: 10.1016/S0273-1223 (97)00575-1

[13] Brito RS, Almeida MC, Ribeiro A. Enhancing hydraulic data reliability in sewer. Water Practice & Technology. 2021;**17**(1):434-444. DOI: 10.2166/wpt.2021.093

[14] Tamasanu F. Contributions to the hydraulic and functional optimization of water intakes. [Thesis]. Romania: "Gheorghe Asachi" Technical University of Iasi; 2012. (in Romanian)

[15] Luca M, Luca Al L, Scripcariu CF, Chirica S. Contributions to simulation non-permanent movement in sewerage collectors. IOP Conference Series: Materials Science and Engineering. 2019; **586**:012037, IOP Publishing. DOI: 10.1088/1757-899X/586/1/012037

[16] Luca M, Tamasanu F, Luca Al L. Flow Modelling with Free Level in the Polycentric Channels. Ovidius University Annals, Year XII, Issue 14, Civil Engineering Series, Constantza, Romania: "Ovidius" University Press Constantza; 2012. p. 155-162

[17] Luca M, Cretu FC, Scripcariu CF. Consideration on topographic parameters of rehabilitation sewer collectors. RevCad Journal of Geodesy and Cadastre. 2017;**22**:237-242. "1 December" University of Alba-Iulia, Aeternitas Publishing House Alba-Iulia, Romania. ISSN 1583-2279

[18] Popescu S. Computer Applications In The Hydraulics Of Hydrotechnical Systems. Iasi, Romania: Cermi Publishing House; 1999. pp. 81-124. (in Romanian)

[19] Scripcariu CF. Contributions to the Optimization of Sewage Network in Order to Increase Operational Safety [Thesis]. Romania: "Gheorghe Asachi" Technical University of Iasi; 2018. (in Romanian)

Chapter 8

Achieving Sustainable Development Goal Related to Water and Sanitation through Proper Sewage Management

Aditi Agarwal, Amanpreet Kaur, Sonika Saxena and Sudipti Arora

Abstract

Due to urbanization, there is an increasing need for infrastructure and services, leading to pressure on the sewage system. As a result, water contamination and sewage-related illnesses are emerging. On-site sanitary facilities are insufficient, and current sewage systems are outdated, causing freshwater contamination and diseases such as typhoid, malaria, etc. Untreated domestic sewage/wastewater, mining waste, industrial wastewater, agricultural waste, and other contaminants are polluting most aquatic ecosystems worldwide, leading to harm to surface water bodies, sewage drainage systems, surface water, and groundwater. Various sewage disposal methods are discussed, but they are not sustainable. The UN proposed Sustainable Development Goals (SDGs) in response to the need for sustainability and the effects of pollution and population growth. SDG 6 aims to ensure equitable access to safe and affordable drinking water, sanitation, and hygiene. It also includes goals to enhance water quality, increase water usage efficiency, develop integrated water resource management, and restore aquatic ecosystems. Efficient sewage disposal is crucial to reduce detrimental effects on the environment and public health. It is necessary to emphasize SDGs to protect the environment sustainably. It is crucial for the international community to work together to find effective and sustainable solutions to the problem of sewage management.

Keywords: sewage disposal, sustainable development goals, sanitation, eutrophication, contamination

1. Introduction

Water scarcity is a major global issue. It is so relevant that United Nation has given priority to water management across the globe. Similarly, India also is dealing with water crisis and there is a growing need for freshwater at household and commercial level. As per the latest reports by the committee of the Ministry of Water Resources (2000), India is overall consuming 1093 billion m^3 and this consumption will increase to 1447 billion m^3 by the next 30 years. The common man is enraged and agitated about the pollution issues in India and worried about it getting worse each passing day. Negligent disposal of industrial and domestic wastewater is polluting the water-bodies of India. In Indian culture, the water bodies are not only the source of life but are considered very sacred. Indians are religiously sentimental about the water bodies of the nation are getting serious about their detriment [1].

Population wise India wears an inglorious crown as a second-highest population on the entire planet. India is known as a land of agriculture, thus 93% of freshwater water goes in agricultural practices and remaining 7% is consumed by industrial and domestic processes. Due to consistent utilization of surface freshwater, currently in India, the three fourth of freshwater bodies are now contaminated by the discharge of untreated sewage wastewater. Owing to regrettable sanitation practices, 80% of the total water consumption of Indian subcontinent is returned as wastewater, polluting the waterbodies and soils.

As per recent reports by environmentalists and environmental scientists, many major Indian cities and suburbs are facing persistent issues in the matter of solid liquid waste management including sewage management, especially in rural areas. Current situation of rural India paints a really bad picture of sewage and sanitation management. Villages of the country are still lagging behind in the matter of proper sanitation practices and better sewer lines in domestic surroundings. Mixing of sewage waste in open freshwater is a major cause of pathogenic contamination in domestic water supplies and diseases related to it. Water borne pathogens are responsible for approximately 3.5 million human deaths as per recent reports by UNDP [2]. According to a report of WHO-UNICEF, Indian villages contributes to more than 50% open defecation practices responsible for many health concerns including high infant mortality, spread of water-borne diseases and child stunting [2]. Freshwater bodies such as lakes, rivers, ponds and groundwater are getting contaminated by sewage due to ignorant open defecation practices in their catchments.

Similar situation is recognized in urban settlements as well, according to some recent reports, urban domestic sewage is a primary cause of decline in aquatic freshwater ecosystem of India. On account of inadequate treatment systems in most the Indian cities, untreated or half treated sewage run offs are fusing with the natural water resources, also contaminating the rivers of cities with effective treatment plants as well. Domestic sewage contributes more than 70% of total pollutants in water as per reported in Water Aid (2016a) [1].

Domestic sewage from major cities and towns of north India alone contributes for the 1528 million m^3 sewage waste along the banks of the river Ganga every day. Unfortunately, due to ever growing population and urbanization along with the languid implementation of Government policies, the sanitation practices are still not on the path of any improvement [3].

In developing nations, sewage or domestic wastewater is main source of water supply due to growing water scarcity but to avail that water, it needs to be properly treated and analyzed before use. Owing to lack of awareness and improper treatment methods, maximum population is consuming this contaminated water on daily basis which is posing serious threat to the community as a whole. The gray wastewater constitutes the washes coming out from households, institutions, industries, business establishments etc. However, the main constituent in sewage wastewater is fecal matter of human and animal activities. Due to the presence of fecal matter, sewage water is rich in organic compounds and pathogenic enterobacteria [4].

Organic wastes, mainly in the form of sewage sludge is recently been considered an important resource available to meet the ever-increasing demand of renewable energy across the globe. Organic compounds can be employed as a resourceful substrate to generate energy in the form of heat and can be utilized in advanced new age technologies as well. In addition, the organic contaminants in sewage can be used as a source of fertilizers for soils and bioremediation of infertile lands [5]. But most of the time these organic contaminants are considered pollutants and cause serious harm to the ecosystem.

In lieu of changing environment and challenges posed by water scarcity and growing demand of clean and hygienic water for consumption, United Nations (UN) general assembly reinforced the Agenda 2030 for sustainable development in September, 2015. Out of 17 Sustainable Development Goals (SDGs) defined for different societal issues, SDG 6 was dedicated completely to the access of clean and adequate water to every individual. The United Nations Environmental Program (UNEP) adopted SDG 6 as clean water and sanitation, which is subdivided into targets to be achieved by the end of 2030. The main elements of the SDG 6 include the availability of quality water to each individual, treatment of wastewater, efficient use of available freshwater, integrated water resource management and sustainable ecosystem. In totality, the SDG 6 aims to ensure the sustainable management and availability of proper sanitation and clean water for one and all by the end of this decade.

These diverse set of objectives is a reflection of the growing realization that a lot of problems with water management, adaptability, and administration must be rectified if the humanity ought to be experiencing overall growth in sustainable manner [6]. In order to achieve Sustainable Development Goal 6 (SDG 6), it is essential to carefully monitor and evaluate social and fiscal water demand at the regional level in a way that does not harm the environment on global scale. India holds a crucial position among developing nations with a complex interrelated web of a rapidly expanding population, geophysical pressure, social insecurity, and economic disparity pertaining to the usage, availability, and access to water and basic sanitation systems [7].

2. Evolution of sewerage systems

"The history of men is reflected in the history of sewers" famously quoted by Victor Hugo in 1892 in his equally famous literature "Les Misérables". He claimed that "it has been a sepulchre, it has served as asylum, crime, cleverness, social protest, the liberty of conscience, thought, theft, all that the human law persecutes or have persecuted is hidden in that hole."

It is well acknowledged that people's interactions with sanitation and drinking water have changed significantly over time as a consequence of societal, cultural, and religious factors. Urban communities have been conscious of the need for potable drinking water since ancient times. However, modern communities did not entirely comprehend the significance of basic sanitation for the safeguarding of public health until the 19th century. Historically, for both individuals and governments, managing wastewater had always presented political challenges and substantial need for technological advancement. The history of waste and sewage management is a reflection of both human brilliance and human frailty. The progress of sewage management over time has been addressed by many economists and scholars, but it is common that they typically lack an engineering insight on the issue.

Evolution of sanitation and sewage management over time is divided into few significant eras to understand the development of sanitation methods chronologically:

1. **Ancient Civilization**: The early human populations were distributed over vast areas, and the waste they produced was sent back into the environment where it was decomposed by organic processes. Disposal problems were essentially non-existent since they were small groups of wandering hunters and subsistence farmers. Raw sewage was discharged through holes made in the ground and closed after being used, in accordance with the Mosaic Law of Sanitation, until the initial developed civilization emerged [8, 9].

2. **The Time of Romans**: The Romans carried on the Assyrian engineering endeavors and put their concepts into practice to create a substantial infrastructure system to serve the entire community. The Romans built two networks to collect spring water and remove storm and effluent, thus creating the first integrated water service and controlling the water cycle from collection to disposal. The Romans recognized that surface water could be utilized for other things outside human consumption and that spring water was of far higher quality than surface water. They also recycled spa effluent by using it to flush bathrooms before disposing of the waste in sewers, which eventually led to the Tiber River. Ancient Rome had a very complex sewer system that was made up of multiple smaller sewers. Although, sewer and water pipes were already present in other Eastern civilizations, Romans did not create them, but they probably made significant improvements.

3. **The Era of Sanitary Downfall**: While the magnificent water distribution systems which would have established the Romans famous for generations were neglected, the huge baths were robbed of all of their belongings. Water started to be collected from streams and aquifers and dumped without even being treated, which was a historically unprecedented step in the wrong direction and caused disease to proliferate. Several thousand years passed during the unsanitary dark ages (476–1800).

4. **The Industrial Age:** The industrial revolution and the subsequent high pace of industrial growth during the 18th century led to an understanding of the significance of waste and sewage disposal. The experiment with organized action to enhance urban environmental conditions started in Britain, one of the first nations to do so. The Bazalgette sewerage system in England, which was built between 1858 and 1865, is a perfect demonstration of this concept. Wastewater was transported from the streets and then released into the Thames by a network of collection sewers and reservoirs. No one understood the river's capability for assimilation or the necessity of removing pollutants before releasing them into the stream.

In Frankfurt, the development of a system in 1867 marked the beginning of the widespread installation of sewers in German cities. The reconstruction of Paris began in June, 1853, when George-Eugene Haussmann took his oath of office as prefect of the Seine. Eugène Belgrand was given the task by Haussmann of conducting a significant restructuring of the city's existing sewerage system beginning in 1854.

The Pugliese Aqueduct, which brought water from the Sele River to Bari, was the last of the major public works projects in Italy that were built between 1870 and 1915. These projects included the aqueducts of Serino and Selino. Almost invariably, building aqueducts for "aristocratic" drinking water was favored over building sewers to collect effluent, partly because the cost of aqueducts was relatively low due to the reusing of old Roman pipeline infrastructure and the construction & maintenance were funded by overseas companies.

The construction of "sewers" utilizing hollowed-out logs began in the 1700s in large towns like Boston and Chicago. The first "water pollution control" law was implemented in Massachusetts, a British colony, in 1647.

5. **The Era of Strict Environmental Regulations**: The twentieth century ultimately saw a breakthrough in environmental research, sewage treatment, and people's sentiments towards pollution. Throughout the century, scientific advancement, societal values, and government actions changed, starting with unregulated pollution and concluding with efforts to strengthen control (10).

The 8th Report of the Royal Commission on Sewer Systems in 1912, established rules and Biochemical tests to be performed on sewage and sewage sludge and introduced the idea of biochemical oxygen demand (BOD), was a defining moment and was already imitated by many other nations.

Scientists were capable of predicting the permissible limits of BOD loads to surface freshwaters owing to the creation of aeration/deaeration models by Streeter and Phelps (1925) and later developed by Imhoff and Mahr (1932) [10, 11]. Meanwhile, authorities made waste management mandatory. Wastewater treatment plants were built in the major cities of Europe prior to the World War I, but warfare contributed in their construction to be suspended [12, 13].

However, political ideologies in several nations impeded the management of sewage. For instance, when the national socialist party was elected to power in Germany, they changed the way wastewater was handled. The "Blood and Soil" philosophy emphasized agricultural exploitation over eliminating the contaminants before use, resulting in massive irrigation of sewage wastewater for agriculture practices.

Later on, the World War II also slowed down the sewage treatment systems till 1948, resulting in increased water pollution. Furthermore, several wastewater treatment facilities suffered damage during the conflict and were never repaired [12].

The United Kingdom and the United States saw tremendous advancements in sewage waste management after the war, but not Europe [13]. Pollution discussions by 1950 were centered on stream use categorization and freshwater quality standards, which were prerequisites to the development of a strategy for sewage management [14]. The general relationship between industrial water pollution and toxicity was established as early as the first decade of the 20th century [15].

In the late 1970s, publicly available gas chromatography and atomic absorption spectrophotometry techniques contributed to the further advancement in our knowledge of environmental pollution [16, 17]. This made possible to identify environmental contaminants accurately. In the early 21st century, a roadmap for the advancement of analytical techniques was established [18].

3. Access to sanitation: current Indian scenario

The extent to which modern sanitation systems are utilized is significantly influenced by the availability of water [1]. In metropolitan areas, over 90% of individuals have safe drinking water, and therefore more than two-thirds have access to basic sanitation facilities, based on research by World Bank in 2011. Access to reliable, cost-effective, and sustainable water supply and sanitation (WSS) services, however, is still a challenge. None of the Indian cities have constant piped water access. Flowing water is never delivered for more than a few hours on daily basis, regardless of the quantity supplied. Untreated sewage regularly spills in exposed sewers. Those who live in cities make up only about half of the population. Between 30 and 70 percent of the water delivered is thought to be non-revenue water, which is caused by leakage, faulty connections, inefficient billing and collection practices, etc. Less than 30–40% of maintenance and operational expenses are covered by user fees. The bulk of urban businesses depend heavily on grants for operating and capital to survive [19].

4. The untold story of pits in urban India

A flushable toilet has come to symbolize modern urban population living in India and throughout the world. More than 80% of Indians living in metropolitan areas have access to solitary or communal bathrooms. Instead of being linked to a sewer network, most of these toilets are pour-flush and joined to a pit or septic systems. In the absence of pipes to remove fecal waste from pits and tankers, it ought to be transferred by non-waterborne methods. Until about the mid-2000s, most pit cleaning work was done by hand. Waste is discarded distant from households in the caste-based practice of manual "scrounging" after being scraped out of dried, unsewered lavatories. Due to recurring sanitation employee fatalities as well as the vile aspect of manual scouring, the Indian government approved the "Prohibition of Employment as Manual Scavengers and their Rehabilitation Act" in 2013. For collection, a water tank with a sturdy, leak-proof design is a requisite [20].

The basic treatment system, or septic tank, lowers the biochemical oxygen demand (BOD) by 30 to 50 percent on average while minimally affecting the pathogens and nutrient content. The wastewater from the tank therefore has to be remedied further, either with underground dispersion systems or impenetrable secondary and tertiary treatment methods. Conventional septic tank systems, also known as "septic systems," include a septic tank and a subsurface dispersal mechanism, like a soak pit and dispersion trench. The soak pit plan has obstacles due to costs, close communication with the septic tank, and associated activities.

Due to a shortage of consistent desludging solutions, families opt for bigger septic tanks which needs less frequent desludging. However, 67% of pit latrines larger than 2000 L only have one compartment, rendering them inefficient for their intended use. The construction of a subsurface dispersion system in addition to the onsite repair of the basic unit would be required in order to handle wastewater through soak pits under these circumstances [21].

5. Deteriorating health status

Water availability is essential for human life. According to the report published in 2021 on World Water Development by UNESCO, water usage has increased exponentially in the previous century and it has increased by about 1% each year since the early 1990s. As a consequence of increased water consumption, water quality is experiencing serious crises. Human wellbeing and long-term socioeconomic development are eventually harmed by environmental pollution and degradation forced by urbanization, subsistent agriculture, and rapid industrialization, which have detrimental effects on the rivers and streams that are critical for health of ecosystem [22]. In recent reports of 2022, WHO estimates that 8,29,000 individuals annually die from acute diarrhea as a consequence of inadequate hygiene practices, sanitation, and potable water. However, diarrhea is generally avoidable, and by addressing these health risks, 2,97,000 newborn fatalities under the age of 5 might be avoided annually [23].

In India, the pathogens mostly responsible for water-borne infections include the bacteria *Escherichia coli*, *Vibrio cholerae* and *Shigella*. Other than that, parasites such as hookworm, *Entamoeba histolytica* and *Giardia* as well as the viruses Hepatitis A, poliovirus, and rotavirus are also responsible for the diseases. The presence of chemicals in the water are known to cause health issues as well. Pesticides which are washed into streams as well as other sources of fresh water can also harm the nervous, endocrine, and reproductive organs.

Phosphorus, organophosphorus, and related compounds can all cause cancer. Infants who are consuming milk are usually diagnosed with blue baby syndrome if nitrate toxicity in drinking water is present. Similarly, Lead contamination affects the central nervous system. Malignant melanoma and complications related to it are both brought on by arsenic poising. Fluoride contamination can damage the nervous system and lighten teeth color. The use of petrochemical products can cause cancer even in little dosages [1]. Skin problems such as melanosis and keratosis are connected to the high arsenic levels in water supply, main cause of water poisoning in Bangalore, as per reported by Kazi *et al.* in 2009 [24]. According to a different study carried out in Bangladesh, stream pollution is a significant contributor to the prevalence of scabies there [25]. Nitrate was the leading contaminant in India in 2019, and it was discovered in 387 districts. India has been ranked 120 out of 122 nations in regards to water quality, having 70% of the country's fresh water being assessed to be highly polluted, as per Niti Aayog [26].

6. Economic and environmental sustainability

The injudicious utilization of freshwater and common wastes from residential, farming, and manufacturing sector in natural lakes and rivers has exacerbated the challenges related to purity, cleanliness and availability, of water supplies. It has been determined that contaminants such as oxygen-demanding substances, diseases causing microbes, micronutrients, minerals, and synthetic organic substances are possible contaminants of municipal effluents.

Ingredients like ammonia in freshwater sources that need oxygen may endanger the aquatic ecosystems. Pathogens are introduced into the subsurface waters via sewerage from municipal wastes, storm water runoffs and industrial effluents. Substantial levels of nutrients including carbonates, nitrate, and phosphate are present in farm effluent. High nutrient levels, particularly those of phosphate and nitrate, if improperly managed, can lead to enrichment, which encourages the proliferation of algae and ultimately result in eutrophication of waterbodies. Hence, prior to getting dumped into the water system, effluent must always be monitored and treated to prevent harm to aquatic life and natural reservoirs [27].

The easy accessibility to sewage treatment plants is already present in urban areas, however, even there in the first, the amount of sewage generated as well as the amount handled varied significantly. The nation's expanding economic and population growth are surpassing the development of water infrastructure in India's key areas. Treatment systems for wastewater should be also be ecologically acceptable for wholesome development. But environmental sustainability and efficiency of wastewater treatment systems is significantly hampered by commercial feasibility. Elevated levels of wastewater treatment may boost the cost of the infrastructure without necessarily improving the benefits, especially the immediate monetary rewards, that would be counterproductive to the facility's capacity to generate a profit. There are trade-offs between wastewater treatment technologies' economic and ecological efficacy. High levels of sewage treatment, which seem to be expensive, could not produce comparable economic advantages by minimizing the impacts of treating wastewater just on ecosystem [1]. India today has a fast-growing population, an unproductive infrastructure, ecological degradation, and social inequality. The concept of sustainable development, which is described as "development that addresses present needs without jeopardizing the ability of future generations to fulfill their own needs," was developed to minimize the above-mentioned effects [7].

7. Sustainable development goals

The United Nations adopted seventeen Sustainable Development Objectives in addition to 169 other goals in 2015. The Sustainable Development Goals (SDGs) cover three main areas of holistic development such as the environment including climate change, aquatic ecosystem, land ecosystems, etc.; society including poverty, malnutrition, non- discrimination, equality and security, institutional mechanisms, etc. and economic growth including reduced injustices, steady work, and economic expansion, etc. [28]. Out of these 17 goals, Sustainable Development Goal 6 is dedicated to access to clean water and sanitation for all.

8. Sustainable development goal 6

Access to proper sanitation and safe drinking water is the major health intervention that offers the greatest economic benefits as well. Nearly half of all the instances of malnourishment are caused by lack of safe drinking water, unhygienic environment, and inadequate sanitation. A further issue is the continuing decline in water purity and growing scarcity. By 2050, it has been predicted that at least one fourth of total world population would live in a country with persistent or long-term freshwater scarcity, affecting over 2 billion individuals currently. It is vital to put measures in place that will ensure human utilization of water is sustainable and environmentally efficient. In light of this, SDG 6 includes far greater array of water related targets than just improving accessibility [29] (**Figure 1**).

SDG 6 is essential since it also affects some other SDGs directly, such as SDG 1 (No Poverty), SDG 2 (Zero Hunger), SGD 3 (Good Health and Wellbeing), SDG 11 (Sustainable Cities and Communities), SDG 14 (Life Below Water) and SDG 15 (Life on Land) [18]. SDG 6 contains eight targets, all of which deals with different challenges regarding water [6].

As per the recent reports by Integrated Water Resource Development, India would likely require more than 900 billion m^3 of water in a low load situation and more than 1000 billion m^3 in a peak load situation by the end of the year 2050. The nation's present water supply is 695 billion m^3, whereas the current sustainable groundwater availability has been estimated to be 1137 billion m^3.

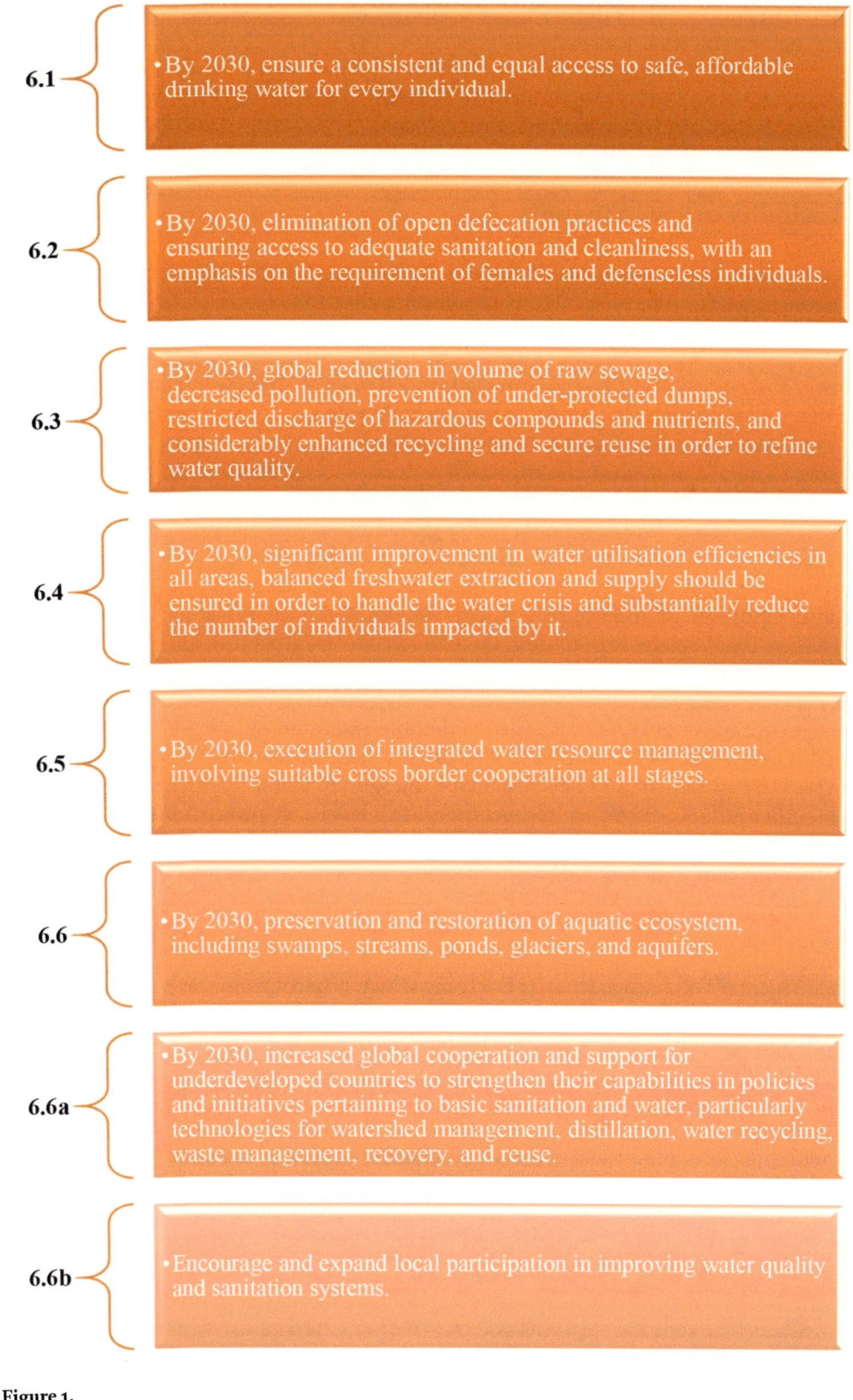

Figure 1.
Targets of SDG 6 to be achieved by 2030.

6 billion population in India endure mild to severe shortage of water, and fluoride poisoning of aquifers has already been found in 184 regions, based on a recent study by the Green Governance Initiative [6]. It is crucial for India to start advancing towards achieving SDG 6 tragets as early as possible to combat the growing issues of water scarcity and poor sanitation among others.

9. Feasibility of SDG 6 in India

Innovation, technology, and materials engineering advancements offer tremendous opportunities to accelerate the implementation of SDG 6. Science solely, however, cannot tackle all of the country's water-related issues. To accomplish SDG 6, three important pillars have been devised which includes sustainable management, science, and commerce. The economic approach towards freshwater scarcity, opportunities, and pollution cost has been given a little value. Due to its scarcity, water must be regarded as a non - renewable resource and its utilization should be controlled keeping in mind the sustainable approach.

In addition, the traditional approach to water management and architecture has to be reassessed. India's water technologists must design strategies which reuse sewage and other wastewater and differentiate between the water resource coming from different sources, expenditures, qualities, and dependability, each utilized for specific requirements and objectives. This necessitates a variety of local suppliers as well as the use of strategic planning, circularity, decentralization, and potential for eco-friendly alternatives.

In order to take connections, unpredictability, and transitions into account, management of water resources must be flexible and interconnected. Using integrated strategies makes it simpler to identify market, mitigate them, and comprehend unexpected consequences. They also enable integrated water infrastructure by drawing together numerous sectors and partners at all levels, spanning between regional to cross-border [30].

10. Conclusions

Management of sewage runoffs is a topic which is becoming ever more relevant. Developed economies are now stretching the purview of wastewater management to incorporate the elimination of trace nutrients and organic compounds in addition to the elimination of carbon - containing contaminants. Eventually, treatment facilities must provide discharge that would be suitable for direct water consumption. Considering the potential of recent technological advances, this seems to be feasible and will therefore spread across the globe. It's indeed crucial to comprehend the physical and chemical properties of wastewater in order to devise an adequate sewage treatment system, select a suitable approach, establish minimum standards for the remnants, ascertain the degree of assessment required to verify the system, and choose the byproducts to be assessed based on the level of toxicity.

As a nation with growing water issues and ever-growing population, India must reconsider its current assumptions and how we perceive and use water in order to meet the targets regarding SDG 6. We cannot continue to think of safe drinking water as something of an excessively abundant and affordable resource. We will be required to employ a considerable measure of creativity, analysis, and ingenuity to come up with solutions that preserve and design water resources environmentally, while also attempting to utilize water efficiently and equally. This same outcome of Sustainable Development Goal 6 (SDG 6) calls for sustainable improvement in the fields of clean water supply and basic sanitation, that further necessitates proper scrutiny and understanding of societal and economic need of potable water at the nation's level in a manner that ought not to propagate adverse environmental consequences on the regional and global levels.

Acknowledgements

I hereby, acknowledge the resources provided by Dr. B. Lal Institute of Biotechnology, Jaipur for the completion of this chapter.

Conflict of interest

The authors declare no conflict of interest.

Author details

Aditi Agarwal, Amanpreet Kaur, Sonika Saxena* and Sudipti Arora
Dr. B. Lal Institute of Biotechnology, Jaipur, India

*Address all correspondence to: sonika@drblal.com

References

[1] Kumar MD, Tortajada C. Growth of treatment plants and reuse of treated wastewater. In: Assessing Wastewater Management in India. Singapore: Springer; 2020. pp. 53-58

[2] Chandana N, Rao B. A critical review on sludge management from onsite sanitation systems: A knowledge to be revised in the current situation. Environmental Research. 2022;**203**:111812

[3] Saxena R, Frost S. Sewage management in some Asian countries. Environmental Management and Health. 1992;**3**(1):18-26

[4] Topare NS, Attar SJ, Manfe MM. Sewage/wastewater treatment technologies: A review. Scientific Reviews and Chemistry Communications. 2011;**1**(1):18-24

[5] Kacprzak M, Neczaj E, Fijałkowski K, Grobelak A, Grosser A, Worwag M, et al. Sewage sludge disposal strategies for sustainable development. Environmental Research. 2017;**156**:39-46

[6] Sadoff CW, Borgomeo E, Uhlenbrook S. Rethinking water for SDG 6. Nature Sustainability. 2020;**3**(5):346-347

[7] Roy A, Pramanick K. Analysing progress of sustainable development goal 6 in India: Past, present, and future. Journal of Environmental Management. 2019;**232**:1049-1065

[8] Vuorinen HS, Juuti PS, Katko TS. History of water and health from ancient civilizations to modern times. Water Science and Technology: Water Supply. 2007;**7**(1):49-57

[9] Larsen Ø. History of Public Health in the Ancient World. 2017

[10] Streeter HW, Phelps EB. A Study of the Pollution and Natural Purification of the Ohio River. Washington, DC: United State Public Health Service; US Department of Health, Education, & Welfare; 1958

[11] Imhoff K, Mahr G, Fair GM. Required degree of treatment of sewage prior to its disposal by dilution. Sewage Works Journal. 1932;**4**(5):892-898

[12] Seeger H. The history of German waste water treatment. European Water Management. 1999;**2**:51-56

[13] Cooper PF. Historical Aspects of Wastewater Treatment. Tokyo, Japan: Decentralised Sanitation and Reuse; 2001

[14] Shifrin NS. Pollution management in twentieth century. J Environ Eng-Asce. 2005;**131**:676-691

[15] Shelford VE. An experimental study of the effects of gas waste upon fishes with especial reference to stream pollution. Illinois Natural History Survey Bulletin. 1917;**011**(04):50-421

[16] Ettinger MB. Developments in detection of trace organic contaminants. Journal of American Water Works Association. 1965;**57**:453-457

[17] Sugar JW, Conway RA. Gas-liquid chromatographic techniques for petrochemical wastewater analysis. Journal (Water Pollution Control Federation). 1968;**40**(9):1622-1631

[18] Lofrano G, Brown J. Wastewater management through the ages: A

history of mankind. Science of the Total Environment. 2010;**408**(22):5254-5264

[19] http://web.worldbank.org/archive/website01291/WEB/0__CO-23.HTM [Internet]

[20] Prasad CS, Ray I. When the pits fill up:(in) visible flows of waste in urban India. Journal of Water, Sanitation and Hygiene for Development. 2019;**9**(2):338-347

[21] Dasgupta S, Agarwal N, Mukherjee A. Moving up the on-site sanitation ladder in urban India through better systems and standards. Journal of Environmental Management. 2021;**280**:111656

[22] Xu X, Yang H, Li C. Theoretical model and actual characteristics of air pollution affecting health cost: A review. International Journal of Environmental Research and Public Health. 2022;**19**(6):3532

[23] https://www.who.int/news-room/fact-sheets/detail/drinking-water [Internet]

[24] Kazi TG, Arain MB, Baig JA, Jamali MK, Afridi HI, Jalbani N, et al. The correlation of arsenic levels in drinking water with the biological samples of skin disorders. Science of the Total Environment. 2009;**407**(3):1019-1026

[25] Hanif MA, Miah R, Islam MA, Marzia S. Impact of Kapotaksha river water pollution on human health and environment. Progressive Agriculture. 2020;**31**(1):1-9

[26] https://www.statista.com/statistics/1111747/india-number-of-districts-with-contaminated-water-by-contamination-type/ [Internet]

[27] Yenkie KM. Integrating the three E's in wastewater treatment: Efficient design, economic viability, and environmental sustainability. Current Opinion in Chemical Engineering. 2019;**26**:131-138

[28] https://ggiindia.in/goal-6-clean-water-and-sanitation/ [Internet]

[29] Biswas S, Dandapat B, Alam A, Satpati L. India's achievement towards sustainable development goal 6 (ensure availability and sustainable management of water and sanitation for all) in the 2030 agenda. BMC Public Health. 2022;**22**(1):1-16

[30] Lewis K, editor. Water governance for poverty reduction: Key issues and the UNDP response to Millenium [i.e. millennium] development goals. In: Water Governance Programme. New York City: Bureau for Development Policy, UNDP; 2004

Chapter 9

Sewage Treatment Using Nanoparticles

Heena Tabassum, Allika Sailaja, Huda Afreen and Minal Wani

Abstract

This chapter provides a brief overview of nanomaterials, including classification, shape and structure, nanomaterial types, and applications in the degradation of recalcitrant organic contaminants. With the rapid advancement of nanotechnology science, the use of nanomaterials in environmental applications, particularly water treatment, has piqued the scientific community's interest in recent decades. Nanomaterials have unique properties such as surface-to-volume ratio, quantum effect, low band-gap energy, and so on, which enhance catalytic performance. Wastewater treatment is a critical task of the twenty-first century since it protects the health of our environment and living beings. Because of its ability to affect both living and nonliving organisms, wastewater is always viewed as a serious source of environmental contamination. Many physical, biological, and chemical modes of treatment are implied to comply with wastewater discharge standards set by competent national agencies for environmental protection.

Keywords: nanomaterials, synthetic, lipid based nanoparticle, wastage, sewage

1. Introduction

Sewage treatment is a large component of water enterprise that safeguards public health, natural surroundings and economic development. The rapid population growth in distinctly urbanized and industrialized societies has resulted in the production of enormous volumes of Sewage, which require power and cost-effective remedies to discharge Sewage into receiving water bodies. To meet the discharge limits, present Sewage treatment facilities make use of energy-extensive strategies despite the current medical knowledge which aims in offering known strategies for power saving and restoration of flora. This offers a brief overview of a novel technology that has the potential to reduce the power needs of existing, standard Sewage treatment facilities, both by using replaceable remedies at some point of treatment to lessen the environmental footprint and gain electricity-efficient treatment facilities.

Water and Sewage control are enormously critical and interdependent requirements that can strongly affect a human being. Untreated Sewage contains huge amounts of organic and inorganic material from domestic, industrial and also public facilities, poisonous compounds, and many pathogenic microorganisms. If left untreated, Sewage can pollute surface and floor water reservoirs, consequently posing critical threats to public health and the surroundings. Subsequently, the objective of Sewage management is to offer reliable safety and adequate discharge of Sewage into the aquatic environment. But rapid and localized populace growth has brought about massive volumes of clean water being consumed every day and respectively huge volumes of Sewage being produced, which stresses, even more on the present Sewage management centres. On top of this a rapid deterioration of the high-quality water reservoirs, especially because of the elevated urbanization, industrialization and farming activities, is found. This is evident by the excess of natural pollutants and vitamin (N and P) masses in aquatic bodies. All of the above implies that more intensive Sewage treatment technologies, which are associated with high working volumes need to be adopted to protect public health and the natural environment. Sewage treatment incorporates diverse physical, chemical and organic processes, as well as their combination, that allows you to produce an effluent which can be safely disposed of into the surroundings without causing any sort of long-term unfavorable results to human beings or other living beings. Nevertheless, to meet Sewage discharge, high working volume demands are required, leading to high operational prices and making Sewage management unsustainable. Therefore, additional green and energy-efficient treatment constructions, that require a decrease to zero external quantities of strength to perform and therefore lower operational costs, have to be delivered on a huge scale. Sewage management is usually carried out in five stages as explained below:

1. Pre-treatment consists of bar screens which remove large objects like tree branches, fruit and vegetable peels, plastic wrappers, sheets and papers etc. Then the resulting water is sent into a flow equalization tank followed by a grit elimination channel.

2. Primary Treatment includes a primary sedimentation tank where suspended solids are settled out via gravity.

3. Secondary treatment consists of an activated sludge system where microorganisms are used for the conversion of organic matter into carbon and nitrogen sources.

4. Tertiary treatment includes disinfection of Sewage using UV irradiation and an activated carbon filter.

5. Sludge treatment generally is carried out for sludge thickening and later incineration of the final waste and finally waste disposal is carried out.

2. Current technologies of sewage treatment

2.1 Anaerobic digestion (AD)

Anaerobic digestion (AD) is a familiar method in which biodegradable constituents of Sewage are broken down with the help of micro-organisms in the absence of oxygen. This can be carried out in a single stage or multiple stages. This not only leads to decreased organic loads but also will simultaneously produce biogas. The process takes place under temperatures of about 30 to 38°C, where mesophilic digestion can take place or about 49–57°C where thermophilic digestion takes place [1]. Anaerobic Digestion is usually carried out in three stages [2]:

1. Hydrolysis: In this stage, microorganisms degrade organic matter present in Sewage into smaller constituents in the presence of water

2. Fermentation of volatile acids: Acidogenesis is carried out thus producing acetic acid, carbon dioxide, and hydrogen as end products.

3. Methane formation: At this stage acetic acid, carbon dioxide, and hydrogen which were produced in the previous stage get converted to methane and carbon dioxide.

Sewage s, as well as the Sludge coming out from the last stage generally is found to be rich in organic matter and thus can be used to produce energy in the form of biogas. Biogas obtained from anaerobic digestion plants can be utilized for various purposes like heating, production of electricity etc. This energy utilization makes the anaerobic digesters a versatile piece of equipment as it sustains their energy and heat requirements without the need for looking out for other sources of external energy requirements [3].

2.2 Photo-fermentation

In this method anaerobic micro-organisms which are photosynthetic like Rhodobacter and *Rhodopseudomonas* act as catalysts and convert organic acids, such as acetic and butyric acids, and sugars into glucose, fructose and sucrose in the presence of sunlight. There is a similar called as dark fermentation which is mostly carried out in the absence of sunlight where anaerobic bacteria, such as *Clostridium* and *Enterobacter*, convert glucose, sucrose, starch and cellulosic materials to H_2 [4]. But for the above-mentioned processes Sewage s containing huge amounts of carbohydrate content is required to produce adequate amounts of hydrogen typically 15% of the maximum theoretical potential. Due to this constraint, these processes have yet to emerge on a large industrial scale.

2.3 Microbial fuel cells

Microbial fuel cells (MFCs) are showing promising results in Sewage treatment in recent years as these are sustainable technologies which can accomplish the removal of organic pollutants with the help of micro-organisms and also produce electricity, water, CO_2 and other inorganic residues as by-products [5–7]. MFCs come under the category of bioreactors and are usually operated under anaerobic conditions. A positively charged ion membrane separates both the anode and cathode. Reactions at the anode include oxidation of organic materials present in Sewage by microorganisms, therefore generating CO_2, electrons and protons. These electrons are then relocated to the cathode compartment via an external electric circuit, producing electricity, whereas protons are moved to the cathode through the membrane. During this process, water is also formed as electrons and protons combine with oxygen in the cathode [8, 9]. MFCs offer the benefits of low cost, as they utilize inexpensive catalysts, that is, microorganisms present in Sewage s, huge energy efficiencies, and low volume of solid disposal.

2.4 Pyrolysis

Pyrolysis is an age-old technique and also a Green technology in comparison to existing Sewage technologies as it is carried out in absence of oxygen resulting in toxic-free by-products. Pyrolysis thermally breaks down the sewage sludge into biogas, bio-oil, and biochar anaerobically [10]. If seen from a thermodynamic perspective, Pyrolysis is an endothermic process (100 kJ kg^{-1}) which is carried out at temperatures fluctuating between 350–1000°C [11]. For carrying out pyrolysis effectively, Sewage management industries would require complex and expensive equipment. Thus the large-scale applications of this process are limited.

3. Sewage treatment using nanoparticles

Nanoparticles are materials in which the structural components lie within sizes ranging from 1 and 100 nm in at least one dimension [12]. Because of this special nature, nanoparticles highly differ in their properties mechanical, electrical, optical, and magnetic in comparison with other materials. In recent years, nanoparticles have found applications in many fields, such as catalysis [13], medicine [14], sensing [15], and biology [16]. Specifically, due to their nano sizes and huge surface areas available for chemical and biochemical reactions, High mobility of nanoparticles in solutions, nanoparticles are being used in water and Sewage treatment extensively. Sewage s contain toxic metal ions like [Hg(II), Pb(II), Cr(III), Cr(VI), Ni(II), Co(II), Cu(II), Cd(II), Ag(I), As(V), and As(III)] [17]. Numerous Conventional chemical and physical techniques like adsorption, precipitation, ion exchange, reverse osmosis, electrochemical treatments, membrane filtration, evaporation, flotation, oxidation and biosorption processes are being expansively used for the removal of such toxins. These conventional techniques do offer good amounts of toxin removal but to treat huge volumes of Sewage, there is an emerging need to search for new alternatives. In the current scenario, widely used nanoparticles for water and Sewage treatment largely comprise zero-valent metal nanoparticles, metal oxide nanoparticles, carbon nanotubes (CNTs), and nanocomposites.

Current research trends show the use of various zero-valent metal nanoparticles, viz.; Fe, Zn, Al, and Ni, in sewage treatment. Silver nanoparticles (Ag NPs) which come under the category of Zero-Valent Metal Nanoparticles are extremely lethal to microorganisms and hence show a strong antibacterial effect against various microorganisms, including viruses [18], bacteria [19], and fungi [20]. These can adhere to the bacterial cell wall of the microorganisms and increase the permeability of the cell walls [21]. They can easily penetrate through the cell walls resulting in structural changes in the cell membrane leading to the death of the cells [22]. Further, when Ag NPs come in contact with bacteria, they generate free radicals. They can damage the cell membrane and are considered to cause the death of cells. Magnetic nanoparticles can also be utilized in water treatment to remove heavy metals, sediments, chemical effluents, charged particles, bacteria and other pathogens. Lower operating costs, Lower energy requirements, lesser sludge discharge, Reduction in the number of pesticides and VOCs (organic chemicals), and Reduction of heavy metals, nitrates and sulfates, color, tannins, and turbidity are some of the many advantages of Sewage treatment using nanoparticles.

3.1 Carbon nanomaterials

Carbon nanomaterials (CNMs) are sheets made of graphene which are rolled up in the form of cylinders having diameters as minor as 1 nm. Carbon nanomaterials possess exceptional structures, and a great capacity to adsorb a wide range of contaminants like dichlorobenzene, ethyl benzene, Zn2+, Pb2+, Cu2+, and Cd2+, dyes, electrical properties, fast kinetics, large surface area, rich porous structures [23]. All these contribute to their diverse applications in advantages in sewage treatment processes.

The nanocomposite is an emerging field of nanomaterials. Nanocomposites are prepared through the chemical deposition of nZVI on CNTs. The resulting adsorbent has huge potential for rapid and effective removal of nitrate components presenting sewage water. These nanocomposites also have good magnetic properties due to which, the adsorbent can be easily separated from the solution by using a magnetic field [24]. Many such nanocomposites can be fabricated creating a network on polyimide supports. These offer advantages of being nontoxic, long-term stable and low-price materials. In concept, ideal composites for physical applications should be continuous, bulk immobile materials in which the nano reactivity is acquired by impregnating a parent material structure with nanoparticles [25]. Still, much research is underway for creating nanocomposites which can serve sewage treatment in a cost-efficient manner.

3.2 Metallic nanoparticles

Metallic nanoparticles have a great deal of potential for many forms of environmental cleanup because of their adaptability. One, two, or three metals and their oxides can be found in a variety of nanostructures. The shape-controlled, stable, and monodispersed properties of metal-based nanomaterials have been thoroughly studied using physical and chemical methods. Water treatments such as adsorption, photodegradation, membrane separation, and chemical disinfection can all benefit from these positive qualities. They have been utilized for self-cleaning surfaces, air purification, and water disinfection because of their noteworthy antibacterial, antifungal, and antiviral activity, which is employed as water disinfectants [26, 27].

3.3 Nanoparticles based on oxidation

Inorganic oxide-based nanoparticles are often created by combining metals and non-metals. The removal of harmful contaminants from wastewater makes considerable use of these nanoparticles. There are also ferric oxides titanium oxides [28], titanium oxide/dendrimers composites [28–30], zinc oxides [31], magnesium oxides, manganese oxides [32, 33] and ferric-oxide [34]. High BET surface area, less environmental impact, less solubility, and lack of secondary pollutants are the characteristics of oxide-based nanoparticles [35].

3.4 Silver based nanoparticles

Due to its low toxicity, water-based microbial inactivation, and well-documented antibacterial action, silver is the substance that is utilized the most frequently.

Silver salts like silver nitrate and silver chloride are used to make silver nanoparticles, which are known to be excellent biocides. Smaller Ag nanoparticles (8 nm) were the most effective, despite the fact that the antibacterial action is size dependant; increasing particle size (11–23 nm) resulting in reduced bactericidal activity.

Additionally, truncated triangular silver nanoplates outperformed spherical and rod-shaped nanoparticles in terms of antibacterial efficacy, demonstrating the importance of shape.

Ag nanoparticles' bactericidal effects can occur through a variety of methods, such as the production of free radicals that damage bacterial membranes, interactions with DNA, adherence to cell surfaces that change the characteristics of the membranes, and damage to enzymes.

Immobilized nanoparticles have become more significant because of their potent antibacterial properties. Gram-positive and Gram negative bacteria have been observed to be particularly sensitive to embedded Ag nanoparticles. In a study, cellulose acetate fibers that had been directly electrospun with silver nanoparticles incorporated in them were demonstrated to be efficient against both types of bacteria. Ag nanoparticles are also added to several kinds of polymers to create antibacterial nanocomposites and nanofibers. In a study, antimicrobial nanofilters made of poly (−caprolactone) based polyurethane nanofiber mats with Ag nanoparticles were created.

Ag nanoparticles are included in several types of nanofibers that have been manufactured for antimicrobial applications and have shown excellent antibacterial capabilities [36–39].

Ag nanofiber-coated polyurethane foam water filters have demonstrated effective antibacterial activities against *Escherichia coli* (*E. coli*). Other examples of inexpensive drinkable microfilters made with Ag nanoparticles exist and can be applied in rural areas of developing nations [40].

Ag nanoparticles are also used in water filtration membranes, such as polysulfone membranes, where they are effective in reducing biofouling and are effective against a wide range of bacteria and viruses. These membranes with Ag nanoparticles demonstrated effective antibacterial properties against *E. coli*, Pseudomonas, and other microbes. The efficiency of Ag nanocatalyst alone and in combination with carbon covered in alumina for the destruction of microbiological pollutants in water has been proven.

Although Ag nanoparticles have been successfully employed to destroy bacteria, viruses, and reduce membrane biofouling, their long-term effectiveness against membrane biofouling has not been observed. This is mostly because silver ions are lost over time.

Therefore, additional efforts to stop this loss of silver ions are needed in order to permanently control membrane biofouling. As an alternative, the problem can be resolved by doping Ag nanoparticles with other metallic nanoparticles or its composites with metal-oxide nanoparticles. This may also result in the simultaneous removal of inorganic and organic substances from water and wastewater [41].

3.5 Titanium based nanoparticles

TiO2 nanoparticles have mostly been utilized as a catalyst in organic reaction and wastewater breakdown. Because of their special characteristics and lower tendency to aggregate due to the existence of greater repulsive forces, microorganisms (such as bacteria) are often used in the biosynthesis of TiO2 nanoparticles. The morphological structures of TiO2 nanoparticles, such as spherical titania (TNP), titania nanotubes

(TNT), and titania nanosheets, have been extensively researched in the field of water purification along with development and improvement. High specific surface area, pore volume, and pore size are typical characteristics of TNP. As a result, TNP offers an active site that is very easy to reach and can be employed for organic pollutant adsorption. The typical hydrothermal method for producing TNT from TNP uses a potassium hydroxide/sodium hydroxide (KOH/NaOH) solution. TNT is anticipated to be more advantageous than TNP because of its tubular structure, which also provides bigger pore volume and a higher interfacial charge carrier transfer rate. TNT also offers the high hydrophilicity qualities of TiO2 [42].

One of the most significant chemically stable nanoparticles, titanium dioxide (TiO2), has attracted the most interest for its use as a photocatalyst and adsorbent in the removal of contaminants from wastewater. Some microorganisms, such as bacteria, can biosynthesize these nanoparticles as reducing agents for nanofactories in order to create nontoxic, environmentally friendly ways to generate nanoparticles [43].

3.6 Iron based nanoparticles

Ferric oxide is a cheap substance for the adsorption of toxic metals due to the natural abundance of iron and its straightforward production technique. It is an environmentally benign substance that can be utilized in contaminated environments without raising the risk of secondary contamination. The pH, temperature, amount of adsorbent, and incubation period all play a role in how different heavy metals bind to Fe2O3 nanoparticles. Different researchers modified the surface of Fe2O3 to boost its adsorption capability. Recently, there have been multiple publications on the use of magnetic oxides, particularly Fe3O4, as nanoadsorbents to remove hazardous metal ions such Ni2+, Cr3+, Cu2+, Cd2+, Co2+, Hg2+, Pb2+, and As3+ from wastewater. For instance, Shen et al. [44] have found that pH, temperature, the quantity of the adsorbent, and the incubation duration all have a significant impact on the adsorption efficiency of Ni2+, Cu2+, Cd2+, and Cr6+ ions by Fe3O4 nanoparticles. According to Palimi et al. [45], 3 aminopropyltrimethoxysilane was used to modify the Fe2O3 nanoparticles' surfaces.

When it comes to simultaneously removing many contaminants from wastewater, such as Cr3+, Co2+, Ni2+, Cu2+, Cd2+, Pb2+, and As3+, nano-adsorbents have strong affinity.

3.7 Manganese oxide based nanoparticles (MnO)

Due to their high BET surface area and polymorphism structure, manganese oxide (MnOs) nanoparticles have remarkable adsorption capacity [46]. It has been frequently utilized to remove several heavy metals from wastewater, including arsenic [47]. Hydrous manganese oxide (HMO) and nanoporous/nanotunnel manganese oxides are the most commonly used modified MnOs. HMO was created by mixing MnSO4H2O into a NaClO solution. The inner-sphere is usually responsible for the adsorption of numerous heavy metals on HMOs, including Pb (II), Cd (II), and Zn (II). Divalent metals did, however, adsorb on the surface of HMOs in two stages. Metal ions first adsorb on the outside surface of HMOs, and then intraparticle diffusion occurs .

3.8 Zinc oxide (ZnO) based nanoparticles

For the adsorption of heavy metals, zinc oxide (ZnO) possesses a porous micro/nanostructure with a high BET surface area. For the removal of heavy metals from wastewater, nano assemblies, nano-plates, microspheres with nano-sheets, and hierarchical ZnO nano rods are frequently utilized as nano-adsorbents In comparison to commercial ZnO, the aforementioned modified forms of ZnO nano-adsorbent exhibit a high removal effectiveness of heavy metals. ZnO nano-plates and porous nano-sheets were utilized to remove Cu (II) from wastewater. These modified ZnO nano-adsorbents exhibit greater Cu (II) removal effectiveness compared to commercial ZnO because of their distinctive micro/nanostructure. Additionally, nano-assemblies were employed to eliminate a variety of heavy metals, including Co2+, Ni2+, Cu2+, Cd2+, Pb2+, Hg2+, and As3+. Due to their electropositive character, microporous nano-assemblies exhibit strong affinity for the adsorption of Pb2+, Hg2+, and As3+. According to Kumar et al. [48] mesoporous hierarchical ZnO nano-rods have a good removal efficiency for Pb (II) and Cd (II) from wastewater.

Singh et al. [49] reported the removal of numerous harmful metal ions from wastewater using porous ZnO nano-assemblies, including Co2+, Ni2+, Cu2+, Cd2+, Pb2+, Hg2+, and As3+. It has been claimed that Hg2+, Pb2+, and As3+ demonstrate superior removal efficiency (63.5% Hg2+, 100% Pb2+, and 100% As3+) because of their stronger attraction to ZnO nano-assemblies due to their high electronegativity.

3.9 Magnesium oxide based nanoparticles (MgO)

Magnesium oxide (MgO) and iron oxide (Fe2O3) are potential metal oxide nanoparticles (NPs) for the adsorption of textile and tannery wastes. Due to their nanostructure and numerous active sites, these NPs have large surface areas and a high capacity for the adsorption of heavy metals. Ecosystem harm from NP biotreatment is nonexistent. In earlier research, for the purification of textile colors and the eradication of particular heavy metals magnesium oxide (MgO) is utilized. MgO microspheres are a new structure that can increase the removal of heavy metals' adsorption affinity. The shape of NPs has undergone many forms of alteration to boost the adsorption capability of MgO. These include nanorods, nanobelts, fishbone fractal nanostructures, nanowires, nanotubes, nanocubes, and three-dimensional things. Kiran et al. [50], reported that the remediation of Reactive Brown 9 dye was then carried out using MgO-NPs once the reaction's key variables (dye concentration, nanoparticle concentration, pH, and temperature) had been optimized. The highest degree of decolorization (95.8%) was achieved at 0.02% dye concentration, 0.003 mg/L MgO-NP concentration, pH 4, and 40°C. The mineralization of the examined dye samples was evaluated using TOC and COD, and their values were found to be 88.56% and 85.34%, respectively. Other troublesome colors could also be treated using the magnesium oxide nanoparticles in stages. It is imperative to get rid of these harmful colors because they ruin the aquatic environment and spread several diseases.

3.10 Al2O3 based nanoparticles

Aluminum oxide nanoparticles (ANPs) are used in a variety of industrial and personal care products. *E. coli* has been examined for the growth-inhibitory effect of alumina nanoparticles over a broad concentration range (10–1000 g/mL). These metal oxides' antibacterial properties are ascribed to the production of reactive oxygen species (ROS), which results in cell wall breakdown and eventual cell death. However, alumina nanoparticles might neutralize free radicals. The ability of these

NPs to protect cells from oxidative stress-induced cell death appears to depend on the particle's structure but is unrelated to its size between 61,000 nm [51].

3.11 Copper based nanoparticles

A variety of bacterial pathogens responsible for hospital acquired infections were successfully eliminated by CuO nanoparticles (CuO NPs). However, a significant portion of CuO NPs are scavengers. The ability of these NPs to protect cells from oxidative stress-induced cell death appears to depend on the particle's structure rather than its size between 61,000 nm. It is necessary to produce a bactericidal action [52].

3.12 Curcumin-loaded nanocarriers for hospital wastewater treatment

Water contamination with a wide range of chemical, microbiological, and toxic substances is a growing environmental concern. In a study by (Mozhgan et al), the heated high-speed homogenization process was used to create eco-friendly curcumin-loaded nanostructured lipid carriers (NLC-curcumin). NLC-curcumin had an average particle size of 137.9 3.21 nm and a zeta potential of −23.36 3.5 mV. The nanoparticles' morphology, thermal behavior, antioxidant properties, and infrared spectroscopy were also studied. The potential of NLC-curcumin on bacterial growth reduction in the actual environment of hospital wastewater was evaluated using colony forming unit per milliliter (CFU/ml) analysis. The results show that 0.125 M NLC-curcumin in Mueller-Hinton agar media significantly lowers the proportion of wild bacteria strains in autoclaved wastewater at 37°C. NLC-Curcumin (0.125 M) significantly reduced the percentage of the microbial total count at 25°C in the original hospital wastewater treatment as shown in **Table 1** [53].

4. Type of nanomaterials in wastewater treatment

A substantial amount of study has been conducted on the use of nanotechnology for wastewater treatment. Based on the materials used, nanotechnology may be divided into three categories: Nanoadsorbents, nanocatalysts, and nanomembranes present in **Table 1** and **Figure 1**.

4.1 Nano-adsorbents

Adsorbent nanoparticles are nanoparticles composed of organic or inorganic materials with a high affinity for adsorbing chemicals. This implies they can remove a large amount of pollution. Because of their key qualities, such as catalytic potential, small size, high reactivity, and increased surface energy, these nanoparticles may be used to remove many types of contaminants. Metallic nanoparticles, mixed oxide nanostructures, magnetic nanoparticles, and metal oxide nanoparticles all have distinct adsorption mechanisms. Example is Zeolites comprise aluminosilicate minerals having a surface morphology filled with electrostatic pores populated by cations and water molecules. Cations and water molecules have a wide range of movement options, allowing for ion exchange and reversible dehydration. CNTs feature a large surface area, a large number of adsorption sites, and variable surface chemistry as shown in **Table 1**.

S. No	Nanoparticle	Type	Load	Method of action	References
1	Curcumin-Loaded nanocarriers	Nanostructured lipid carriers (NLC)	Curcumin	Antimicrobial properties of curcumin are the main feature of this nanoformulation.	[53]
2	Zeolites	Nano-adsorbents	—	Zeolites comprise aluminosilicate minerals having a surface morphology filled with electrostatic pores populated by cations and water molecules. Cations and water molecules have a wide range of movement options, allowing for ion exchange and reversible dehydration.	[54]
3	CNTs Nanoadsorbents	Nano-adsorbents	—	CNTs feature a large surface area, a large number of adsorption sites, and variable surface chemistry. CNTs are used for pollution cleanup, determining the pre-concentrate, and revealing contaminants. The engagement of carbon nanotubes with metal cations via electrostatic attraction and chemical bonding explained the process of oil contamination in wastewater utilizing carbon nanotubes.	[54]
4	Silica, Clay, Titanium dioxide, Iron, Aluminum Nano-catalysts	Nano-catalyst	Manganese Ferrous Oxides, Silica Oxides and Silver ($MnFe2O4-SiO2-Ag$)	Nanoparticles provide a wide surface area for chemical interaction that directly affects the reaction rate, providing impressive catalytic performance. Thus its used in catalyzing reactions on removal/transformation of pollutants in wastewater	[55]
5	CNTs Nano membrane	Nano-membrane	—	Carbon nanotubes are employed to create a nanomembrane that can be reused and is capable of eliminating pathogens and pollutants.	[56]

S. No	Nanoparticle	Type	Load	Method of action	References
6	Eugenol NLC	Nanostructured lipid carriers (NLC)	Eugenol	Eugenol induced cell wall breakdown leading to cell death is the primary goal of this NLC. It aims to diminish wild bacterial strains in sterilized wastewater and hospital wastewater.	[57]
7	Algal nanoparticles	Simple Nanoparticles	Silver, Gold	*Euglena gracilis* and other microalgae were used in biosynthesis of metallic nanoparticles, proving to be a potential source for in vivo and in vitro biosynthesis of Ag/Au NPs.	[58]

Table 1.
Varied instances of the use of nanocarriers in the treatment of wastewater.

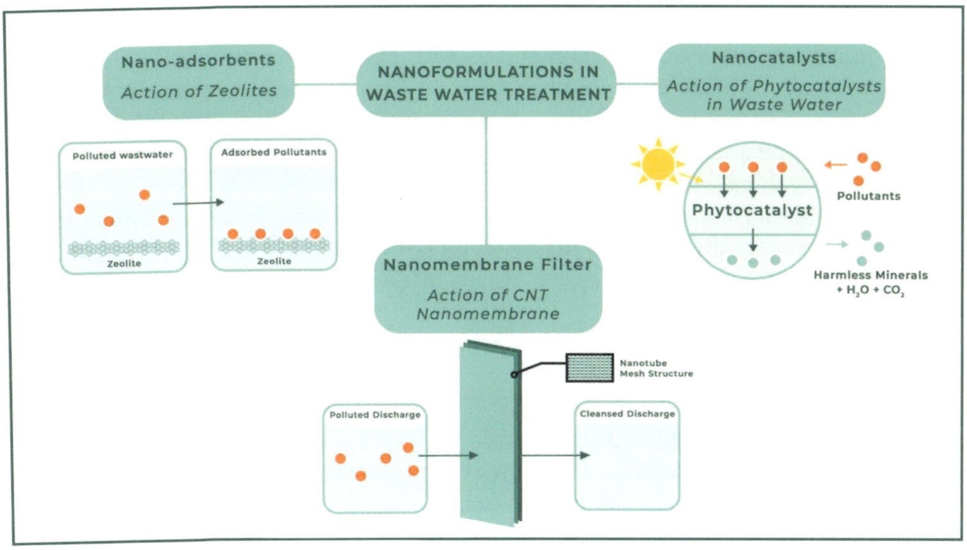

Figure 1.
Schematic explanation of the method of action followed by different types of nanoformulations used in wastewater treatment currently.

4.2 Nano-catalyst

Light energy interacts with metallic nanoparticles in nano-catalysis, resulting in high and broad photocatalytic activity. Because of its great and broad photocatalytic activity, this therapy is gaining favor. Bacteria and enzymes participate in a photocatalytic reaction. Manganese Ferrous Oxides, Silica Oxides and Silver (MnFe2O4-SiO2-Ag) act as catalyst and its mode of action shown in **Table 1**.

4.3 Nano-membrane

The separation of particles from wastewater is the responsibility of a nano-membrane. These filters are extremely good in filtering out dyes, heavy metals, and other contaminants. Nanomaterials utilized as nano-membranes include nanotubes, nanoribbons, and nanofibers.

Nanoparticles integrated into membranes are more convenient and beneficial than nano-adsorbents, nano-catalysts, and nano-membranes because this procedure not only has a powerful physical treatment, but it also contains nanoparticles to increase the quality of the treatment.

To inhibit bacterial development, several eco-friendly nanostructured lipid carriers (NLCs) were created as delivery agents to appropriately carry an antibacterial agent (eugenol) into hospital wastewater. Hot high-speed homogenization was used to create eugenol-loaded nanostructured lipid carriers. The nanocarriers were then analyzed using several methods including transmission electron microscopy, Fourier transforms infrared and dynamic scanning calorimetry. The ability of the produced eugenol-loaded nanostructured lipid carriers to reduce the bacterial growth rate in culture media and hospital wastewater was determined using the turbidity test and colony counting technique, respectively. NLC-mean eugenol's size and zeta potentials were 78.12 6.1 nm and 29.43 2.21 mV, respectively. The maximum inhibitory impact of NLC-eugenol in culture medium was reported in standard and wild *Staphylococcus aureus* strains (43.42% and 26.41%, respectively) at 0.125 M concentration. The antibacterial activity of NLC-eugenol in sterile wastewater on wild bacteria strains revealed that 0.125 M was the most effective concentration to reduce bacterial quantities on wild *S. aureus* and *Enterococcus faecalis* strains (38% and 33.47%, respectively) at 37°C. At 25°C, NLC-eugenol at 0.125 M had the best impact, lowering total microbiological agents by 28.66% in hospital wastewater as shown in **Table 1** [57].

4.4 Microalgal

Nanoparticles can be utilized for a variety of applications, including medical treatment, solar and fuel cells for efficient energy generation, water, and air filters to minimize pollution, and as catalysts in existing industrial processes to remove the usage of harmful ingredients. Wet techniques are the most traditional and widely utilized ways of producing nanoparticles (physical and chemical). These strategies are classified into two approaches: top-down and bottom-up. Growing nanoparticles in a liquid medium containing reducing and stabilizing agents such as potassium bitartrate, sodium dodecyl benzyl sulfate, methoxypolyethylene glycol, polyvinyl pyrrolidone, or sodium borohydride are how nanoparticles are created chemically. In addition, physical processes include pyrolysis and attrition. Unfortunately, the physical and chemical procedures utilized in both approaches have some implications due to their poor environmental effect, lengthy manufacturing methodology, and prohibitively high cost. According to research, nanoparticles attract atoms and molecules owing to their high surface energy, modifying their surface characteristics. As a result, they are unable to live in their natural habitat in their naked state. Nanoparticles are not suitable for therapeutic use due to these environmental interactions. This mechanism of action has raised awareness about the importance of developing nontoxic and environmentally friendly procedures for the assembly and synthesis of nanoparticles.

Physical processes also include pyrolysis and attrition. Unfortunately, both systems' physical and chemical procedures have some issues because of their bad environmental impact, delayed production methodology, and excessively expensive cost. Nanoparticles, according to study, attract atoms and molecules due to their high surface energy, changing their surface features. As a result, they are unable to survive in their native habitat while nude. Because of these environmental interactions, nanoparticles are not suited for therapeutic usage. This mechanism of action has increased awareness of the importance of developing nontoxic and environmentally friendly procedures for nanoparticle assembly and synthesis and mode of action as shown in **Table 1** [59].

I. Direct exploitation of extracted biomolecules from disrupted cells of microalgae

II. Exploitation of cell-free supernatants made of microalgae culture media

III. Biosynthesis of nanoparticles of different natures from whole cells of microalgae

IV. Using living cells of microalgae

To make gold nanoplates, the microalgal biomass is lyophilized and then exposed to RP-HPLC, or reverse-phase high-performance liquid chromatography, until the gold shape-directing protein (GSP), which controls the shape of nanoparticles, is separated. Furthermore, this protein is exposed to an aqueous HAuCl4 solution, resulting in the formation of gold nanoparticles of various forms. In the case of silver nanoparticles, PLW (proteins with low molecular weight) and PHW (proteins with high molecular weight) found in microalgae biomass is responsible for converting silver ions into their metallic counterparts. Tang et al. biosynthesized silver and gold nanoparticles using a fine powder of *Spirogyra insignis* (Charophyta) [58].

Many studies on the fate of NPs and their effects on biological wastewater treatment have been conducted, and many successes have been reported [60].

1. At certain concentrations, most NPs, including Ag, ZnO, CuO, Al2O3, SiO2, CNTs, and magnetic NPs, can cause varied degrees of harm to microorganisms. When compared to other NPs, Ag, Cu, and ZnO NPs have significantly large hazardous effects at similar exposure doses. TiO2 NPs, in instance, do not exhibit high toxicity to microorganisms in both short-term (even at 500 mg L1) and long-term (50 mg L1) exposures.

2. By interacting with biomass, WWTPs utilizing activated sludge have the capacity to remove most forms of NPs, including Ag, Cu, ZnO, CuO, and TiO2, but not SiO2. Under specific circumstances, Ag NPs, Cu NPs, and ZnO NPs can be partly converted into Ag+, Cu2+, and Zn2+, respectively. The low removal of SiO2 NPs is attributed to their strong colloidal stability in wastewater and their limited biosorption proclivity.

3. Most NPs have varying degrees of influence on the efficacy of biological wastewater treatment, including the removal of nitrogen, phosphorus, and

organic contaminants. Under most situations, the effects are dose-dependent, and exposure length (short-term or long-term) also plays a role in unfavorable consequences.

a. At low concentrations, Ag, CuO, and ZnO NPs have small or moderate effects on TN removal; TiO2 NPs have some inhibitory effects on nitrifying and denitrifying bacteria, as well as AMO and NOR; and Al2O3 and SiO2 NPs have negative effects on nitrification and denitrification.

b. Al2O3, TiO2, and SiO2 NPs have no significant adverse effects on phosphorus removal, but ZnO NPs can result in net phosphorus removal failure at certain concentrations.

c. Ag NPs can reduce COD removal when it comes to organics removal. During anaerobic biological wastewater processes, ZnO NPs and CuO NPs can suppress methane formation. Under most situations, the effects are dose-dependent. Some of the effects are caused by the NPs themselves, whereas others are caused by liberated ions such as Ag+, Cu2+, and Zn2 +.

5. Discussion

Sixth of the world's population, on average, lacks access to clean drinking water. The availability of freshwater supplies is limited, making it difficult for the world to meet growing demands for clean water. Depletion as a result of (i) protracted droughts, (ii) rising population, (iii) stricter health-based laws, (iv) conflicting demands from various users, and (v) water contamination. Therefore, it is vital to take action to create an inventive technology that can supply clean, affordable water to suit human needs. A healthy existence requires access to clean drinking water. 80% of diseases in nations like India are caused by water, especially drinking water. The World Health Organization (WHO) advises that any water meant for drinking should have zero fecal and total coliform levels in each 100 mL sample. Today, a variety of methods, including chemical and physical processes, are employed to clean water. These methods include the treatment of chlorine and its derivatives, ultraviolet light, boiling, low-frequency ultrasonic irradiation, distillation, reverse osmosis, and water sediment filters [61].

The three primary categories of wastewater treatment techniques are physical, chemical, and biological. Filtration is a key component of solid-liquid separations, which are the main focus of physical wastewater treatment. There are two broad categories that conventional and unconventional filtration methods fall under. Applications involving water purification rely on this technique. The treatment process is just one component of a conventional water treatment system, which offers a variety of technology and equipment alternatives based on the treatment's objectives. Understanding the function of filtration in water purification in comparison to other technologies and the goals of various unit processes is crucial. This cost-effective method can eliminate wastewater microorganisms and suspended particulates in specific circumstances, such as when membranes are used. However, it is unable to reduce the wastewater's organic pollution and heavy metal levels on its own, which are dangerous when reused in the home or industrial settings. One of the most prevalent instances of this technique is membrane filtration, whose structure may be easily adjusted utilizing cutting-edge technology like nanoparticles as well as used with other types of treatment.

Chemical techniques of treatment depend on chemical reactions between the contaminants and the person using the chemical agent, and they help either completely remove contaminants from water or neutralize any negative effects they may have. Chemical treatment techniques can be used both alone and in conjunction with physical processes to treat a variety of problems. By using this pricey process, the wastewater's organics will be removed, but new compounds, some of which may be dangerous, will be introduced. For instance, activated carbon adsorption is frequently used in home and industrial treatments to eliminate turbidity and the smell of water without causing any negative side effects.

Although the biological treatment of wastewater appears straightforward because it depends on natural processes to aid in the breakdown of organic chemicals, it is actually complicated, poorly understood, and occurs where biology and biochemistry meet. Organic materials, such as rubbish, organic wastes, partially digested foods, heavy metals, and poisons, can be found in wastewater. Organic debris is typically broken down by bacteria, nematodes, and other tiny organisms in biological treatments. Worldwide application of biological treatment is possible due to its adaptability, affordability, and environmental friendliness. Many mechanical or chemical techniques fall short of biological treatments in terms of effectiveness or efficiency. A notable example of this is the conventional activated sludge (CAS) procedure. These systems frequently consist of an aeration tank that serves as a biological degrading agent and a secondary clarifier to separate treated wastewater sludge [62].

The primary purpose of a wastewater treatment system is to remove primary pollutants such as suspended particles, biochemical oxygen demand (BOD), nutrients (organic and inorganic), toxicity, and coliform bacteria. The sedimentation process is used in a traditional wastewater system to remove dissolved organic matter and suspended particulates. Sewage preliminary treatment removes 60% of large solid materials through a well-designed sedimentation tank and approximately 35% of BOD delivered by sewers responsible for obstructing flow through the plant or damaging equipment. Heavy grit particles, rags, fecal materials, and wood can be removed from sewage by passing it through screen bars. The secondary treatment method seeks to minimize suspended particles and BOD by lowering organic matter by 85 percent. This is mostly accomplished by a diverse population of heterotrophic bacteria capable of exploiting the organic ingredient for energy and development. Some of the secondary wastewater treatment operating systems include fixed film and suspended growth reactors. Tertiary treatment techniques strive to eliminate 95% of organic ions. It can be done either biologically or chemically, but it is a costly process. Chemical precipitation, reverse osmosis, carbon adsorption, and ozonation are examples of advanced treatment methods based on technologically complex techniques. These methods remove nutrients like phosphorus and nitrogen, which can cause eutrophication in surface water. To remove tiny particles on a small scale, technologies such as land application, filtration, and lagoon storage are utilized. Several primary and secondary treatment plants have been installed in a variety of locations to remove settled materials and oxidize organic material from wastewater. Furthermore, even after tertiary treatment, complete removal of the incoming waste load is not possible, and as a result, many organisms remain in the water bodies.

6. Conclusion

Waste water treatment could greatly benefit from the usage of nanoparticles. Its distinctive quality of having a large surface area can be effectively employed to remove harmful metal ions, microbial pathogens, organic, and inorganic solutes from water. Numerous kinds of nanomaterials, including zeolites, dendrimers, carbonaceous nanomaterials, and metal-containing nanoparticles, have also been shown to be effective for treating water. The chapter discusses recent developments on various nanomaterials, including molecularly imprinted polymers (MIPs), bioactive nanoparticles, molecularly structured catalytic membranes, nanosorbents, nanocatalysts, and nanosorbents with applications in waste water treatment. On both a small- scale and a large-scale, waste water may now be treated in a variety of effective ways with the help of nanotechnology.

Author details

Heena Tabassum*, Allika Sailaja, Huda Afreen and Minal Wani
Dr. D.Y. Patil Biotechnology and Bioinformatics Institute, Dr. D.Y. Pati Vidyapeeth, Pune, Maharashtra, India

*Address all correspondence to: heena.tabassum@dpu.edu

References

[1] Reith JH, Wijffels RH, Barten H. Bio-Methane & bio-Hydrogen: Status and Perspectives of Biological Methane and Hydrogen Production. The Netherlands: Dutch Biological Hydrogen Foundation; 2003. ISBN: 90-9017165-7

[2] POST. Energy and Sewage, Postnote Number 282. Parliamentary Office of Science and Technology. Vol. 7. London, SW1P 3JA, UK: Millbank; 2007. ISBN 9781780405025

[3] Cao Y, Pawłowski A. Sewage sludge-to-energy approaches based on anaerobic digestion and pyrolysis: Brief overview and energy efficiency assessment. Renewable and Sustainable Energy Reviews. 2012;**16**:1657-1665. DOI: 10.1016/j.rser.2011.12.014

[4] Su H, Cheng J, Zhou J, Song W, Cen K. Hydrogen production from water hyacinth through dark- and photo-fermentation. International Journal of Hydrogen Energy. 2010;**35**:8929-8937. DOI: 10.1016/j.ijhydene.2010.06.035

[5] Ahn Y, Hatzell MC, Zhang F, Logan BE. Different electrode configurations to optimize performance of multi-electrode microbial fuel cells for generating power or treating domestic waste water. Journal of Power Sources. 2014;**249**:440-445. DOI: 10.1016/j.jpowsour.2013.10.081

[6] Barua PK, Deka D. Electricity generation from biowaste based microbial fuel cells. International Journal of Energy, Information and Communications. 2010;**1**(1):77-92

[7] Du Z, Li H, Gu T. A state of the art review on microbial fuel cells a promising technology for Wastw water treatment and bioenergy. Biotechnology Advances. 2007;**25**:464-482. DOI: 10.1016/j.biotechadv.2007.05.004

[8] Oh SE, Logan BE. Hydrogen and electricity production from a food processing waste water using fermentation and microbial fuel cell technologies. Water Research. 2005;**39**:4673-4682. DOI: 10.1016/j.watres.2005.09.019

[9] Rabaey K, Verstraete W. Microbial fuel cells: Novel biotechnology for energy generation. Trends Biotechnology. 2005;**23**(6):291-298. DOI: 10.1016/j.tibtech.2005.04.008

[10] Agrafioti E, Bouras G, Kalderis D, Diamadopoulos E. Biochar production by sewage sludge pyrolysis. Journal of Analytical and Applied Pyrolysis. 2013;**101**:72-78. DOI: 10.1016/j.jaap.2013.02.010

[11] Samolada MC, Zabaniotou AA. Comparative assessment of municipal sewage sludge incineration, gasification and pyrolysis for a sustainable sludge-to energy management in Greece. Waste Management. 2014;**34**:411-420. DOI: 10.1016/j.wasman.2013.11.003

[12] Buzea C, Pacheco II, Robbie K. Nanomaterials and nanoparticles: Sources and toxicity. Biointerphases. 2007;**2**(4):MR17-MR71. DOI: 10.1116/1.2815690

[13] Parmon V. Nanomaterials in catalysis. Materials Research Innovations. 2008;**12**(2):60-61. DOI: 10.1179/143307508x304228

[14] Liang X-J, Kumar A, Shi D, Cui D. Nanostructures for medicine and pharmaceuticals. Journal of Nanomaterials. 2012;**2012**:2. Article ID 921897. DOI: 10.1155/2012/921897

[15] Kusior A, Klich-Kafel J, Trenczek-ZajacA,SwierczekK,RadeckaM, Zakrzewska K. TiO2–SnO2 nanomaterials for gas sensing and photocatalysis. Journal of the EuropeanCeramic Society. 2013;**33**(12):2285-2290. DOI: 10.1016/j.jeurceramsoc.2013.01.022

[16] Bujoli B, Roussi'ere H, Montavon G, et al. Novel phosphate– Phosphonate hybrid nanomaterials applied to biology. Progress in Solid State Chemistry. 2006;**34**(2-4):257-266. DOI: 10.1016/j.progsolidstchem.2005.11.039

[17] Tambe Patil BB. Waste water treatment using nanoparticles. Journal of Advanced Chemical Engineering. 2015;**5**(3):10-4172. DOI: 10.4172/2090-4568.1000131

[18] Borrego B, Lorenzo G, Mota-Morales JD, et al. Potential application of silver nanoparticles to control the infectivity of Rift Valley fever virus in vitro and in vivo. Nanomedicine: Nanotechnology, Biology and Medicine. 2016;**12**(5):1185-1192. DOI: 10.1016/j.nano.2016.01.021

[19] Kalhapure RS, Sonawane SJ, Sikwal DR, et al. Solid lipid nanoparticles of clotrimazole silver complex: An efficient advances in materials science and engineering 7 nano antibacterial against Staphylococcus aureus and MRSA. Colloids and Surfaces B: Biointerfaces. 2015;**136**:651-658. DOI: 10.1016/j.colsurfb.2015.10.003

[20] Krishnaraj C, Ramachandran R, Mohan K, Kalaichelvan PT. Optimization for rapid synthesis of silver nanoparticles and its effect on phytopathogenic fungi. Spectrochimica Acta—Part A: Molecular and Biomolecular Spectroscopy. 2012;**93**:95-99. DOI: 10.1016/j.saa.2012.03.002

[21] Sondi I, Salopek-Sondi B. Silver nanoparticles as antimicrobial agent: A case study on *E. coli* as a model for Gramnegative bacteria. Journal of Colloid and Interface Science. 2004;**275**(1):177-182. DOI: 10.1016/j.jcis.2004.02.012

[22] Danilczuk M, Lund A, Sadlo J, Yamada H, Michalik J. Conduction electron spin resonance of small silver particles. Spectrochimica Acta—Part A:Molecular and Biomolecular Spectroscopy. 2006;**63**(1):189-191. DOI: 10.1016/j.saa.2005.05.002

[23] Khin MM, Nair AS, Babu VJ, Murugan R, Ramakrishna S. A review on nanomaterials for environmental remediation. Energy & Environmental Science. 2012;**5**(8):8075-8109. DOI: 10.1039/c2ee21818f

[24] Azari A, Babaei A-A, Kalantary RR, Esrafili A, Moazzen M, Kakavandi B. Nitrate removal from aqueous solution using carbon nanotubes magnetized by nano zero-valent iron. Journal of Mazandaran University of Medical Sciences. 2014;**23**(2):14-27. DOI: 10.1186/s40201-014-0115-5

[25] Tesh SJ, Scott TB. Nano-composites for water remediation: A review. Advanced Materials. 2014;**26**(35):6056-6068. DOI: 10.1002/adma.201401376

[26] Gao C, Zhang W, Li H, Lang L, Xu Z. Controllable fabrication of mesoporous MgO with various morphologies and their absorption performance for toxic pollutants in water. Crystal Growth and Design. 2008;**8**(10):3785-3790. DOI: 10.1021/cg8004147

[27] Guerra FD, Attia MF, Whitehead DC, Alexis F. Nanotechnology for environmental remediation: Materials and applications. Molecules. 2018;**23**(7):1760. DOI: 10.3390/molecules23071760

[28] Barakat MA, Al-Hutailah RI, Hashim MH, Qayyum E, Kuhn JN. Titania-supported silver-based bimetallic nanoparticles as photocatalysts. Environmental Science and Pollution Research. 2013;**20**(6):3751-3759. DOI: 10.1007/s11356-012-1306-1

[29] Barakat MA, Ramadan MH, Alghamdi MA, Algarny SS, Woodcock HL, Kuhn JN. Remediation of Cu (II), Ni (II), and Cr (III) ions from simulated wastewater by dendrimer/titania composites. Journal of Environmental Management. 2013;**117**:50-57. DOI: 10.1016/j.jenvman.2012.12.025

[30] Tuzen M, Soylak M. Multiwalled carbon nanotubes for speciation of chromium in environmental samples. Journal of Hazardous Materials. 2007;**147**(1-2):219-225. DOI: 10.1016/j.jhazmat.2006.12.069

[31] Gupta K, Bhattacharya S, Chattopadhyay D, Mukhopadhyay A, Biswas H, Dutta J, et al. Ceria associated manganese oxide nanoparticles: Synthesis, characterization and arsenic (V) sorption behavior. Chemical Engineering Journal. 2011;**172**(1):219-229. DOI: 10.1016/j.cej.2011.05.092

[32] Qu X, Alvarez PJ, Li Q. Applications of nanotechnology in water and wastewater treatment. Water Research. 2013;**47**(12):3931-3946. DOI: 10.1016/j.watres.2012.09.058

[33] Xu P, Zeng GM, Huang DL, Feng CL, Hu S, Zhao MH, et al. Use of iron oxide nanomaterials in wastewater treatment: A review. Science of the Total Environment. 2012;**424**:1-10. DOI: 10.1016/j.scitotenv.2012.02.023

[34] Tyagi I, Gupta VK, Sadegh H, Ghoshekandi RS, Makhlouf AH. Nanoparticles as adsorbent; a positive approach for removal of noxious metal ions: A review. Science Technology and Development. 2017;**34**(3):195-214

[35] Najafpoor A, Norouzian-Ostad R, Alidadi H, Rohani-Bastami T, Davoudi M, Barjasteh-Askari F, et al. Effect of magnetic nanoparticles and silver-loaded magnetic nanoparticles on advanced wastewater treatment and disinfection. Journal of Molecular Liquids. 2020;**303**:112640. DOI: 10.1016/j.molliq.2020.112640

[36] Saikia J, Gogoi A, Baruah S. Nanotechnology for water remediation. In: Environmental Nanotechnology. Cham: Springer; 2019. pp. 195-211. DOI: 10.1007/978-3-319-98708-8_7

[37] Thamilselvi V, Radha KV. Silver nanoparticle loaded silica adsorbent for wastewater treatment. Korean Journal of Chemical Engineering. 2017;**34**(6):1801-1812. DOI: 10.1007/s11814-017-0075-4

[38] Ma R, Levard C, Judy JD, Unrine JM, Durenkamp M, Martin B, et al. Fate of zinc oxide and silver nanoparticles in a pilot wastewater treatment plant and in processed biosolids. Environmental Science & Technology. 2014;**48**(1):104-112. DOI: 10.1021/es403646x

[39] Sheng Z, Van Nostrand JD, Zhou J, Liu Y. Contradictory effects of silver nanoparticles on activated sludge wastewater treatment. Journal of Hazardous Materials. 2018;**341**:448-456. DOI: 10.1016/j.jhazmat.2017.07.051

[40] Amin MT, Alazba AA, Manzoor U. A review of removal of pollutants from water/wastewater using different types of nanomaterials. Advances in Materials Science and Engineering. 2014;**2014**:1-24. DOI: 10.1155/2014/825910

[41] Tetteh EK, Rathilal S, Asante-Sackey D, Chollom MN. Prospects

of synthesized magnetic TiO2-based membranes for wastewater treatment: A review. Materials. 2021;**14**(13):3524. DOI: 10.3390/ma14133534

[42] Qader MAHA, Al-Rawi AM, Al-Ahmedy KK. Designing Wastewater Treatment Plant by Using Biosynthesized Titanium Oxide Nanoparticles. 2021;**2021**:1-147. DOI: 10.53730/ijhs.v6nS1.4899

[43] Shen YF, Tang J, Nie ZH, Wang YD, Ren Y, Zuo L. Preparation and application of magnetic Fe3O4 nanoparticles for wastewater purification. Separation and Purification Technology. 2009;**68**(3):312-319. DOI: 10.1016/j.seppur.2009.05.020

[44] Palimi MJ, Rostami M, Mahdavian M, Ramezanzadeh B. Surface modification of Fe2O3 nanoparticles with 3-aminopropyltrimethoxysilane (APTMS): An attempt to investigate surface treatment on surface chemistry and mechanical properties of polyurethane/Fe2O3 nanocomposites. Applied Surface Science. 2014;**320**:60-72. DOI: 10.1016/j.apsusc.2014.09.026

[45] Luo T, Cui J, Hu S, Huang Y, Jing C. Arsenic removal and recovery from copper smelting wastewater using TiO2. Environmental Science & Technology. 2010;**44**(23):9094-9098. DOI: 10.1021/es1024355

[46] Vo TT, Nguyen TTN, Huynh TTT, Vo TTT, Nguyen TTN, Nguyen DT, et al. Biosynthesis of silver and gold nanoparticles using aqueous extract from Crinum latifolium leaf and their applications forward antibacterial effect and wastewater treatment. Journal of Nanomaterials. 2019;**2019**:1-14. DOI: 10.1155/2019/8385935

[47] Kumar SS, Venkateswarlu P, Rao VR, Rao GN. Synthesis, characterization and optical properties of zinc oxide nanoparticles. International Nano Letters. 2013;**3**(1):1-6. DOI: 10.1186/2228-5326-3-30

[48] Kiser MA, Westerhoff P, Benn T, Wang Y, Perez-Rivera J, Hristovski K. Titanium nanomaterial removal and release from wastewater treatment plants. Environmental Science & Technology. 2009;**43**(17):6757-6763. DOI: 10.1021/es901102n

[49] Kiran S, Ashraf A, Afzal G, Abrar S, Rahmat M, Asif S. Green synthesis of magnesium oxide nanoparticles using leaves of Iresine herbstii for remediation of reactive brown 9 dye. Global NEST Journal. 2022;**24**(2):291-296. DOI: 10.30955/gnj.003929

[50] Mohammad G, Mishra VK, Pandey HP. Antioxidant properties of some nanoparticle may enhance wound healing in T2DM patient. Digest Journal of Nanomaterials and Biostructures. 2008;**3**(4):159-162

[51] Ren G, Hu D, Cheng EW, Vargas-Reus MA, Reip P, Allaker RP. Characterisation of copper oxide nanoparticles for antimicrobial applications. International Journal of Antimicrobial Agents. 2009;**33**(6):587-590. DOI: 10.1016/j.ijantimicag.2008.12.004

[52] Mahapatra O, Bhagat M, Gopalakrishnan C, Arunachalam KD. Ultrafine dispersed CuO nanoparticles and their antibacterial activity. Journal of Experimental Nanoscience. 2008;**3**(3):185-193. DOI: 10.1080/17458080802395460

[53] Mozhgan SM, Shajari KR, Rostamizadeh K, Reza SR, Shapouri LT, Taghavi L. Eco-friendly curcumin-loaded nanostructured lipid carrier as an efficient antibacterial for hospital

wastewater treatment. Environmental Technology & Innovation. 2020;**18**(2020):100703. DOI: 10.1016/j.eti.2020.100703

[54] El-Sayed MEA. Nanoadsorbents for water and wastewater remediation. The Science of the Total Environment. 2020;**739**:139903. DOI: 10.1016/j.scitotenv.2020.139903

[55] Eskandarinezhad S, Khosravi R, Amarzadeh M, Mondal P, Filho M, Jorge F. Application of different Nanocatalysts in industrial effluent treatment: A review. Journal of Composites and Compounds. 2021;**2**::43-56. DOI: 10.52547/jcc.3.1.5

[56] Liang HW, Wang L, Chen PY, Lin HT, Chen LF, He D, et al. Carbonaceous nanofiber membranes for selective filtration and separation of nanoparticles. Advanced Materials (Deerfield Beach, Fla.). 2010;**22**(42):4691-4695. DOI: 10.1002/adma.201001863

[57] Shajari M, Ahmadi N, Zamani M, Rostamizadeh K, Shapouri R. Hospital wastewater treatment using eco-friendly eugenol nanostructured lipid carriers: Formulation, optimization, and in-vitro study for antibacterial and antioxidant properties. Water Environment Research: A Research Publication of the Water Environment Federation. 2022;**94**(7):e10751. DOI: 10.1002/wer.10751

[58] Li Y, Tang S, Song W, et al. Biosynthesis of silver nanoparticles using Euglena gracilis, Euglena intermedia and their extract. IET Nanobiotechnology. 2015;**9**:19-26

[59] Agarwal P, Gupta R, Agarwal N. Advances in synthesis and applications of microalgal nanoparticles for wastewater treatment. Journal of Nanotechnology. 2019;**2019**:1-9. DOI: 10.1155/2019/7392713

[60] Wang S, Liu Z, Wang W, You H. Fate and transformation of nanoparticles (NPs) in municipal wastewater treatment systems and effects of NPs on the biological treatment of wastewater: A review. The Royal Society of Chemistry. 2017;**7**:37065-37075. DOI: 10.1039/C7RA05690G

[61] Samanta HS, Das R, Bhattacharjee C. Influence of Nanoparticles for Wastewater Treatment—A Short Review. Austin Chemical Engineering. 2016;**3**(3):1036

[62] Oghyanous FA. Nanoparticles in wastewater treatment. In: Eyvaz M, Albahnasawi A, Gürbulak E, Yüksel E, editors. Water Conservation—Inevitable Strategy. London: IntechOpen; 2021. DOI: 10.5772/intechopen.100239

Section 2

Treatment of Sewage Sludge

Chapter 10

Remediation and Management of Sewage Sludge

Farzana Malik, Muhammad Yousuf, Zeeshan Umer, Wajid Malik, Abdul Majid and Sajid Iqbal

Abstract

In recent times, along with urbanization, the population of the city is also increasing rapidly. In this regard, the discharge of municipal sewage is increasing year by year, which is a worrying situation for living beings as well as the environment. In fact, wastewater is an important by-product of modern industry and contributes significantly to polluting the aquatic environment. Its sources are based on many industries and anthropogenic pollutants. The nature of wastewater is organic and inorganic. Many harmful pollutants especially heavy metals are present in sewage sludge and wastewater. Phytoremediation has become a significant experimental and practical strategy to use plants to remove heavy metals from sewage waters, sludges, spillage sites, and polluted places.

Keywords: sewage, sludge, contamination, phytoremediation, wastewater

1. Introduction

A vital requirement for human survival is water. However, this problem will only get worse over time because we are still far from fulfilling the world's standards for clean water. Due to declining water quality, global climate change, and population growth, there is a growing need for clean drinking water [1]. In recent times, the city's population has been increasing rapidly along with urbanization. In this regard, the discharge of municipal sewage is increasing year by year, which is a worrying situation for living organisms as well as for the environment [2]. High concentrations of microplastics (MP) have been observed in municipal sewage sludge. In various countries, it is used as a fertilizer for agricultural land. The application of microplastics to land can contaminate other uncontaminated areas, leading to uncontrolled pollution in those areas [3]. Municipal sewage is an accumulation of commercial water, domestic water of citizens, and municipal unit water. Municipal wastewater must be properly treated. Otherwise, it will be a serious threat to the ecological environment [2].

In fact, wastewater is an important by-product of modern industry and contributes significantly to polluting the aquatic environment. Based on the many industries and pollutants, there are different types of sewage sludge. Each industry sector releases its own unique set of contaminants. In general, there seem to be two types of wastewaters released from industries, chemical wastewater and biological wastewater [4].

IntechOpen

2. Types and sources of wastewater

2.1 Chemical (inorganic) wastewater with sources

Among them, the electroplating sector is a significant source of pollution. Compounds containing chromium, cadmium, nickel, lead, iron, zinc, and titanium are discharged by the metal working industries. Solvent waste is produced by dry cleaning and auto repair industries, Silver is released by photo processing industries; ink and dye are released by printing industries. Petroleum products and phenolic compounds are widely discharged by the petrochemical industry. Additionally, pulp and paper mill effluents comprise suspended particles, organic wastes, and chloride organics and dioxins because the pulp and paper sectors extensively depend on chlorine-based compounds. Wastewater from factories that produce food atoms includes a significant number of suspended particles and organic matter (**Table 1**) [4].

2.2 Biological (organic) wastewater with sources

Biological industrial wastewater is the by-product of large-scale chemical plants and other industrial operations that primarily use organic materials for chemical reactions. The organic components in the wastewaters have diverse properties and originate from various sources. After the wastewater has undergone the proper pretreatment, these can only be eradicated through biological treatment. Numerous organic industrial wastewaters are produced by the following sectors and locations:

i. Factories that produce soap, synthetic detergents, insecticides, herbicides, cosmetics, organic dyes, glue, and adhesives

ii. Paper and cellulose manufacturing factories

iii. Fermentation and brewery and factories

iv. Oil refinery manufacturing industry

Industries	Contaminants
Steel and iron	COD, BOD, oil, cyanide, phenols, metals, and acids
Refineries and petrochemicals	COD, BOD, mineral oils, chemicals COD, heavy metals, chromium, organic chemicals, phenols, and cyanide
Leather and textiles	Sulfates, solids and chromium, and BOD
Paper and pulp	Chlorinated organic compounds COD, solids, and BOD
Microelectronics	Organic chemicals and COD
Non-ferrous metals	SS and fluorine
Mining	Acids, SS, salts, and metals

Industrial wastewater contaminant.

Table 1.
Sources with contaminants.

v. Leather and tanneries

vi. Textile industries [4].

3. Types of sewage contaminants

3.1 Organic sewage contaminants

Various sources, such as urban runoff, domestic wastewater, industrial discharges, and wet and dry atmospheric deposition, can contribute organic contaminants to wastewater. Sewage sludge typically contains large quantities of many kinds of organic contaminants because they tend to sorb on the suspended solids in wastewater. Sewage sludge contains more than 300 chemicals, representing several different chemical groups, according to the type of sewage that the household produces, whether it's municipal or industrial. The range of their concentrations is between pg. kg-1 and g kg-1. Phthalic acid esters (PAEs), organochlorinated pesticides, monocyclic aromatics, polychlorinated biphenyls (PCBs), phenols, polycyclic aromatic hydrocarbons (PAHs), chlorobenzenes (CBs), amines, nitrosamines, and polychlorinated dibenzo-p-dioxins and furans (PCDD/Fs) are the contaminants that are most frequently found [5].

3.2 Inorganic sewage contaminants

The dangerous inorganic chemicals found in sewage sludge are generally heavy metals, having concentrations ranging from cadmium 1–3410, arsenic 3–230, copper 80–2300, zinc 101–49,000, nickel 2–179, lead 13–465 to chromium 10–990,000. Environmental risks are not eliminated by thermal procedures, which often transport, and pollutants can build up in both solid and liquid states. Using sewage that has not been treated, which is a source of numerous hazardous and lethal substances, it may be particularly harmful [6].

4. Role of heavy metals in sewage pollution

There have been many harmful substances introduced into the agro-ecosystem, but heavy metals are particularly problematic because they do not degrade and are bio toxic. In addition to lowering crop yield and quality, heavy metals in soil can pose a hazard to the ecosystem's security and public health [7]. For the most part, because of their toxicity, bioaccumulation, and environmental endurance, heavy metals are well-known inorganic contaminants. Along with other contaminants, these inorganic ones slowly migrate to the nearby water and soil. Waterways can become contaminated by heavy metals in several ways. In places with such contamination, severe cases of diseases caused by heavy metals in living systems have been documented. In some countries, wastewater can be discharged into lakes, rivers, and other bodies of water. Such practices contaminate freshwater, harming the environment. Most the earth's natural species are severely affected by heavy metal contamination. The earth's crust contains heavy metals, but due to human activity, there are geochemical and biochemical imbalances involving naturally occurring compounds [8]. Because they may cause cancer and mutagenesis, heavy metal ions from the paper, leather, and textile industries

Source	Heavy metal	Affect
Use of fertilizers	As, Pb, Hg, Cu	water, soil, and crops
Use of pesticides	As, Cd, Cr, Pb, Ni, Hg	Water, soil, and crops
Reclamation of land	Cr, As, Cu, Cd	Agricultural soil
Dyeing and tanning	Zn, Wood preservation, As	Water and soil
Manufacturing of batteries	Hg	Water and soil
Coating and plating	Cr	Water and soil
Metal refining	Zn, Ni, Fe,	Water and soil
Mining activities	Cd, Ni, Cu, As, Hg, Pb, Zn	Water and soil
Smelting operations	Cd, Co, Cu, Cr, As, Ni, Sb, Hg, Pb	Water and Soil

Table 2.
Sewage sludge is one of the sources of heavy metals in water and soil.

are a severe issue. Some contaminants are poisonous, damaging, and carcinogenic to ecosystems and people. Impurities in some heavy metals are extremely hazardous. Arsenic has long been recognized as a deadly substance. The very poisonous heavy metals chromium, lead, copper, mercury, cadmium, zinc, and nickel, among others, can have detrimental toxic effects. Additionally, the following nitrates, fluoride, selenides, chlorides, phosphates, and chromates have a high-level dangerous effect (**Table 2**) [1].

5. Entrance of heavy metals into our food chain

Compared to other contaminants, heavy metals are considered to pose the main threats to food safety [7]. The natural ecosystem services are degraded by heavy metals being present in contaminated soils, which ultimately impact human health through the food chain. Due to their abundance of advantageous and necessary nutrients and minerals, vegetables are a significant component of the human diet. Unfortunately, plants can absorb and accumulate excessive amounts of heavy metals in both edible and inedible sections of their bodies. Heavy metal concentrations in vegetables (fruit, leafy vegetables, and root vegetables) have been found to be high in recent years. Finally, the accumulation of heavy metals in vegetables and edible crop components in sewage sludge-contaminated soils is a cause for considerable worry due to the negative and permanent consequences that metals have on both human and animal health overall through the food chain [9].

5.1 Sewage wastewater used in agriculture

Nitrogen and phosphorous are found in sewage sludge, particularly because of the nitrification and denitrification phases of the wastewater treatment process. As a result, sludge has special nutritional properties because it contains nutrients that are necessary for plant growth. Sludge may also contain other substances, therefore, that can be hazardous if they get into the human food chain [10].

6. Types of sewage and wastewater treatments

The accumulation of wastewater through urban wastewater treatment systems is a rising environmental problem. A crucial component of sewage management is manure sludge. Through the treatment of wastewater, sludge is created. Wastewater is made up of liquid or water-borne wastes that are separated from institutions, homes, industries, and organizations, as well as groundwater, stormwater, and surface water [11].

It is organic because it contains carbon compounds like paper, human waste, vegetable matter, and so on. It is composed of 99.9% water and 0.1% solids. In the area, there is industrial effluent in addition to sewage from the local communities. In the same manner that sewage is handled physically, chemically, and/or by microorganisms, many industrial wastes also have an organic nature. Using either physical–chemical processes or microorganisms, in order to treat sewage and wastewater, complex organic compounds must be broken down into less unstable, odorless substances. The contaminants in wastewater are removed by physical, chemical, and biological means [12].

In basic, advanced, and secondary wastewater treatment methods, sludge is produced. It is categorized as basic, advanced, and secondary sludge produced in innovative wastewater treatment, respectively. Secondary sludge is made up of biological solids as well as additional settable solids, while primary sludge is made up of settable solids transported in the untreated wastewater. Advanced wastewater can produce sludge that contains highly resilient viruses, heavy metals, nitrogen, or phosphorus.

6.1 Sludge disposal and treatment

i. Initial treatment (screening, comminuting)

ii. The first thickening (belt, centrifuges, gravity, flotation, drainage)

iii. Stabilizing liquid sludge (lime addition, aerobic digestion, anaerobic Digestion)

iv. Secondary thickening (gravity, flotation, drainage, belt, centrifuges)

v. Conditioning (elutriation, chemical, thermal)

vi. Dewatering (belt press, plate press, drying bed, centrifuge)

vii. Storage (dry sludge, compost, liquid sludge, ash)

viii. Transportation (pipeline, road, sea)

ix. Destination (agriculture/horticulture, landfill, reclaimed land forest, land building, other uses).

x. Final treatment (pyrolysis, wet oxidation, drying, incineration, composting, and line addition) [10].

6.2 Disposal of municipal sludge

Sewage sludge is now primarily disposed of via agricultural usage; 11% is burned, 40% is dumped, 37% of the produced sludge is used in agriculture, and 12% is used in other sectors, including forestry, silviculture, land restoration, and so on. Significant scientific interest has been produced by the most recent developments in management of sewage, including co-combustion of sewage sludge with other substances for future use as a source of energy, moist oxidation, decomposition, gasification, and combustion of sludge.

6.3 Incineration technique

The most attractive disposal technique is still incineration. In terms of procedure engineering, fuel efficiency, and plant efficiency, incineration technology has made significant progress in recent years. Advanced fluid bed incinerators are gaining popularity due to their lower capital and operating cost. Sludge volume is massively diminished; after combustion, it is only about 10% of what it was after mechanical dewatering.

i. Burning hazardous organic compounds to dust

ii. Since sewage sludge has roughly the same calorific value as brown coal, incineration presents the opportunity to recover that energy content.

iii. Reduction of odor production [10].

6.4 Thermal processing

To meet the ever-stricter regulations, sewage sludge is thermally processed, which makes use of the energy that has been stored in the sludge while also minimizing any negative environmental effects. Sludge is widely recognized for having high moisture levels. To decrease the amount of moisture, most of the energy released during thermal operations is used. Various contemporary technologies have recently been established, providing an alternate tendency to the disposal of sewage sludge, particularly with the declining availability and rising cost of land for landfilling. The primary representatives of the thermal processing are pyrolysis, gasification, moist oxidation, and combustion. These technologies can all be categorized under the heading of thermal exploitation of sewage sludge [10].

6.5 Combustion

For the warm air processing of sewage, several techniques have been developed and are available on the market. The most well-established method involves the one-track- and co-combustion of sewage sludge, with mono-burning (combustion) being considerably more prevalent. For the warm air processing of sewage sludge, various fluidized bed and hearth technologies have been developed. The extremely well-recognized ones include the one-track- and co-burning of sewage sludge, with being significantly more common. As opposed to single hearth furnaces, multiple hearth furnaces typically consume mechanically wet (dewatered) sludge. Whereas, fluidized bed furnaces may burn sludge with a dry matter composition of 41–65 wt% that is both wet and semi-dried. The release and combustion of volatiles, drying of sludge, and the burning of the extreme residue content left over as char are the dominant factors that might possibly alter the general combustion procedure of sewage sludge. According to the makeup of sewage sludge, burning of the sludge might be viewed as a cause of possible contaminants; thus, caution must be used while disposing of it.

i. Furans and dioxins, HCl, N2O, HF, SO2, as well as NOx, and Cx Hy emissions

ii. Processing solid waste products, such as bed and filter ash

iii. Metals that are released [10]

6.6 Pyrolysis

The process of pyrolysis involves the thermal decomposition of organic materials within an oxygen-off environment at temperatures between 300 and 900°C. To put it another way, pyrolysis is the process of heating sewage sludge within an inactive atmosphere to make available organic stuff that may then be recycled. This process focuses the heavy metals around a solid carbonaceous deposit, not necessarily as much as in debris from burning, which makes it look less polluted Compared to standard procedures (combustion, incineration).

i. The gaseous component of this non-condensable gas (NCG) is mostly composed of hydrogen, carbon monoxide, methane, and carbon dioxide, with negligible volumes of numerous additional gases.

ii. The liquid part of the stream is made up of tar and oil, which also includes acetic acid, acetone, and methanol.

iii. The portion of the solids is mostly made up of char, which is typically just uncontaminated carbon mixed along with trace volumes of inert substances [10].

Pyrolysis consists of the following steps:

i. Volatile compounds vaporizing

ii. Solid char is the primary byproduct of the breakdown of non-volatile components. In addition to the char, fumes and tar are also created.

iii. The char could go through a higher-level pyrolysis process. Many hydrocarbons and aromatic chemicals are present in this stage's final volatile phase [10].

6.7 Wet oxidation

The thermal process category includes the wet oxidation of sewage sludge. It utilizes pure or ambient oxygen and occurs in an aqueous phase, and its temperatures lie between 150 and 330°C and pressures of 1 to 22 MPa. The procedure requires a high temperature to avoid boiling at the necessary temperatures. The procedure involves heat degradation, hydrolysis, oxidation, and conversion of the organic substance in the sewage sludge to water, nitrogen, and carbon dioxide. Two separate regimes operate during the entire operation.

i. The initial happens at temperatures between higher than or below 374°C and 10 MPa of pressure.

ii. A pressure of 21.8 MPa and supercritical temperatures below 374°C for the second.

6.8 Gasification

Technologies that make it possible to use garbage as fuel are extremely important. Theoretically, nearly all biological wastes along with a humidity content of 5–30% are able to be successfully gasified, yet not all biofuels can do this. Gasification is known to be influenced by fuel characteristics like surface, size, and moisture in addition to shape, carbon content, and volatile matter. Sludge and other less-priced materials might be useful as the feedstock for gasification. Numerous variables, including the input fuel, reactor type, and others, affect the gasification process' ability to create gas with a given amount of energy. To produce a suitable gas for power production, essential research into the impacts of sludge on gasification is crucial. Gasification technology can be used to both alleviate the environmental issue and turn the sewage sludge into a usable energy source.

7. Phytoremediation of heavy metal polluted sewage and wastewater

Pesticides, oils, colors, phenol, cyanides, hazardous organics, phosphorus, suspended particles, and heavy metals can all be found in untreated industrial and home wastewater. Among these harmful compounds, heavy metals are easily collected in the environment. To reduce the risk to the environment and human health, it is crucial to remove harmful contaminants from the environment. The fact that heavy metals can take on various chemical forms makes it challenging to remove them from wastewater. Most metals are not biodegradable, and they can easily move between trophic levels to accumulate chronically in the biota. High pollutant concentrations in wastewater are extremely harmful to both human health and aquatic ecosystems.

The development of several heavy metals, including zinc, cadmium, lead, nickel, and so on, is a result of wastewater irrigation. Some of these metals, for instance Ni, Zn, Cd, Cu and Pb, are frequently noticed in the soil's subsurface when untreated wastewater is used to irrigate the soil. Long-term wastewater irrigation raises the level of hazardous heavy metal content in the soil [13].

Phytoremediation has become a significant experimental and practical strategy to use plants to remove heavy metals from sludges, sewage waters, spillage sites, and polluted places. When compared to other remediation methods, phytoremediation offers several benefits:

i. It can be carried out with little impact on the environment.

ii. It is applicable to an inclusive range of toxins, as well as many metals for which there are only some other possibilities.

iii. It may produce a lesser amount of resultant air and water wastes than conventional techniques.

iv. Organic contaminants can degrade to carbon dioxide and water, eliminating their environmental injuriousness.

v. For large quantities of water with low pollutant concentrations, it is economical; taking up contaminated groundwater by plants can stop migration off-site [14].

The idea of using metal accumulator plants in phytoremediation for the removal of heavy metals and several more-than toxins was originally proposed in 1983, but it has been around for 300 years. Numerous terms, including agro-remediation, green remediation, vegetative remediation, botanic remedy, and organic (green) technology, are used to refer to phytoremediation. For the elimination of potentially hazardous metals from the environment, phytoremediation is regarded as an efficient incredibly attractive, economically advantageous, and environmentally beneficial technology. In phytoremediation, plants collect pollutants through their roots and then move them to their aboveground parts of the body [13].

To degrade, remove, or immobilize the pollutants, phytoremediation uses a variety of mechanisms, including accumulation (phytoextraction, rhizofiltration,), degradation (rhizo-degradation, phytodegradation,), immobilization (Phyto stabilization and hydraulic control), and dissipation (phytovolatilization). Plants use a variety of these methods to lower the amounts of pollutants in soil and water, depending on the contaminants. For instance, plants absorb and store HMs in their tissues, and they breakdown organic contaminants to lessen their toxicity in the soil and water. Depending on the types, forms, and mediums of the contaminants, various plants use various strategies or combinations of them to remediate soil and water. Rhizo-filtration, phytodegradation, phytovolatilization, rhizo-degradation, and phytodegradation are all methods for cleaning up contaminated groundwater. Rhizo-filtration, phytodegradation, and rhizo-degradation are three treatment options for surface and wastewater contamination. Through phytodegradation, phytovolatilization, phytoextraction, and rhizo-degradation, contamination caused by soil, sediments, or sludge is remedied [15].

7.1 Phytoextraction

Phytoextraction can be used to eliminate heavy metals [15]. By translocation in the sections of roots that can be harvested, pollutants are retained in the shoot tissue [16]. One of the environmentally safe, long-lasting alternatives for soil cleanup is phytoextraction, which also offers the potential for reusing the metals that are extracted through phytomining. The hyperaccumulating plants' tolerance levels and development restrictions frequently place a limit on phytoextraction. Through the introduction of moderate stress cues that lead to acclimatization in the plant, priming can affect the tolerance of plants to stress. Utilizing various types of chemicals that are added exogenously to plant organs (such as roots, leaves, etc.), plant priming's capacity to increase abiotic stress tolerance has been thoroughly studied [9].

7.2 Phytodegradation

Enzymatic activity breaks down organic pollutants [16]. Organic substances can undergo phytodegradation either inside the plant or in the rhizosphere. This technique can be used to remove a wide variety of substances and classes of compounds from the environment, including as solvents in groundwater, aromatic compounds in soils, volatile compounds in the air, and petroleum [8]. Petroleum is a pollutant that can remain in the environment for a very long time before vegetation fully recovers, and its persistence can be attributed to the hydrocarbons' slow biodegradation. The effects of petroleum on plants might be direct when they encounter oil or indirect when biotic and abiotic changes related to plant development occur. Because of their potential to degrade contaminants, several species of Caesalpiniacae, Mimosaceae, Fabaceae, and Poaceae have been investigated. When compared to soils with no vegetation, the decomposition of petroleum and its derivatives in soils with Juncus roemerianus Scheele, Sorghum bicolor L. Moench, Vigna sinensis (L.) Endl. ex Hassk., Panicum maximum Jacq, Medicago sativa L., Brachiaria brizantha. Stapf., and Festuca arundinacea Vill. The Poaceae family of plants are most important as compared to other plant families because they encourage the removal of pollutants. By altering the physical and chemical properties of the soil, plants and their roots have a direct impact on the breakdown of contaminants [10].

7.3 Phytovolatilization

Gas exchange at stomata on leaf surfaces aids in pollutant removal. Microbiological processes connected to plant roots may enhance/speed up the transformation of organic pollutants like pesticides and hydrocarbons into harmless forms. By combining metal(loid)-immobilizing soil additives with plants that can withstand high levels [16].

7.4 Phyto stabilization

Contaminants are immobilized inside the root zone by an adsorption mechanism working in conjunction with the accumulation and precipitation phenomenon [16]. Microbiological processes connected to plant roots may enhance/speed up the transformation of organic pollutants like pesticides and hydrocarbons into harmless forms. By combining metal(loid)-immobilizing soil additives with plants that can withstand high levels, Phyto stabilization can be improved [10].

7.5 Rhizo-degradation

It is the microbial breakdown of pollutants in plant root zones [16]. Plants influence a site's water balance, alter the pH and redox potential of the soil, and promote microbial activity. These unintended effects could increase root zone destruction or lessen chemical leakage into groundwater. Upon entering plants, substances may be metabolized, retained, or volatilized into the atmosphere [12].

7.6 Rhizo-filtration

It is the storage of contaminants by plants following root absorption from an aqueous growth media [16]. Rhizo-filtration is a phytoremediation technology that can reduce groundwater contamination by using the root systems of plants to remove a variety of contaminants (both organic and inorganic). The chemicals from the polluted water are absorbed and deposited in the root systems of the plants. The bioavailability of pollutants in the food chain can be greatly reduced by the interaction

DOI: http://dx.doi.org/10.5772/intechopen.109408

between the roots of plants and pollutants in contaminated groundwater. The plants chosen for rhizo-filtration should be easy to cultivate, have a low maintenance cost, and produce few wastes when disposed of once the roots of the plant have become completely saturated with the pollutants. The effectiveness of rhizo-filtration was influenced by a variety of variables, including plant species, groundwater quality (pH and temperature), and the chemical properties of organic pollutants. The influencing elements placed a cap on how well the plants could execute rhizo-filtration. The elements could be considered to maximize rhizo-filtration's effectiveness, hence speeding up the removal of organic contaminants from groundwater [2].

7.7 Phyto desalination

Plant species primarily in saline soils remove salts [16]. Phytoremediation using halophytes is preferable since it can be carried out without these issues and is very simple to do. Different halophyte species, such as grasses, shrubs, and trees, can remove the salt from salt-affected problematic soils by salt excluding, excreting, or accumulating through their morphological, anatomical, and physiological adaptation at the organelle and cellular levels. Meeting the fundamental needs of people in salt-affected areas can also be accomplished by utilizing halophytes to reduce salinity [9].

8. Conclusion

Water is a vital requirement for living organisms. However, this problem will only get worse over time because we are still far from fulfilling the world's standards for clean water. The discharge of municipal sewage is increasing year by year, which is a worrying situation for living organisms as well as for the environment. Heavy metals in soil and water can pose a hazard to the ecosystem's security and public health. Most metals are not biodegradable, and they can easily move between trophic levels to accumulate chronically in the biota. High pollutant concentrations in wastewater are extremely harmful to both human health and aquatic ecosystems. To degrade, remove, or immobilize the pollutants, phytoremediation uses a variety of mechanisms, including degradation (phytodegradation, rhizo-degradation), accumulation (rhizo-filtration, phytoextraction), dissipation (phytovolatilization), and immobilization (hydraulic control and Phyto stabilization). Plants use a variety of these methods to lower the amounts of pollutants in soil and water, depending on the contaminants. Municipal sewage is an accumulation of commercial water, domestic water of citizens, and municipal unit water. Municipal wastewater must be properly treated. Otherwise, it will be a serious threat to the ecological environment.

Author details

Farzana Malik[1*], Muhammad Yousuf[1], Zeeshan Umer[2], Wajid Malik[3], Abdul Majid[4] and Sajid Iqbal[5]

1 Department of Environmental Science, University of Agriculture, Faisalabad Pakistan

2 Department Society and Culture, University of Calabria, Italy

3 Department of Epidemiology and Public Health, University of Agriculture Faisalabad, Pakistan

4 Department of Textile Engineering, Hamdard University Karachi, Pakistan

5 Department of Livestock Management, University of Veterinary and Animal Sciences Lahore, Pakistan

*Address all correspondence to: farzanaawan0115@gmail.com

References

[1] Ji GD, Yang YS, Zhou Q, Sun T, Ni JR. Phytodegradation of extra heavy oil-based drill cuttings using mature reed wetland: An in situ pilot study. Environment International. 2004;**30**(4):509-517

[2] Kumar S, Prasad S, Yadav KK, Shrivastava M, Gupta N, Nagar S, et al. Hazardous heavy metals contamination of vegetables and food chain: Role of sustainable remediation approaches-a review. Environmental Research. 2019;**179**:108792

[3] Kristanti RA, Ngu WJ, Yuniarto A, Hadibarata T. Rhizofiltration for removal of inorganic and organic pollutants in groundwater: A review. Biointerafce Res. Appl. Chem. 2021;**4**:12326-12347

[4] de Farias V, Maranho LT, de Vasconcelos EC, da Silva Carvalho Filho MA, Lacerda LG, Azevedo JAM, et al. Phytodegradation potential of Erythrina crista-galli L., Fabaceae, in petroleum-contaminated soil. Applied Biochemistry and Biotechnology. 2009;**157**(1):10-22

[5] Bolan NS, Park JH, Robinson B, Naidu R, Huh KY. Phytostabilization: A green approach to contaminant containment. Advances in Agronomy. 2011;**112**:145-204

[6] Naseem T, Waseem M. A comprehensive review on the role of some important nanocomposites for antimicrobial and wastewater applications. International journal of Environmental Science and Technology. 2022;**19**(3):2221-2246

[7] Hanchang SHI. Industrial wastewater-types, amounts and effects. Point sources of pollution: Local effects and their control. 2009;**2**:191

[8] Karalija E, Selović A, Bešta-Gajević R, Šamec D. Thinking for the future: Phytoextraction of cadmium using primed plants for sustainable soil clean-up. Physiologia Plantarum. 2022;**174**(4):e13739

[9] Hasanuzzaman M, Nahar K, Alam M, Bhowmik PC, Hossain M, Rahman MM, et al. Potential use of halophytes to remediate saline soils. BioMed Research International. 2014;**2014**:598341

[10] Cai QY, Mo CH, Wu QT, Zeng QY, Katsoyiannis A. Occurrence of organic contaminants in sewage sludges from eleven wastewater treatment plants, China. Chemosphere. 2007;**68**(9):1751-1762

[11] Adamcová D, Vaverková MD, Břoušková E. The toxicity of two types of sewage sludge from wastewater treatment plant for plants. Journal of Ecological Engineering. 2016;**17**(2): 33-37

[12] Liu B, Ma X, Ai S, Zhu S, Zhang W, Zhang Y. Spatial distribution and source identification of heavy metals in soils under different land uses in a sewage irrigation region, Northwest China. Journal of Soils and Sediments. 2016;**16**(5):1547-1556

[13] Ali S, Abbas Z, Rizwan M, Zaheer IE, Yavaş İ, Ünay A, et al. Application of floating aquatic plants in phytoremediation of heavy metals polluted water: A review. Sustainability. 2020;**12**(5):1927

[14] Amin M. Phytoremediation of heavy metals from municipal wastewater by Typhadomingensis. African Journal of Microbiology Research. 2012;**6**(3):643-647

[15] Fytili D, Zabaniotou A. Utilization of sewage sludge in EU application of old and new methods—A review. Renewable and Sustainable Energy Reviews. 2008;**12**(1):116-140

[16] Kafle A, Timilsina A, Gautam A, Adhikari K, Bhattarai A, Aryal N. Phytoremediation: Mechanisms, plant selection and enhancement by natural and synthetic agents. Environmental Advances. 2022;**8**:100203

Chapter 11

Investigating Issues and Problems of Using Sewage Effluent in Agriculture

Amir Moradinejad

Abstract

The ever-increasing growth of the population and the rapid development of industries are important factors that have caused an increase in water consumption and wastewater production in communities. On the other hand, in countries located in arid and semiarid regions, available water resources are limited. Therefore, the use of non-conventional water resources (sewage) in these countries is becoming more important day by day. The use of wastewater as a permanent source of water in agriculture, in addition to providing a part of water needs, also saves and sustains water resources. In this research, the effluent of the wastewater treatment plant of Arak city in the central province of Iran was studied in order to check its quality and usability in agriculture. The quality of the effluent was compared with the standards. The results of the research showed that the above wastewater has no restrictions for use in agriculture according to the investigated parameters. This text is compiled based on the results of various research studies conducted in different parts of the world. Finally, the challenges and opportunities of using wastewater in agriculture and providing suitable solutions to reduce the problems have been discussed.

Keywords: irrigation, sewage, sewage treatment plant, soil pollution, water quality

1. Introduction

1.1 What is sewage?

Wastewater is water used for specific consumption that cannot be reused. Because this water is often impure and has an unpleasant smell, it is also called sewage.

Wastewater composition: Wastewater consists of approximately 9.99% water and other solid materials, some of which are organic materials and the other part are solid minerals dissolved or suspended in water. The bad smell of sewage is often due to organic substances in it. The source of wastewater may be domestic, industrial, agricultural, or combined. In terms of physical, chemical, biological, and polluting properties, it has four states: weak, medium, strong, and very strong. In this article, emphasis is placed on the most common type of treated wastewater, which originates from urban wastewater. This wastewater is water that contains human body waste

IntechOpen

(excrement and urine) and wastewater resulting from sanitary measures such as bathing, washing clothes, cooking, and other kitchen uses.

Today, the problem of water shortage and environmental destruction is considered as one of the biggest problems of human societies. In the last century, due to the increase in population growth and the development of the range of human activities in different sectors, the per capita consumption of water has increased greatly. The increase in per capita consumption as well as indiscriminate use of water resources has caused the quantitative and qualitative crisis of water resources to appear in many regions of the world, especially in places that naturally face unfavorable climate and limited water resources. Effluent from urban sewage treatment plants is a great source of water that can be used in agriculture and green space. In this situation, wastewater treatment and recirculation are the most important solution in the development of water resources management, which can play an important role in water scarcity problems [1]. The use of wastewater in agriculture is common in many countries of the world, including the United States of America, Canada, France, Germany, Mexico, Brazil, Egypt, Morocco, Jordan, Saudi Arabia, Qatar, China, etc. [2–8]. In the country of Iran, in recent years, due to the limitation of water resources, population growth, development of urbanization, industries, and agriculture, as well as the development and implementation of numerous plans for the collection and treatment of wastewater, the use of wastewater in agricultural lands has become particularly important and is in the priorities of the program. A study was conducted to identify the state of sewage treatment and the quality of production effluents in Kish Island. The results of this study showed that the quality of wastewater in Kish Island was consistent with the reuse standards of the Environmental Protection Organization for irrigation in all parameters except the total number of coliforms and fecal coliforms. Compared to the standards of the World Health Organization, the said effluent is suitable for drip irrigation and tree irrigation, and it was not found suitable for watering sports fields and green spaces of hotels. Amjad et al. (2006) by examining the facilities and capabilities of reusing urban wastewater in Yazd showed that the quality of the wastewater compared to the standards of the Environmental Protection Organization for reuse in agriculture and irrigation is suitable for irrigation and agriculture [9]. Alaton et al. (2007) studied the effluents of four selected refineries in Turkey and showed that the results are appropriate in terms of common control parameters and heavy metal concentrations. However, the wastewater of selected treatment plants has not been satisfactory in terms of microbes, especially fecal coliforms [1]. Hong Yong and Abbaspour (2007) analyzed the potential of reuse in Beijing city by applying linear programming model. The results of this study evaluated the effective and key factors of reuse potential and provided the basic foundations of this evaluation in other Chinese cities as well [10]. Almas et al. (2006) investigated the performance of sewage stabilization ponds (WSP) in Aden city with experiments on effluents and showed that it is possible to use effluents for limited irrigation [11]. Due to the differences in climatic, plant, social, cultural conditions, soil quality, and the variability of wastewater characteristics from one region to another and even over time in one place, it is wrong to only rely on the application of the instructions provided in other regions of the world, and in the long term, it causes irreparable damage to soil and water resources [12]. Although the use of wastewater in the agricultural sector has many advantages, it is because such waters contain substances such as salts, sodium, chlorine, boron, pathogenic microorganisms, and in some cases, heavy metals or organic and inorganic compounds. Another disadvantage is that their unplanned use can cause very adverse environmental consequences, many of which will not be possible to compensate for, at least in the short term. Salinization of soils, destruction of soil structure, poisoning of plants, and reduction of their performance, pollution of surface and underground water sources, and spread of diseases are prominent examples of these effects. For this reason, in order to prevent short-term and long-term adverse effects of wastewater use, special plans and provisions should be considered. In this article, mainly the challenges associated with the application of wastewater in agriculture are examined and the points that are necessary to be followed in the planning and management of wastewater application plans are discussed, and not paying attention to them will prevent the achievement of the goals of the sustainable development program of agriculture. It creates ambiguity. The general purpose of reuse of wastewater in agriculture is to optimize and preserve the availability of water resources by returning the wastewater flows to the ground and rational use of freshwater resources. Experience has shown that the presence of significant amounts of substances such as phosphate, potassium, and nitrogen in wastewater, which all play a valuable role in the fertility of agricultural land, has been effective in increasing the amount of crops. On the other hand, due to the provision of water for agriculture, new lands can be cultivated, and this will play a key role in controlling the migration of villagers to cities.

2. Location of the area

The location of Iran in a dry and semiarid climate and severe pressures on renewable water sources as a result of recent droughts and the increasing development of urbanization, today the optimal use of available non-conventional water such as urban sewage and Home has been taken into consideration. The reuse of wastewater in the study area of this research is considered important in order to meet the increasing needs of water. At the time of conducting this research, the effluent of Arak sewage treatment plant was randomly used by downstream farmers after being discharged to Mighan desert. Therefore, in order to prevent threats to public health, soil contamination, the entry of pollutants into water sources, and the contamination of agricultural products, it is necessary to reuse it consciously and with the necessary investigation along with the quality control of the effluent at the source. In this research, the wastewater treatment plant of Arak city was studied in order to check its quality and usability in agriculture. The present research is cross-sectional descriptive and sampling of wastewater and conducting tests to determine the quality of wastewater and compare it with the standards.

Arak is one of the cities of Central Province in Iran. The study area of North Arak is an agricultural pole. The water used and needed by different departments is provided exclusively from underground water. Therefore, the limitation of water resources in agriculture and the increasing need for water in the industrial sector have become the concern of the provincial officials. Arak sewage treatment plant is located 10 kilometers north-east of the city (**Figure 1**). This treatment plant with the capacity to treat the wastewater of 105,000 people and receiving wastewater from a number of industries has about 75,000 cubic meters of treated wastewater per day, which is released into the environment and finally flows into the Meghan desert lake. Many lands in the villages around the refinery are barren, and cultivating them can be an effective help in creating employment.

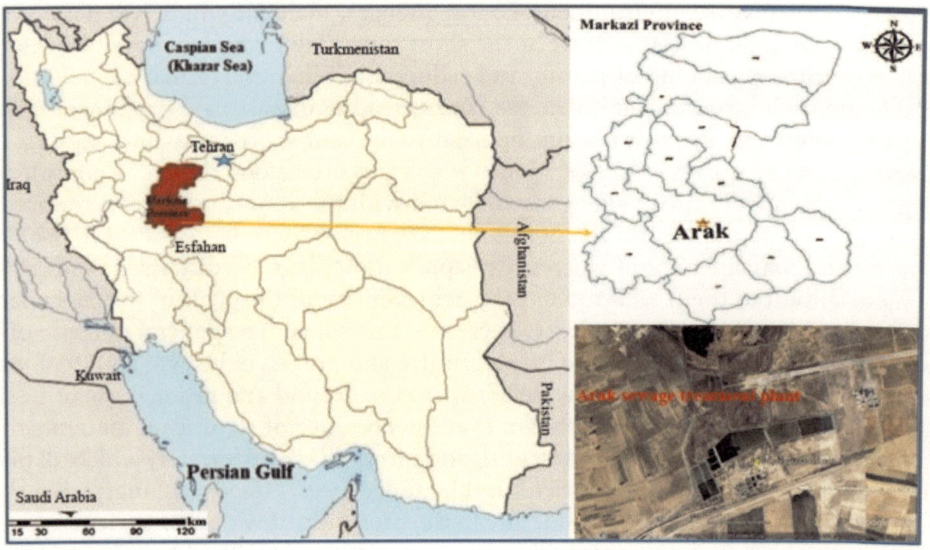

Figure 1.
The location of Markazi Province in Iran.

3. Method of work

This study was a cross-sectional descriptive study. Sampling of wastewater treatment plant effluent in order to determine the required parameters for quality determination in the two cold seasons of autumn and winter 2013 and hot spring and summer 2013 in order to influence the maximum and minimum ambient temperature conditions on the performance of the treatment plant and finally the quality of the effluent. It was done monthly. The samples were prepared in composite form at 6-hour intervals (four times a day) and after the required protection, they were transferred to the laboratory. Samples related to microbial tests were collected in sterile containers and kept at a temperature of four degrees Celsius and transported to the laboratory. The samples related to heavy metal testing were transferred to the laboratory by adding nitric acid and bringing the pH below two. Also, the COD test samples were transferred to the laboratory by adding sulfuric acid and bringing the pH below 2. Some parameters such as dissolved oxygen, temperature, and pH were determined with portable devices on site. All experiments were performed based on the methods recommended in the 2005 standard method book (**Figure 2**). In order to determine health risks, microbial indicators (total coliforms and feces) and heavy metals cadmium, lead, and copper were quantified in terms of their importance. BOD5 and COD parameters were measured to determine the amount of organic substances in the effluent. The analysis of the results was done in the Excel environment, and the statistical indicators including the mean and standard deviation were obtained. It is possible to make a decision about the usability of wastewater in different options based on the results of the tests conducted on the wastewater and comparison with the standards. Various standards for using wastewater in different fields have been provided by international organizations such as EPA, WHO, NAS, and FAO, as well as the quality standards of Jordan for the use of treated domestic wastewater in irrigation. In our country, the standard for the use of wastewater in agriculture and irrigation has been provided by the Environmental Protection Organization. The analysis is based on

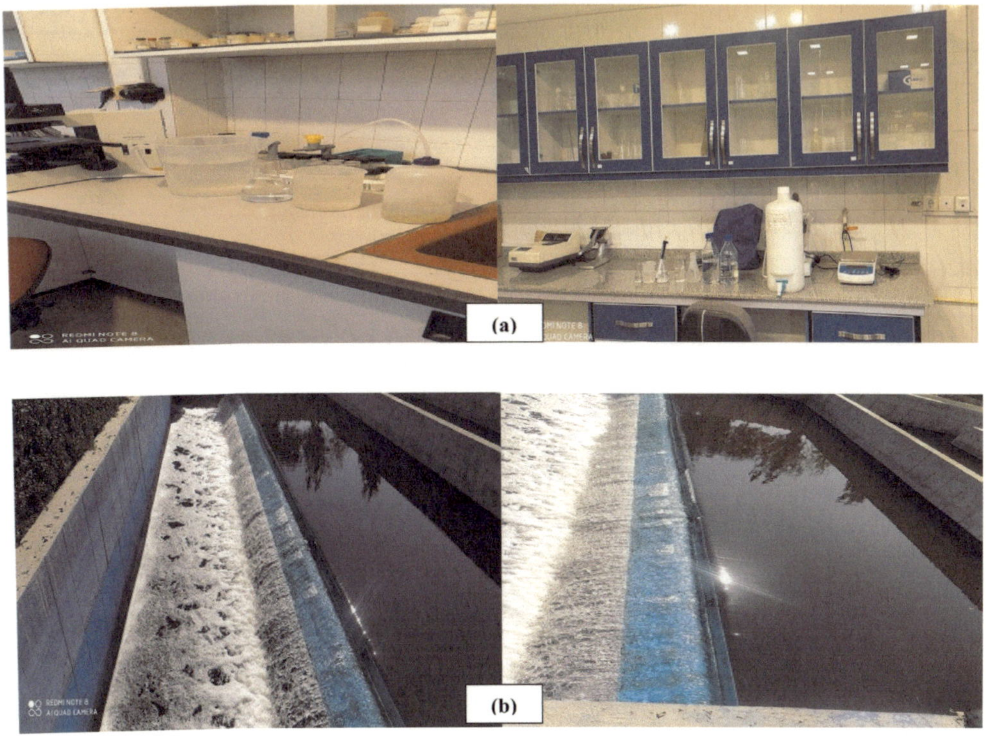

Figure 2.
Sampling and preparation of wastewater samples to send to the laboratory.

existing standards including WHO (World Health Organization), EPA (US Environmental Protection Agency), IRNDOE (Iranian Environmental Protection Organization), NAS (US National Academy of Sciences), FAO (World Food Organization), JORS (Jordanian Environmental Standard), and the vice president of strategic planning and supervision.

4. Results and discussion

Tables 1 and **2** show the comparison of the effluent quality of Arak sewage treatment plant with the standard of Iran Environmental Organization regarding reuse in agriculture. According to **Tables 1** and **2**, the average total and fecal coliforms were 878.9 and 379.6 per 100 ml, respectively, with a standard deviation of 17.2 and 11.4. The average parameters of COD and BOD5 were 49.6 and 23.3 mg/liter, respectively, and the standard deviation was 6.12 and 2.19. Therefore, it is below the recommended limits of 1000 total coliforms and 400 fecal coliforms per 100 ml. Parasite eggs in the samples taken from the sewage are less than the standard with an average of 0.5 and a standard deviation of 0.1134. The average of heavy metals cadmium, copper, and lead were 0.0564, 0.08, and 0.513 mg/liter, respectively, and the standard deviation was 0.0147, 0.0361, and 0.0921. It complies with the standard of the environmental organization in this field. The average values of pH, turbidity, and dissolved oxygen parameters in wastewater were 7.55, 21.31 NTU and 5.14 mg/L with standard deviation of 0.304, 10.3, and 1.2, respectively, which are all in

Parameter	Unit	Standard	Average cold season	Standard deviation cold season	Average warm season	Standard deviation of the hot season	Average for the whole year	Standard deviation of the whole year
COD	(mg/L)	200	48.41	5.20	52.6	7.44	49.6	6.12
BOD5	(mg/L)	100	22.79	2.10	24.4	2.12	23.3	2.2
PH	0	6-8.5	7.63	0.24	7.4	0.36	7.5	0.30
DO	(mg/L)	2	1.94	0.12	1.9	0.16	1.9	0.13
Total coliform	Amount per 100 m	1000	880.58	17.1	874.8	17.60	878.9	17.2
Fecal coliform	Amount per 100 m	400	375.33	9.49	389.7	8.957	379.558	11.3
parasite eggs	Number per liter	<1	0.508	0.121	0.06	0.086	0.524	0.11
Pb	(mg/L)	1	0.5	0.093	0.54	0.086	0.513	0.09
cu	(mg/L)	0.2	0.068	0.02	0.11	0.05	0.08	0.04
Cadmium	(mg/L)	0.05	0.058	0.014	0.05	0.017	0.056	0.015

Table 1.
Comparison of the quality of Arak wastewater treatment plant effluent with the standard of Iran Environmental Organization.

accordance with the standard of the Environment Organization. Life has been consistent in these areas. According to **Table 2**, the values of bar, chromium, and oil in the laboratory samples are lower than the standards, and in terms of these parameters, there is no problem in different uses. Also, the average values of the elements in the treated wastewater have been compared with the existing standards (**Figure 3**). The results show that there is no problem in terms of use for agriculture. In **Figure 4**, the comparison between the average values of heavy elements in treated wastewater in two hot and cold seasons has been done, the results show that cold and heat do not have a great effect on the elements. **Figure 5** compares the average values of the elements in the treated wastewater in different seasons. The results show that the average values of the above eight parameters have the least amount of change in the whole year. The same results can be seen in the two parameters of total coliform and fecal coliform. **Figure 6** shows the comparison of the average values of elements in treated wastewater in different months of the year. Except for cadmium, DO, and pH elements, the results show the highest value in July and the lowest value in February. Of course, in the case of the COD parameter, the highest value happened in September.

5. Discussion

According to **Tables 1, 2** and **Figure 3**, the factors in the effluent are within the appropriate range compared to the existing standards. Therefore, from the chemical point of view, the use of wastewater in agriculture has no special limitations. From the microbial point of view, considering that the products that will be irrigated with wastewater are fodder products or products that cannot be consumed raw by humans,

Standard limit											Observation rate	Parameter mg/L	Row
Planning deputy					Agriculture								
Industrial use	**Aquifer nutrition.**	**Back to surface water.**	**Green space and recreational area**	**agriculture**	**JS**	**FAO**	**NAS**	**IRNDOE**	**EPA**	**WHO**			
0	1	2	0.7	1	1	0.7	0	1	1	0.7	0	(B)	1
10	10	10	0	10	0	0	0	10	0	0	<1	(Oil)	2
0	1	0.5	0	1	0.1	0.1	0.1	1	0.1	0.1	<0.1	(Cr)	3

Table 2.
Comparison of observed values of quality parameters of wastewater with existing standards.

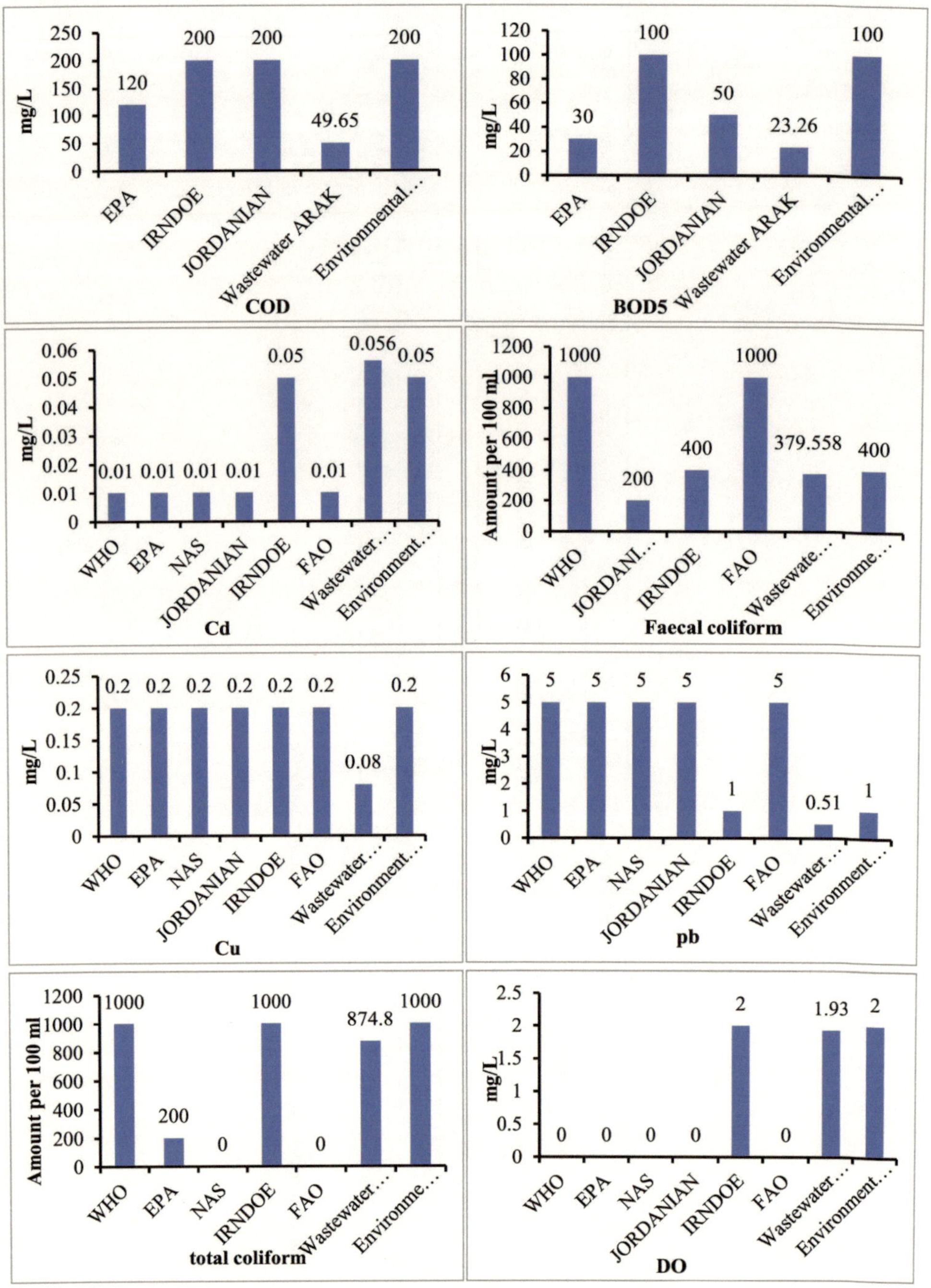

Figure 3.
Comparison of the average values of the elements in the treated wastewater with the existing standards.

it can be said that the transmission of pathogenic bacteria is very weak and there is a limit for the use of wastewater in agriculture, green space, and artificial feeding of underground aquifers does not. Considering that the daily flow rate of the effluent from the treatment plant is 1000 liters per second, and the average hydromodulus of

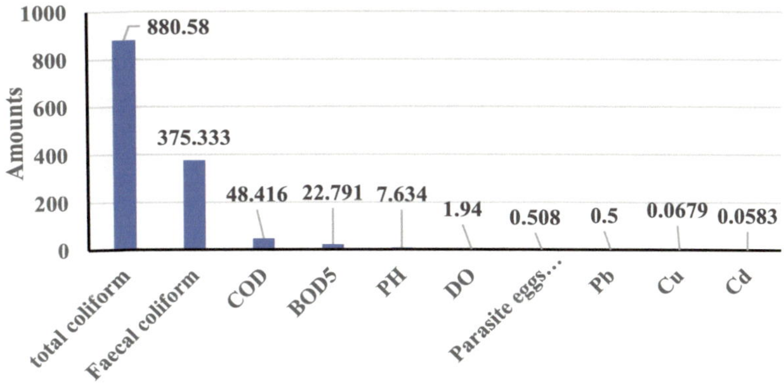

Figure 4.
Comparison of the average values of the elements in the treated wastewater in two hot and cold seasons.

crops in the region is 1.4 liters per second per hectare in the month of August, it is possible to use this water in the cropping season for about 714 hectares of land. He allocated the lands that are barren due to lack of water in the villages around the treatment plant to agriculture and green spaces, and in the non-agricultural season, he used this water for the artificial feeding of the underground aquifers or to the Miqan desert. By artificially feeding the underground aquifers around the desert, the advance of salty water to the underground aquifers of the surrounding villages is prevented. If the effluent is allocated to the irrigation of only one crop, with about 86,400 cubic meters of treated wastewater per day, according to **Table 3**, approximately 4198 hectares of wheat or 1998 hectares of alfalfa or 3679 hectares of clover or He irrigated 16,898 hectares of fodder corn or 18,814 hectares of sunflowers or 30,115 hectares of spruce or 30,586 hectares of walnut trees.

In the application of wastewater for agriculture, the selection of plants should be in accordance with the principles that do not cause contamination of the irrigated crops with pathogenic agents and transfer to the consumer. Therefore, according to the information obtained from the quality of wastewater, compared to the standards of the World Health Organization, the said wastewater is suitable for drip irrigation and tree irrigation, but not suitable for watering sports fields and green spaces of hotels. Also, watering vegetables and products that are consumed raw are not recommended at all. Irrigation of root crops such as potatoes and sugar beets is also not recommended due to direct soil contact. Due to the fact that wheat, barley, and legumes are not consumed directly, they can be cultivated with the effluent of this refinery. It is important that in all cases of application, irrigation with wastewater should be stopped at least 2 weeks before harvest (WHO publication, 2006) [13]. Irrigation of fruit gardens is recommended, but irrigation of fruits that are consumed fresh is not recommended due to the possibility of soil contamination. Trees whose products are consumed as dry fruits such as walnuts and almonds can be suggested. Timber trees such as cypress, pine, elm, and fir can be irrigated with wastewater without restrictions. Also, irrigation of industrial crops such as cotton has no restrictions. Although in this study, the concentration of heavy metals was lower than the recommended standards of the Environmental Protection Organization, but due to the cumulative effect of these elements, the first priority is to irrigate non-edible industrial plants such as cotton and wood trees (WHO publication, 1989) [14].

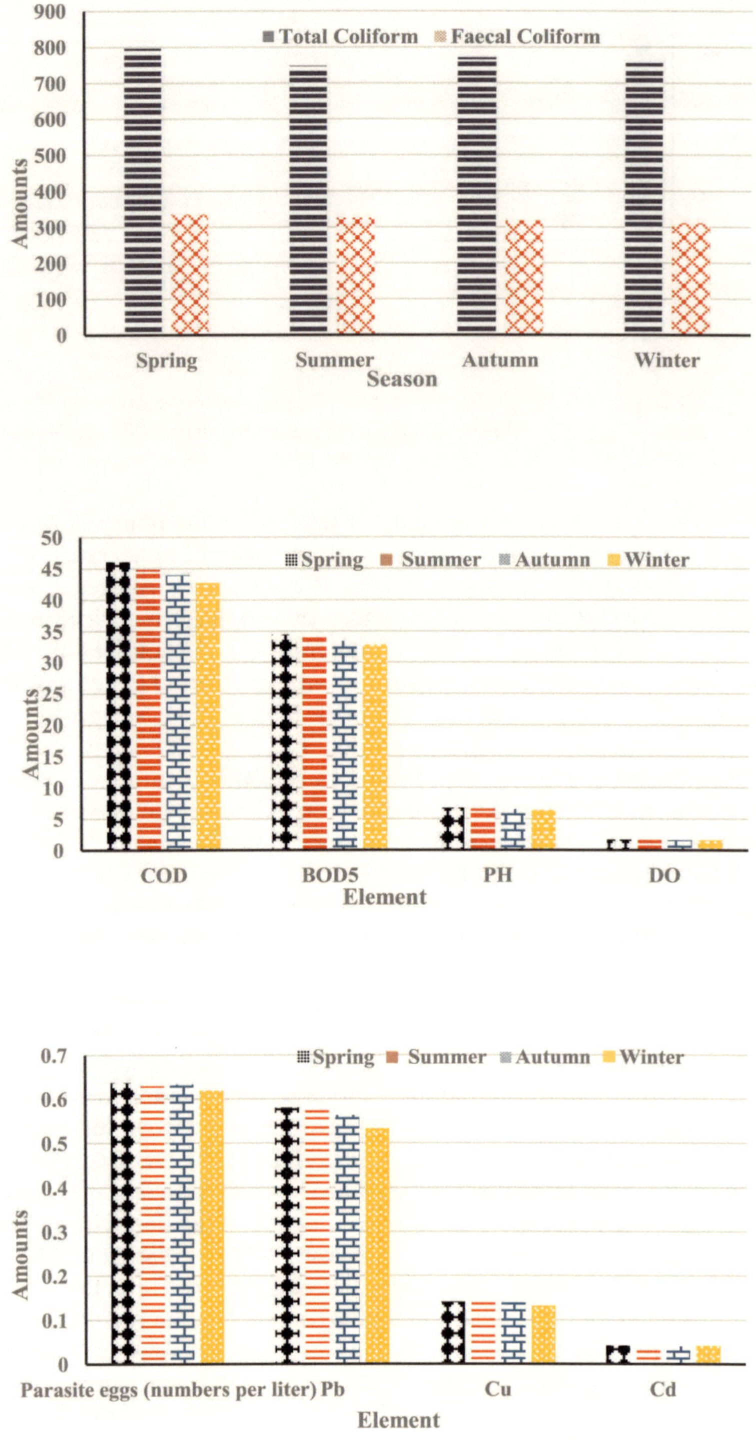

Figure 5.
Comparison of the average values of the elements in the treated wastewater in different seasons.

DOI: http://dx.doi.org/10.5772/intechopen.108636

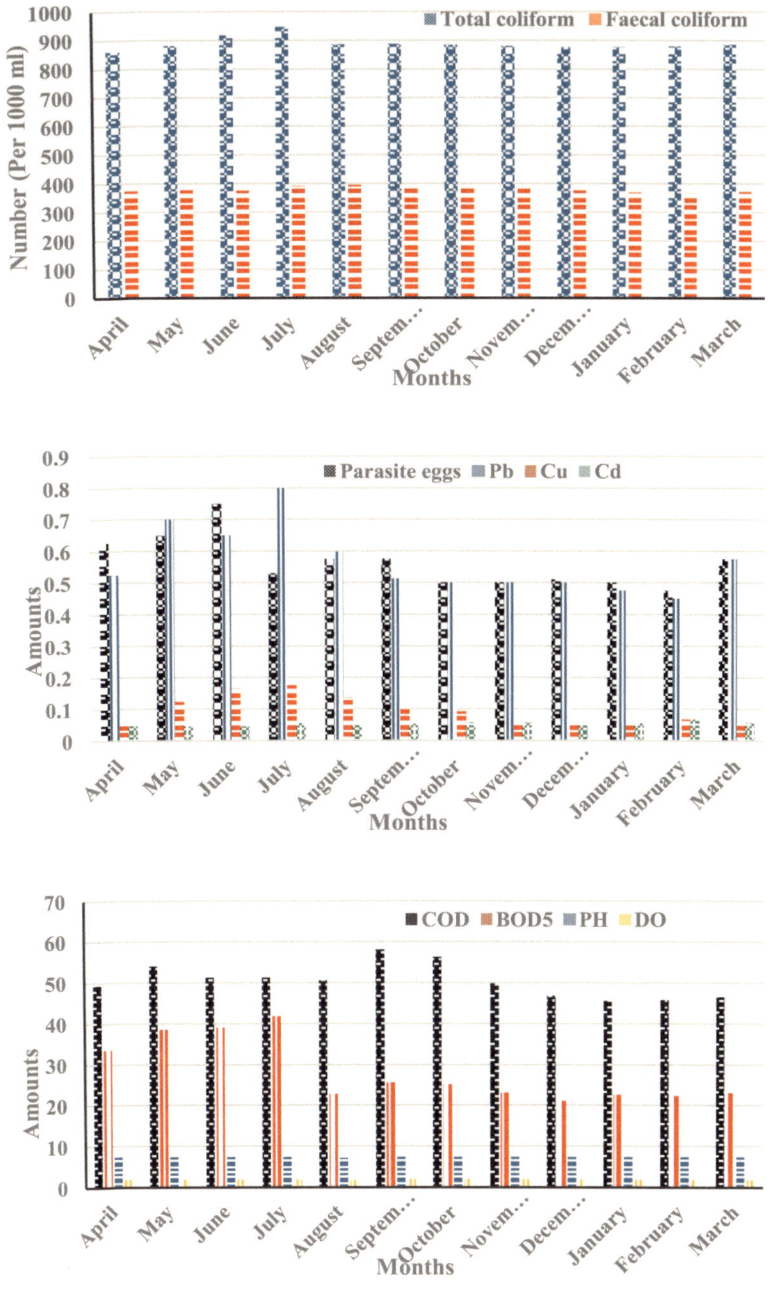

Figure 6.
Comparison of the average values of the elements in the treated wastewater in different months.

5.1 Mixing the wastewater of this treatment plant with primary water

Suitable quality water or primary water can be used directly for product production. On the other hand, raw water mixed with wastewater can be reused. This

Cultivable area (ha)	Pure water required (m^3/ha)	Growth period (Day)	Product name
4198	6520	240	Wheat
1998	13,700	260	Wheat
3679	7440	270	Clover
16,898	6080	110	Fodder corn
18,814	6770	130	Sunflower
30,115	10,830	—	Walnut tree
30,586	11,000	—	Fir tree
714	Average crop yield with hydromodule is 1.3 liters per second per hectare		

Table 3.
Water required for the production of some agricultural products.

combination is possible in two ways: periodic use and mixing. In intermittent use, two water sources are used alternately during the growing season (intra-seasonal intermittent use) or both water sources are used separately during the seasons for different plants (inter-seasonal intermittent use). Choosing a safe option for reuse depends on several important factors, which include: the quality of wastewater, plant resistance to the amount of materials in the wastewater, and the amount of access to freshwater sources. For example, in a place where wastewater is supposed to be used as irrigation water in a cropping season, the important issue is whether the wastewater is used directly or intermittently. The direct use of wastewater is usually done at the farm level without mixing with suitable primary water. The results of the research conducted in India, Pakistan, Central Asia, and Egypt show that surface irrigation with the direct use of wastewater is possible without reducing the yield if the salinity of the wastewater does not exceed the tolerance threshold for the desired plants and the conditions. Drainage should be in good condition. Since plants are more sensitive to salinity in the early stages of growth, according to the research conducted in India, pre-irrigation with water of appropriate quality is of particular importance. In order to obtain a higher yield, it is necessary to pre-irrigate with suitable water and to use wastewater in the subsequent periods of irrigation. Under such conditions, it is possible to use wastewater with a salinity level higher than the plant's tolerance threshold for salinity while preserving the product. Mixing wastewater with water of a suitable quality in such a ratio that the substances in the resulting irrigation water are less than the tolerance threshold of the plant is an acceptable practical method and has been used by many. The option of mixing primary water with wastewater is another management and practical method. It is easy to use because in this method a tank is not needed to mix water from two sources. In addition, many scientists have used good-quality water during the critical stage of plant growth and low-quality water in other growth stages. When the effluent quality is higher than the threshold tolerance value for optimal product production, it can be mixed with other available water sources so that it has an acceptable quality for growing the desired plants. When the mixing operation is carried out at the field level, the water quality can be modified as much as the tolerance limit for the quality of each of the plants. Intermittent use, which is also known as intermittent, is a method that provides the possibility of combined use of suitable wastewater. In this method, wastewater replaces primary water in a predetermined cycle. Intermittent use is used in cases where the effluent quality has exceeded the tolerance of the plant threshold. The intermittent method of using wastewater can be used intra-seasonally and inter-seasonally.

6. Challenges of wastewater application

6.1 Adverse effects on soil, plant, and public health

In the reuse of wastewater in agriculture, due to its inherent characteristics and also due to the occurrence of processes such as decomposition of organic substances, ion exchange, oxidation of minerals, sedimentation, filtration, etc., in the soil system, soil properties can be affected and change especially in the long term. Many researchers in different parts of the world have studied and analyzed the effects of wastewater on the physical, chemical, and biological properties of soil. The increase of soil salinity and sodium as a result of irrigation with wastewater, which respectively causes conditions of reduced water availability for plants and destruction of soil structure, has been reported by various researchers [15–17]. Saber during his research on irrigation with wastewater in Cairo showed that with the increase in the years of using wastewater, the amount of dissolved salts in the depth of 0–20 cm of the soil, significantly up to about three times, compared to non-irrigated soils. It has increased [18]. The results of Ebrahimizadeh et al.'s research (1) also indicate that as a result of irrigation with wastewater compared to conventional water, soil salinity in the layers of 20–40 and 40–60 cm and SAR and sodium in the soil at depths of 0–20, 20–40, 40 and 60 cm has had a significant increase [19]. Smart compared the properties of soils in the northern Adelaide area in Australia that were irrigated with water or wastewater, reported that irrigation with wastewater increased the salinity, sodium, and boron content of the soils in the region, although the observed increase was still to a certain extent. It has not affected the performance of agricultural products. But the observed increase in sodium and SAR of the soil is alarming in terms of the destruction of the soil structure and the reduction of its drainage capacity [20]. The study and research conducted on the soils of the Moose Jaw region in the Canadian state of Saskatchewan, which has been irrigated by wastewater in an area of about 1200 hectares since 1982, have shown that the soil salinity has increased significantly so that the average EC Soil has reached from 0.75 to 1.6 in 1997. More salt accumulation is reported in the 1 meter surface layer of the soil. The results of investigations on shallow underground waters in the mentioned area also indicated an increase in sodium, chloride, sulfate, and bicarbonate concentrations. Also, the studies conducted in connection with another big project that started in Swift Current in the same state in 1978 in an area of about 338 hectares indicate that there has been a significant increase in soil salinity in some places. In the recent region, the amount of chlorine, hardness, sodium, sulfate, and manganese in shallow underground water has increased [21]. In his research in Australia, Patterson concluded that the high SAR in the effluent from domestic sewage treatment plants leads to a decrease in the saturated hydraulic conductivity of the soil, so that with the increase of SAR from zero to 3, the hydraulic conductivity The saturation is 50%, and if it increases to 15%, the hydraulic conductivity is reduced by 75% [22]. In Parvan's research report, it has been reported that saturated hydraulic conductivity decreased by 30% in the surface layer of the soil due to long-term irrigation with wastewater [23]. Alizadeh et al. showed in their research that corn irrigation with the treated wastewater of Mashhad city for 2 years resulted in a 156% decrease in the permeability of the soil compared to the time before the beginning of the research [24]. Shadkam and others also observed a significant decrease in the hydraulic conductivity of the studied soils as a result of irrigation with wastewater [25]. Gholamhossein and Al-Saati in the study of the characteristics of production wastes in Saudi Arabia reported that the use of such wastes due to the inappropriate amount of salinity and sodium can increase the salinity of the soil and change the exchangeable sodium percentage of the soil. Therefore, they recommended the use of wastewater only as auxiliary water [26]. Moadad and Hanifeh Lu in the investigation of the quality of the effluent from the wastewater treatment plant west of Ahvaz city for use in agriculture came to the conclusion that in terms of some parameters such as sulfate, chloride, and salinity, the said effluent exceeded the standards of the Environmental Protection Organization of Iran [27]. And especially in terms of salinity, according to Ayers and Westcott irrigation water quality guidelines, it is evaluated as having a very bad outcome. In the reuse of wastewater, the pH of the soil can also be changed. Since the availability of nutrients required by the plant as well as the solubility of many elements and toxic compounds depends on the pH of the soil, changing this parameter can reduce the absorption of the nutrients required by the plant and in this way or by affecting the availability Toxic elements and compounds affect plant growth and performance [28]. The studies of Saber show that the irrigation of the lands of Cairo city with sewage has led to a decrease in the pH of the soil, but in the investigations carried out by Mahida, an increase in the pH in dry and semiarid soils of India due to irrigation with wastewater was reported. Irrigation with wastewater, especially due to the chlorination process in treatment plants, can increase the concentration of this element in the soil and reach the level of toxicity for plants [29]. Agricultural plants and fruit trees are sensitive to chlorine ion, and if the amount of this element in the saturated soil extract reaches about 10, it causes poisoning for many plants [28]. In his research work on the long-term effects of wastewater, Parvan has reported an increase in the amount of chlorine in different depths of the soil. Among the other impurities that are found in the wastewater of treatment plants, especially in industrial areas, stone metals.

7. Standards and guidelines

Standards and guidelines for reuse of wastewater in agriculture are very different in different countries of the world. One of the most important common guidelines regarding the physical and chemical parameters of irrigation water is Ayers and Westcott's guideline, which is summarized in **Table 4**. Because Ayres and Westcott's guidelines are based on many studies and researches and taking into account factors such as leaching percentage, changes in soil permeability due to EC and SAR, plants' tolerable capacity against salinity, sodium, and bar toxicity and other trace elements have been developed, it can be a suitable basis in evaluating the quality of wastewater for use in agriculture [28].

Regarding the microbiological parameters of sewage and effluent due to their importance in public health, guidelines have been provided by various organizations such as WHO and American EPA for use in agriculture, which are shown in **Tables 4–6**, respectively. The purpose of WHO in developing such guidelines is to determine the guiding indicators for design engineers to choose the appropriate technologies for wastewater treatment and for planners to choose the best management options. Comparison of the mentioned tables shows that the American EPA guidelines are more strict than the WHO guidelines [13, 14, 30]. According to the EPA, for the

The amount of restriction in use			Unit	Irrigation problems			
Severe limitation	**Low to medium restriction**	**Unlimited**					
Salinity (affecting the amount of water needed for the plant)							
>3	0.7–3	<0.7	m/dS	ECw			
				or			
>2000	450–2000	0.45<	l/mg	TDS			
Permeability (the effect on the rate of water penetration into the soil, which is evaluated by considering ECw and SAR)							
<0.2	0–7.2	>0.7		=	ECw	0-3 = and	SAR
<0.3	1–2.3	>1.2		=	ECw	3-6=	SAR
<0.5	1–9.5	>1.9		=	ECw	6-12=	SAR
<1.3	2.1–9.3	>2.9		=	ECw	12-20=	SAR
<2.9	2–5.9	>5		=	ECw	20-40=	SAR
Toxicity of certain ions (sensitive plants)							
				Sodium(Na)			
>9	3–9	<3	SAR	Surface irrigation			
	>3	<3	me/l	Rain irrigation			
				Chloror(Cl)			
>10	4–10	<4	me/l	Surface irrigation			
	>3	<3	me/l	Rain irrigation			
>3	0.3–7	<0.7	mg/l	Br			
Other effects (sensitive plants)							
>30	5–30	<5	mg/l	Nitrogen			
				Bicarbonate			
>8.5	1.8–5.5	<1.5	me/l	Rain irrigation			
	Normal range			pH			

Table 4.
Water quality guidelines for irrigation [28].

application of wastewater in the irrigation of plants that are not processed (for example, plants that are eaten raw), no fecal coliforms should be detectable in 100 ml of wastewater samples. Meanwhile, for watering such plants, WHO has considered the total number of fecal forms in 100 ml of wastewater to be less than or equal to 1000.

Investigations show that heavy metals are the most important harmful pollutants in urban wastewater and irrigation with these wastewaters increases its amount in plants, especially vegetables and soil. Of course, the wastewaters of small and non-industrial cities are less contaminated with heavy metals and chemical compounds. The maximum permissible nitrate concentration that can be present in drinking water is set by the US Environmental Protection Agency as 45 mg/liter. Also, according to the standard of Iran's Environmental Protection Organization, the permissible limit of nitrate for discharging into surface water is 50 mg/liter and for discharging into water

Purification methods expected to meet microbiological guidelines.	Overall fecal forms, geometric mean (number in 100 ml[c]	Intestinal nematode[b] Arithmetic mean (number of eggs per liter)[c]	Vulnerable groups	Reuse conditions	Classification
A series of stabilization sheets designed to achieve microbiological index or their equivalent treatment.	≤1000	≤1	Workers, consumers and the general public	Watering plants that are eaten raw, Sports fields and public parks[d]	A
Storage in stabilization sheets for 8 to 10 days or equivalent methods to remove worms and general fecal forms.	The standard is not recommended.	≤1	Workers	Irrigation of grains, industrial plants, fodder plants, pastures and trees[e]	B
The required pretreatment depends on the irrigation method. Minimum: initial settlement.	Not applicable	Not applicable	None	Irrigation of floor B plants provided that workers and the public are not exposed	C

[a]In special cases, epidemiological, social, cultural conditions, and environmental factors should be considered and the guidelines should be modified based on them.[b]Scaris and Nericoris species and hookworms.[c]During the irrigation period.[d]For public green spaces such as hotels, i.e., where direct public contact is possible, a stricter guideline (less than or equal to 200 coliforms per 100 ml) should be considered.[e]In the case of fruit trees, irrigation should be stopped 2 weeks before fruit picking and no fruit should be collected from the ground. In addition, rain irrigation should not be used.

Table 5.
WHO (1989) Guidelines for the Application of Effluent (Treated Sewage) in Agriculture[a] [14].

Buffer distance	Recommended Monitoring	Wastewater quality	Purification needed	Type of application
300 feet from drinking water sources and 100 feet from publicly accessible areas	• pH - weekly • BOD- weekly • Turbidity- Daily • FC Daily • remaining chlorine - Continuous	• 9-6 = pH • mg/l 30 ≤ BOD • mg/l 30 = SS • ml 100/200 ≤ FC • mg/l 1 = Cl2 Remainder	• Secondary purification • Disinfection	Irrigation: • Food plants that are processed commercially • Orchards and vineyards • pastures • Pastures for dairy cattle • Pastures for livestock
50 feet from drinking water supply wells and 100 feet from publicly accessible areas	• pH - weekly • BOD- weekly • Turbidity- Daily • FC Daily • remaining chlorine - Continuous	• 6-9 = pH • mg/l 30 ≤ BOD • NTU1 ≤ turbidity • ml 100/ (0) = FC • mg/l 1 = Cl2 Remainder	• Secondary purification • filteration • Disinfection	Food plants that are not commercially processed.

Table 6.
USEPA guidelines for wastewater reuse in irrigation [28].

Name of metal	FAO standard	Iranian standard	German standard
Pb	5	1	0.2
Ni	0.2	2	0.033
Hexavalent chromium	—	1	0.2
Fe	5	3	—
cu	0.2	0.2	0.2
Cd	0.01	0.05	0.0033
Zn	2	2	0.5

Table 7.
Allowable concentration standard of some heavy metals for irrigation with wastewater.

absorbent wells, it is 10 mg/liter. **Table 7**. It shows the permissible concentration standard of some heavy metals (mg/liter) for irrigation with wastewater.

8. Sewage pollution intensity

The intensity of sewage pollution is called the power of sewage pollution or its concentration, and the more waste materials in the sewage, the stronger it is. Usually, the strength and weakness of wastewater in terms of organic matter in it are measured according to the following indicators:

A. BOD5 (Biological Oxygen Demand)

BOD5 is the amount of oxygen needed by bacteria to decompose organic matter in wastewater. This standard is the most important tool for measuring biochemically degradable organic substances, which is commonly used in wastewater. The higher the oxygen required, the higher the concentration of organic substances in wastewater that can be oxidized by bacteria. It has been proved by experience that the BOD of a sample is different in hours and even in the early days, and today, at the global level, the value of this index within 5 days (BOD5) has been chosen as the standard. BOD5 relates the reduction of oxygen in the sample (which is consumed by aerobic bacteria in 5 days to decompose organic matter) with the concentration of organic matter that can be decomposed by bacteria, and its unit is mg/liter.

B. COD (Chemical Oxygen Demand)

Chemically required oxygen.

In this method, the amount of oxygen proportional to the chemical decomposition and stabilization of organic materials is called chemically required oxygen. This index is in milligrams per liter.

C. TSS (Total Suspended Solids)

Total suspended solids include materials such as organic particles such as plant roots, soil mineral particles, and plastic particles. Suspended solids are another

Indicators	Chemical composition of untreated urban wastewater			
	Iran (Tehran, Refinement)	Pakistan	America	France
BOD	115–300	193–762	110–400	100–400
COD	125–304	83–103	250–1000	300–1000
TSS	110–790	76–658	100–350	150–500

Table 8.
Range of BOD, COD, and TSS changes in some countries (mg/liter).

sign of wastewater quality. This index is expressed in milligrams per liter. The range of changes of the mentioned three indicators is different depending on the type of wastewater and its characteristics. **Table 8** shows the range of changes in BOD, COD, and TSS (mg/liter) in some countries.

In addition to the above guidelines, different steps have been taken in various countries in order to ensure public health and protect the environment, in the planning of wastewater application in agriculture, an example of which is the development of microbiological standards and guidelines. In terms of developing and applying such standards and guidelines, countries can be divided into several groups: (A) In industrialized and advanced countries such as America and France, standards and guidelines have been developed with a conservative point of view and based on advanced technology and high cost, as well as with low risk tolerance. (B) In some other countries, WHO guidelines, which are based on low technology and low cost, are accepted and form the basis of control [31, 32] and (C) In contrast to the above countries, the third group of countries, which mostly include developing countries, have accepted very strict standards without study and planning. Although these standards are accepted by legal authorities and are very good for having an international image, they are practically unacceptable and unenforceable due to economic and technical reasons. For such countries, perhaps the best solution is the step-by-step development of standards, which, if implemented, can effectively prevent health risks. In relation to the country of Iran, the environmental standards for the use of wastewater are very advanced and include a total of more than 50 physical, chemical, and microbiological parameters, which is ideal if implemented. But according to the existing economic and technical conditions, there are major problems as follows that do not make the implementation of the standards practical:

1. The standards have not been developed based on the economic, social, and epidemiological conditions of the country.

2. Some important qualitative parameters that are very important in terms of agriculture and soil quality, such as: EC, TDS, and SAR, are not included in the current standards.

3. It is not economically and technically possible to ensure the quality of the effluent quality parameters to meet the developed standards.

4. At present, the quality monitoring of effluents from treatment plants is only based on limited parameters such as BOD, PH, TSS, COD, total forms

and total fecal forms, and other important parameters in agriculture are not measured. Therefore, such statistics cannot be a suitable basis for evaluating the quality of wastewater for use in agriculture. It should be remembered that strict standards and advanced technology do not necessarily mean reducing environmental risks and risks associated with wastewater. Because in many cases, the lack of proper operation of sewage treatment plants, the lack of sufficient funding for the management of treatment plants and the application of appropriate monitoring systems, or the lack of enforcement of laws, cause advanced technology and developed standards to lose their practical meaning. And as a result, environmental risks increase [31, 32].

9. Lack of coordination between relevant bodies and institutions

In relation to the discussion of wastewater treatment and the use of wastewater in agricultural lands, many organizations, institutions, and groups are involved, each of which has specific responsibilities and duties and pursues specific goals. These groups are mainly:

1. Legislative and law-enforcing institutions and bodies

2. Companies and institutions responsible for supplying water for agriculture and other uses

3. Institutions and companies responsible for wastewater treatment

4. Organizations responsible for ensuring community health

5. Organizations responsible for planning for agricultural sectors, forests and pastures

6. Farmers and consumers of agricultural products

Although each of the above groups does its best within the scope of their legal responsibilities and specific goals, unfortunately, in many countries, they do not pay attention to the fact that projects such as the use of wastewater in agriculture are not projects that are only responsible for be the responsibility of a specific organization or body. These types of plans are multi-dimensional and inter-organizational plans, as can be seen from their nature, for which comprehensive planning requires full coordination between all the abovementioned organizations. It is obvious that the success in formulation, implementation, and long-term continuation of such plans, in which all legal, technical, economic, social, environmental, and health aspects of society should be considered, can be accompanied by success and continue the alliance. There should be opinions, unification of goals, and maximum effort to create coordination among the involved and officials of the relevant organizations. The result of such cooperation and coordination is the formulation and correct implementation of comprehensive and principled plans for the use of wastewater in agriculture, which leads to increasing the positive effects and reducing the environmental risks associated with such plans [28–35].

10. Sociocultural aspects

One of the important and fundamental aspects of the success of wastewater utilization programs in agriculture is the acceptance of wastewater as a source of irrigation water by farmers and the acceptance of the general public in buying and consuming products irrigated by this source [35]. To achieve this goal, it is necessary that farmers and people are fully informed about the importance of such plans and how to implement them. Also, if possible, their opinions will be collected and reviewed and used in developing programs. In line with this, education and promotion as well as gaining the public's trust in terms of ensuring their health and guaranteeing the protection of natural resources (soil, surface water, underground water, plants, etc.) can solve many social-cultural problems.

Inadequacy of statistics, information, and research.

Since the climatic conditions, characteristics of production effluents, types of crops, economic, social, technical, cultural, and health conditions of different countries are different from each other, therefore, each country cannot simply use the results of other countries' studies in their planning. Good luck. The continuity and success of long-term plans for the use of wastewater in agriculture depend on the plans being developed based on comprehensive information and results obtained from numerous short-term and long-term research conducted in local conditions. The aforementioned research should cover various topics such as: location, type of crops and cultivation pattern, environmental and health risks, determining the amount of acceptable risks, determining risk points, risk reduction management methods, costs related to different risk reduction options, localization of guidelines and standards, suitable options for training farmers and people, etc. [34, 36]. The truth is that currently in many countries, especially developing countries, regardless of the need to conduct research and obtain the necessary statistics and information, based on the belief that "because other countries have been successful, we will also be successful." Wastewater application plans are compiled and implemented without thinking and delaying, which unfortunately will affect the future generations.

The use of wastewater in agriculture for economic development purposes will only be viable if long-term conservation and preservation of resources as well as public health protection are possible. Examining the challenges associated with the use of wastewater in agriculture shows that many of these challenges can be solved with basic planning and the application of correct management methods. In such methods, an integrated control system (a set of different methods) is used to prevent, reduce, and compensate for environmental and health risks, which results in reducing costs, not requiring strict standards and ensuring the success of planning. In the integrated management system of wastewater application in agriculture, a set of different options are used, the most important of which are:

1. Using appropriate standards and guidelines,

2. Using optimal purification methods,

3. Application of appropriate cultivation patterns,

4. Using proper planting and irrigation methods,

5. Applying the necessary methods to limit the contact and exposure of workers and the public,

6. Compilation and implementation of the necessary instructions (for various relevant groups such as farmers and controlling executive agents).

Creating and implementing accurate and efficient monitoring systems. It is obvious that the selection of a suitable set of the above solutions depends on conditions such as the availability of available resources, the agricultural and social situation of the region, the prevalence of fecal-origin diseases in the region, and the market demand for products irrigated with wastewater, which must be implemented before implementation. The selected collection should be carefully examined and studied.

11. Conclusions and suggestions

According to the results of this study, chemically, the use of Arak wastewater treatment plant effluent in agriculture has no special limitations in terms of the investigated parameters. From a microbial point of view, considering that the products that will be irrigated with wastewater are fodder products or products that cannot be consumed raw by humans, it can be said that the transmission of pathogenic bacteria is very weak. It is recommended to carry out a comprehensive and accurate research on the elements present in the products harvested with sewage and the results should be provided to the users and officials. It is necessary to carry out planned reuse in a prudent manner along with quality control in the wastewater treatment stage. Finally, from the integrated control management system such as: use of appropriate standards and guidelines, use of optimal purification methods, use of appropriate cultivation patterns, use of appropriate planting and irrigation methods, application of necessary methods in order to limit contact and exposure of workers and the public and formulation and implementation of the necessary instructions (for various relevant groups such as farmers and controlling executive agents) creation and implementation of accurate and efficient monitoring systems to prevent, reduce, and compensate for environmental risks. Environmental and sanitary use. The use of wastewater in agriculture for economic development purposes will only be viable if long-term conservation and preservation of resources as well as public health protection are possible. Examining the challenges associated with the use of wastewater in agriculture shows that many of these challenges can be solved with basic planning and the application of correct management methods. In such methods, an integrated control system (a set of different methods) is used to prevent, reduce, and compensate for environmental and health risks, which results in reducing costs, not requiring strict standards and ensuring the success of planning. In the integrated management system of wastewater application in agriculture, a set of different options are used, the most important of which are: The use of appropriate standards and guidelines, the use of optimal purification methods, the use of appropriate cultivation patterns, the use of appropriate planting and irrigation methods, the application of necessary methods in order to limit the contact and exposure of workers and the public, formulation and implementation of necessary instructions (for various relevant groups such as farmers and controlling executive agents), creating and implementing accurate and efficient monitoring systems. It is obvious that the selection of a suitable set of the above solutions depends on conditions such as the availability of available resources, the agricultural and social situation of the region, the prevalence of fecal-origin diseases in the region, and the market demand for products irrigated with wastewater, which must be implemented before implementation. The selected collection should be carefully examined and studied.

Author details

Amir Moradinejad
Soil Conservation and Watershed Management Research Department, Markazi Agricultural and Natural Resources Research and Education Center, Arak, Agricultural Research Education & Extention Organization (AREEO), Tehran, Iran

*Address all correspondence to: amir_24619@yahoo.com

References

[1] Alaton IA, Tanik A, Ovez S, Iskender G, Gure M, Orhon D. Reuse potential of urban wastewater treatment plant effluents in Turkey: A case study on selected plants. Desalination. 2007; **215**(1):159-165

[2] Neirizi S. Reuse of purified wastewater as a solution for water supply. Journal of Water and Environment. 1999:**34**(2):4-12

[3] Irfani Aghaz A. 2018. Investigating the effectiveness of treated domestic wastewater in the irrigation of lettuce and tomato cultivation. Proceedings of the Conference on Environmental Aspects of Using Wastewater in Irrigation. National Irrigation and Drainage Committee of Iran. Tehran. Pages: 61 to 79

[4] Angelakis AN, Marecos Do Monte MHF, Bontoux L, Asano T. The status of wastewater reuse practices in the mediterranean basin: Need for guidelines. Water Research. 1999;**33**(10):2201-2217

[5] Asano T, Levine AD. Wastewater reclamation, recycling and reuse: Past, present, and future. Water Science and Technology. 1996;**33**(10-11):1-14

[6] Bouwer H. Integrated water management: Emerging issues and challenges. Agricultural Water Management. 2000;**45**:217-228

[7] Thrassyvoulos M, Ioannis KT. Evaluating water resources availability and wastewater reuse importance in the water resources management of small mediterranean municipal districts. Resources, Conservation and Recycling. 2006;**47**:245-259

[8] Cameron DR. Sustainable Effluent Irrigation Phase l: Literature Review. Canada-Saskatchewan, Agriculture Green Plan: International Perspective and Standards. Tech. Rept. Prepared for Irrigation Sustainability Committee; 1996

[9] Amjad M, Salimi SS, Maghsoodloo B. Study the Opportunities of Reusing Municipal Wastewater of the City of Yazd. Isfahan: 9th National Congress on Environmental Health; 2006

[10] Yang H, Abbaspour K. Analysis of wastewater reuses potential in Beijing. Desalination. 2007;**212**:238-250

[11] Almas AAM, Scholz M. Potential for wastewater reuse in irrigation: Case study from Aden (Yemen). International Journal of Environmental Studies. 2006; **63**(2):131-142

[12] Hassan Aghli A, Liaqat A, Mirabzadeh M. Changes in the amount of soil materials as a result of irrigation or domestic sewage and its self-treatment. Journal of Water and Wastewater. 2008;**42**:11-12

[13] WHO. Guidelines for the Safe Use of Wastewater. Excreta and Grey water Wastewater use in Agriculture; 1989

[14] WHO. Health Guidelines for the Use of Wastewater in Agriculture and Aquaculture. 2006. p. 77

[15] Danesh Sh., Alizadeh A. Wastewater application in agriculture, opportunities and challenges. 1st national conference, the role of recycled water, wastewater in water resource management, challenges and solutions. Mashhad, Iran: 2008

[16] Alizadeh A. Using Reclaimed Municipal Wastewater for Irrigation of Corn. ICID-CIID, Seoul, Korea:

International Workshop on Wastewater Reuse Management; 2001. pp. 147-154

[17] Danesh Sh, Alizadeh A. Evaluation of the quality standards for the use of sewage treatment plants in agriculture. 9th International Congress of Civil Engineering. Isfahan, iran: 2012

[18] Saber MSM. Prolonged effect of land disposal of human waste on soil conditions. Water Science and Technology. 1986;**18**:371-374

[19] Ebrahimizadeh MA, Hasan Li AM, Ahmadi Rad Sh. Minimal environmental effects of urban sewage effluent on soil in corn cultivation. 1st national conference on environmental engineering. Tehran, Iran: 2008

[20] Sttevens DP, Mclaughlin MJ, Smart MK. Effects of long-term irrigation with reclaimed water on soils of the Northern Adelaide Plains, South Australia. Australian Journal of Soil Research. 2003;**41**(5): 933-948. DOI: 10.1071/sr02049

[21] Cameron DR. Sustainable Effluent Irrigation Phase ll: Review of Monitoring Data. Canada-Saskatchewan, Agriculture Green Plan: Moose Jaw. Tech. Rept. Prepared for Irrigation Sustainability Committee; 1997

[22] Patterson RA. Soil hydraulic conductivity and domestic wastewater. Water Science and Technology. 1996; **43**(12):103-108

[23] Parvan M. Effects of Irrigation with Treated Wastewater on Soil Properties. Dissertation of Master's Course in Irrigation and Drainage. Iran: Faculty of Agriculture, Ferdowsi University of Mashhad; 2013

[24] Alizadeh A. Using reclaimed municipal wastewater for irrigation of corn. In: International Workshop on Wastewater Reuse Management. Seoul. Korea: ICID-CIID; 2001. pp. 147-154

[25] Shadkam S, Danesh Sh, Alizadeh A, Parvan M. Investigating the reuse of raw sewage and treated sewage on the hydraulic conductivity of different soil textures. 1st national conference on environmental engineering. Tehran, Iran. 2015

[26] Hussain G, Al-Saati JA. Wastewater quality and its reuse in agriculture in Saudi Arabia. Desalination. 1999;**123**: 241-251

[27] Moazed E, Hanifa Lu A. Evaluation of the quality of incoming and outgoing wastewaters of West Ahvaz wastewater treatment plant for reuse in agriculture. In: 1st national conference on management of irrigation and drainage networks. Ahvaz, Iran. 2007

[28] Ayers RS, Westcot DW. Water Quality for Agriculture. Rome, Ithaly: FAO Irrigation and Drainage Paper, No. 29, FAO; 1985

[29] Mahida UN. Water Pollution and Disposal of Wastewater on Land. New Delhi: McGrow-Hill pub.; 1981. p. 323

[30] US. Environmental Protection Agency. Guidelines for Water Reuse. Report No. EPA. 625/R-04-108. 2004

[31] Fatta D, Kythreotou N. Wastewater as Valuable Water Resource - Concerns, Constraints and Requirements Related to Reclamation, Recycling and Reuse. IWA International Conference on Water Economics, Statistics, and Finance. Greece. 2005

[32] Strauss M. Reuse of Wastewater in Urban Agriculture. ETC-RUAF, CTA Pub: Wageningen, Natheriands; 2001

[33] Environmental rules and standards for reuse of return water and wastewater. Publication No. 535 of the Environmental Protection Organization of Iran. Editing and revision. Tehran, Iran. 2010

[34] Haruvy N. Wastewater reuse-regional and economic considerations. Resources, Conservation and Recycling. 1998;**23**:57-66

[35] Salgot M. Water reclamation, recycling and reuse: Implementation issues. Desalination. 2008;**218**:190-197

[36] Kamizoulis G. Setting Health Based Targets for Water Reuse (in agriculture). Desalination. 2008;**218**(1-3):154-163. DOI: 10.1016/j.desal.2006.08.026

Chapter 12

Study on the Impact of Artificial Recharge on Treated Domestic Sewage

Maipady Ramu Dhanraj and A. Ganesha

Abstract

The proposed research will look into, and study the impact of artificial recharge of treated sewage in the unconfined aquifer. In order to enhance the focus on the development and planning for wastewater management, finding a solution for the disposal of treated sewage has become an important priority. As a result, a quantitative study is necessary to bring out the solution for the disposal of treated sewage. There are two types of artificial recharge of treated sewage practiced globally (a) confined aquifer recharge and (b) unconfined aquifer recharge. Artificial recharge in a confined aquifer may not be sufficient to meet the need for pollutant removal, but artificial recharge practice in an unconfined aquifer will meet the needs of pollutant removal as per the review of the literature. At the moment, treated sewage after tertiary treatment is reused in a variety of ways, such as landscapes in educational institutions, cooling towers in industry, and so on. However, the pollution control board insists zero disposal, then the disposal of treated sewage remaining after reusing will become a burning issue to handle and to dispose it. The idea of the recharge will serve the purpose of replenishment and also a solution for disposal in all-time weathering conditions.

Keywords: biochemical oxygen demand, chemical oxygen demand, total dissolved solids, groundwater recharge, pollutant transport

1. Introduction

Water is a vital source of life for all creatures on the planet earth. Indians have developed various lifestyles as a result of the country's diverse culture and because of the constant growth in population, wastewater management has become a time-consuming procedure, especially in urban areas. The demand for water is increasing every day, and so is the quantity of water squandered. Furthermore, due to a lack of adequate wastewater disposal management, the issue has gained momentum in virtually all cities inside the present environment.

The focus is on driving away sustainable development as a result of increased urbanization and the government's smart city program. In the current situation, adopting a systematic approach to achieve sustainability is critical, since pollution, climate change, and deforestation are on the rise on the one hand, while natural existing water supplies are decreasing day by day on the other end. After China, India is second of the leading populated countries in the globe. Currently, 61,754 million liters of water per day (MLD) of sewage is generated per day, 22,963 MLD of sewage is processed, and 62% of sewage is discharged straight into water bodies without treatment, according to the estimates.

Water consumption for business and home use might reach 29.2 billion cubic meters by 2025 and the population is expected to cross 1.5 billion mark by 2050 [1]. Therefore, the aim has to be to utilize the treated sewage as an alternative source to replenish and conserve it for future purposes by employing an artificial recharge technique. This may increase the efficiency in the use of water by employing the conjunctive use of groundwater thereby reducing the demand for freshwater sources. Though the practice of stormwater recharge is adopted, the significant effects are still unknown after recharge for seasonal emerging pollutants. The artificial recharge system of treated sewage is a promising technique that has significant research potential to understand the mechanism with preliminary hydrogeological investigation to evaluate the impact on unconfined aquifer. This study will head into a new form of disposal for all weathering conditions and more importantly, the potential of the land to adopt this method needs to be addressed.

2. Aquifers

Aquifers are bodies of rock and/or sediment that contain groundwater. The term "groundwater" refers to rainwater that penetrated through the soil and accumulated in voids of the subsurface of the soil. Aquifers naturally filter groundwater by forcing it to travel through tiny pores and between sediments, which aids in the removal of contaminants. However, this natural filtration process may not be sufficient to remove all of the pollutants. Aquifers are classified into two types: confined and unconfined.

Confined aquifers are also known as "Artesian aquifers" since they are located mainly above the base of confined rock strata. Water levels in punctured wells derived from artesian aquifers fluctuate owing to pressure changes rather than the quantity of stored water. The ruptured wells function primarily as conduits for water transfer from replenishment regions to natural or artificial end destinations.

Unconfined aquifers, in contrast to restricted aquifers, are typically found near the ground surface above the water table, although sitting comparatively above impervious clay rock strata. The water table is the highest barrier of groundwater inside an unconfined aquifer. Groundwater in an unconfined aquifer is more sensitive to contamination by surface pollution than groundwater in confined aquifers due to simple groundwater penetration by terrestrial contaminants. The level of groundwater fluctuates and is determined by the amount of groundwater stored in the aquifer, which impacts the rise or fall of water levels in wells that draw their water from aquifers (**Figure 1**).

2.1 Artificial recharge

Artificial recharge is a strategy for replenishing an unconfined aquifer by sending additional surface water into the earth by distributing it on the surface, using recharge wells, or modifying natural conditions to enhance penetration. Artificial recharge is also known as planned recharge, which is described as the storage of water

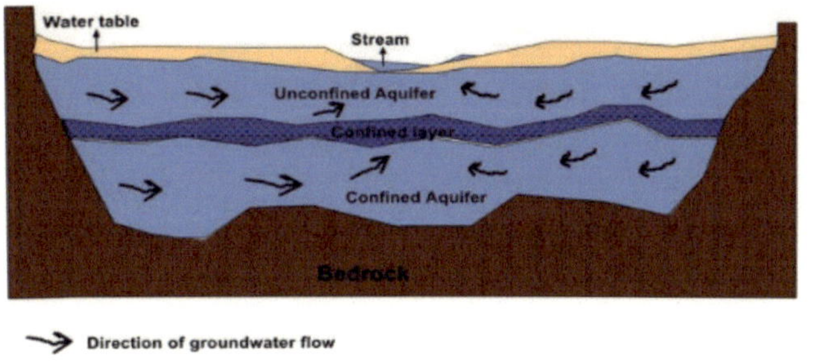

Figure 1.
Unconfined and confined aquifer systems.

underground. During periods of water scarcity, the requirement of additional water is been provided as per demand. Artificial groundwater replenishment with treated wastewater is a crucial and necessary practice with several benefits. It helps to avoid groundwater depletion first and foremost. Second, it maintains water in a certain basin or watershed (i.e., the water is not lost to surface water outflow from the watershed or discharge to the ocean). Third, it may save a lot of money when compared to alternative water sources (**Figure 2**) [3].

With recovered water, three methods of groundwater recharge are typically used: subsurface injection into the vadose zone, surface spreading, and direct injection into the aquifer. **Figure 3** illustrates these three ways.

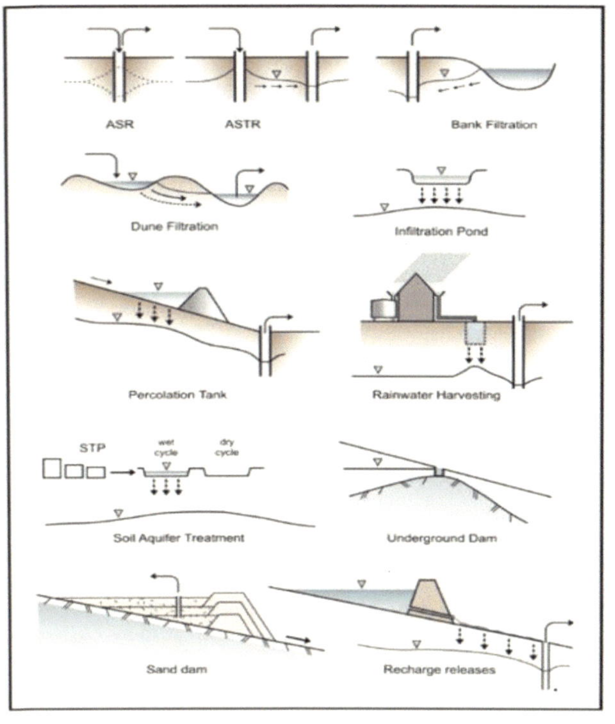

Figure 2.
Different types of aquifer recharge [2].

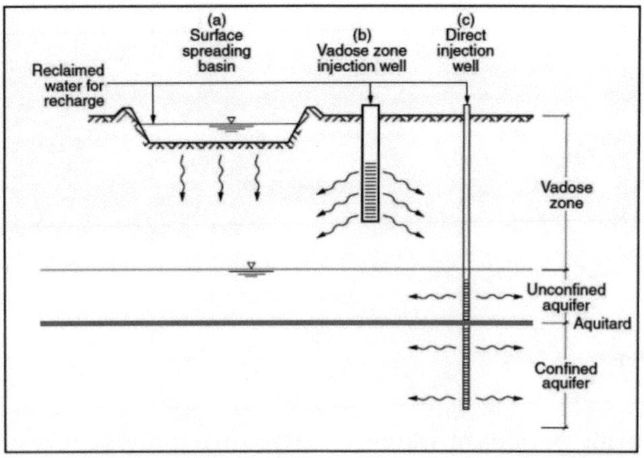

Figure 3.
Different methods in groundwater recharge [4].

2.2 Selection of recharge system

The most frequent technique of groundwater recycling with recharge basins is the surface spread, although it is constrained to aquifers having vadose areas. Injection wells, which have developed recently, are less prevalent in the vadose zone in unconfined aquifers. With confined or unconfined aquifers, direct injection wells may be employed; they may be used to inject a variety of aquifers at various depths (**Table 1**).

2.3 Simulation study

This work is carried out to track contaminant movement in a horizontal direction and is analyzed using a 2D model of pollutant transport in a porous medium with regulated discharge. COMSOL Multiphysics generated a soil matrix for water recharging of treated household sewage. With a 2D analysis, the diffusion is positioned in the center of the recharge as it flows downhill and horizontally through the soil matrix. The soil column dimensions are 1.5 × 1.2 m. The flow pattern in lateral direction was investigated in COMSOL to study the transport of recharge water as well as its pollutant constituents such as completely dissolved solids with adsorption

Characteristic	Recharge basins	Vadose zone (shallow) wells	Direct injection wells
Location where treatment occurs	Vadose zone and saturated zone	Vadose zone and saturated zone	Saturated zone
Aquifer type	Unconfined	Unconfined	Unconfined or confined
Pre-treatment Requirement	Secondary treatment	Secondary treatment & Filtration	Advanced filtration
Maintenance Requirements	Drying and scraping	Drying and Disinfection	Disinfection and flow reversal

Table 1.
Characteristics of aquifer recharge for different [4].

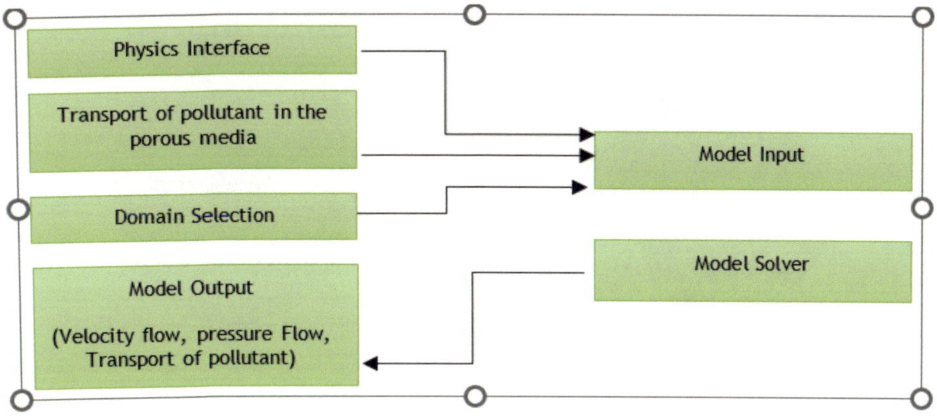

Figure 4.
Flow diagram of pollutant transport modeling.

and desorption characteristics in soil. This research was carried out by developing and running domestic wastewater subsurface infiltration systems. It was discovered that this work discusses the significance of purifying substrates and the types of structures for optimizing diverse operation modes such as HLR, PLR, intermittent operation, aeration, and shunting distribution operation [5]. Metabolomics analysis of a subsurface wastewater infiltration system subjected to organic load fluctuations and current microbiology [6]. Depending on the size of the pores involved, the transport of pollutants in porous media equation was utilized for saturated porous medium. The primary goal of the research is to comprehend the concentration distribution and find the impact after infiltration in the porous medium. Variations in concentration are seen at the outflow based on the study time from the physical model study (**Figure 4**).

Governing equation for the contaminant transport in porous media

$$\frac{\partial\left(\varepsilon_{\mathrm{p}}c_i\right)}{\partial t}+\frac{\partial\left(\rho c_{\mathrm{P}j}\right)}{\partial t}+\nabla\cdot\mathrm{J}_i+\mathrm{u}\cdot\nabla c_i=R_i+S_i \tag{1}$$

$$\mathrm{J}_i=-\left(D_{\mathrm{D}j}+D_{\mathrm{e}j}\right)\nabla c_i \tag{2}$$

$\theta=\varepsilon_{\mathrm{p}}$ Under no flux condition with the associated boundary conditions.

$$\theta=\varepsilon_{\mathrm{p}} \tag{3}$$

$$-\mathrm{n}\cdot\left(\mathrm{J}_i+\mathrm{u}c_i\right)=0 \tag{4}$$

The model for the best results, we must partition the entire model into discrete portions. Depending on the intended effects, this meshing can be further split in a variety of ways. Because the emphasis is on the model's inlet and outflow, a small mesh was chosen based on the geometry of the model. Table 4 shows the statistics used to solve the mesh (**Figure 5**).

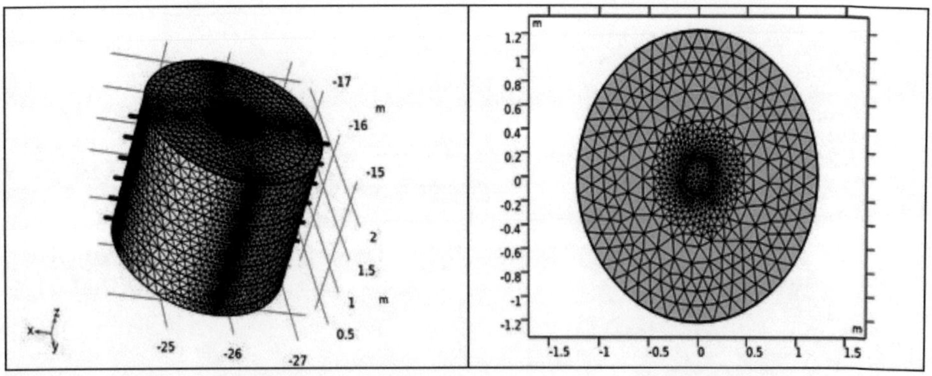

Figure 5.
3D and 2D boundary layer meshing and mesh extrusion make it possible to efficiently discretize.

Property	Minimum element quality	Average element quality	Triangle	Edge element	Vertex element
Value	0.7128	0.9294	1154	72	8

3. Results and observation

The quality of the treated effluent of the sewage treatment plant (STP) is studied and used as an influent (Recharge water) for research work. The samples were collected after the secondary treatment and were tested for parameters such as BOD and COD.

The table below shows the average results of the recharge water used.

Parameter	Raw sewage	Treated effluent
BOD (mg/l)	140	39.24
COD (mg/l)	300	114.00
pH	6.65	6.8

4. Media characteristics

Soil is used as media for the recharge in the physical model developed. The media characteristics such as permeability, porosity, grain size, specific gravity, and density are the parameters determined.

The table indicates the.

Parameters (units)	Soil
Permeability (cm/s)	1.79×10^{-5}
Grain size	1.44
Specific gravity	2.54
Porosity	0.3
Density (gm/cm^3)	1.46

Parameter	Influent	Effluent	% Removal
BOD (mg/l)	53.33	23.85	58.02
COD (mg/l)	103.00	53.61	46.75
TDS (mg/l)	246.10	140.60	58.00
Alkalinity (mg/l)	220.00	105.00	52.06
Acidity (mg/l)	135.00	98.00	28.04
pH	7.90	7.56	—
Chloride (mg/l)	12.50	6.20	49.15

From the results, it was observed that there is a considerable reduction in the pollutant concentration from inlet to outlet which shows the impact of artificial recharge. Further, it is observed that the pollutant removal efficiency is higher in soil media than in sand media. Hence for the prediction analysis, the properties of soil are considered in the simulation model (COMSOL Multiphysics) [7].

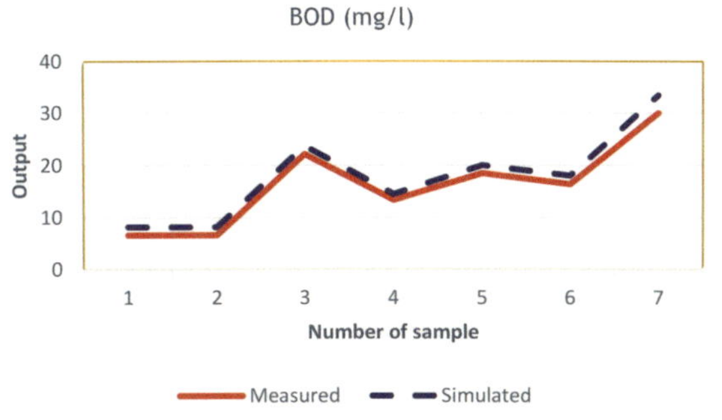

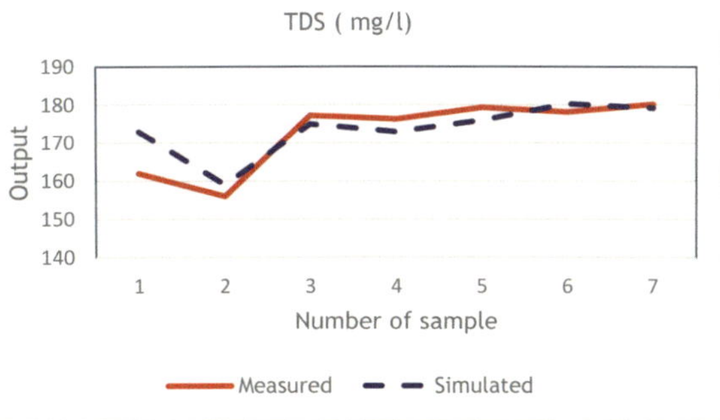

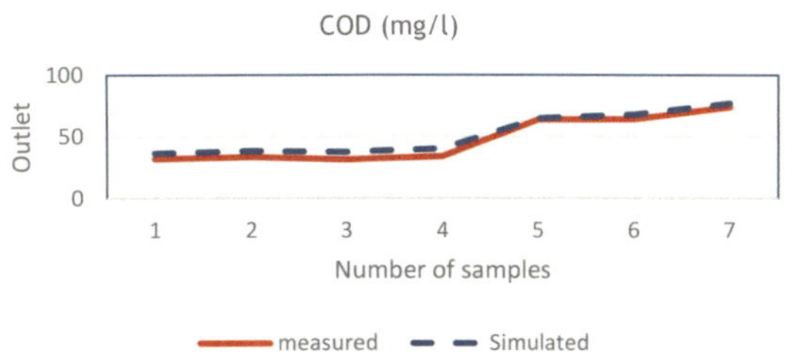

5. Regression analysis

The validation of results from physical and simulation models is further strengthened using linear regression analysis for soil media.

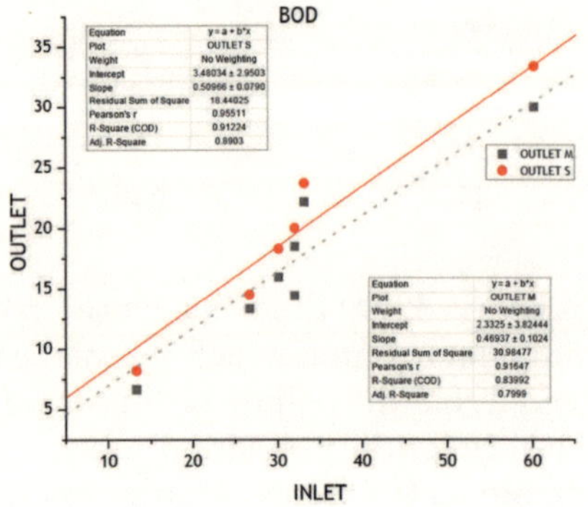

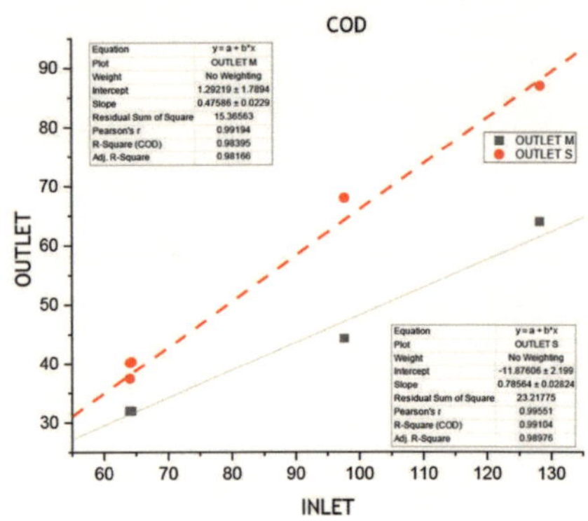

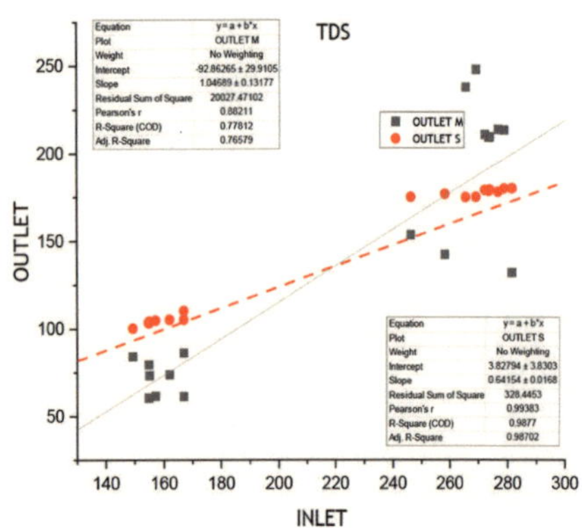

5.1 Prediction analysis of pollutant removal using COMSOL multiphysics

Prediction analysis is carried out using COMSOL Multiphysics in order to determine the pollutant travel distance i.e., (zero concentration at the outlet). For the analysis based on the observation from the physical model used, the radius and time of flow were suitably assumed. By varying the radius of the model and on the basis of the time required from inlet to outlet for the analysis, the parameter BOD is considered with the inlet value of 26.66 and 53.33 mg/l. Following are the boundary conditions used for the prediction analysis.

Infiltration pipe: 0.15 m.

Permeability: 1.79 × 10^{-5} cm/s.

Discharge: 2.5 lpm.

Porosity: 0.3.

Materials used = soil media.

Density = 1.46 gm/cm^3 (**Table 2**)

Radial distance (m)	Time (h)	Inlet BOD =26.66 mg/l	Inlet BOD =53.33 mg/l
		Outlet BOD mg/l	Outlet BOD mg/l
0.52	0.5	11.80	23.00
1.05	1	9.50	19.50
2.11	2	8.50	17.30
4.22	4	5.70	11.70
8.44	8	2.85	5.85
16.86	16	1.90	3.90
25.31	24	0.00	0.90
25.91	25	—	0.00

Table 2.
Simulated model output for pollutant removal (BOD).

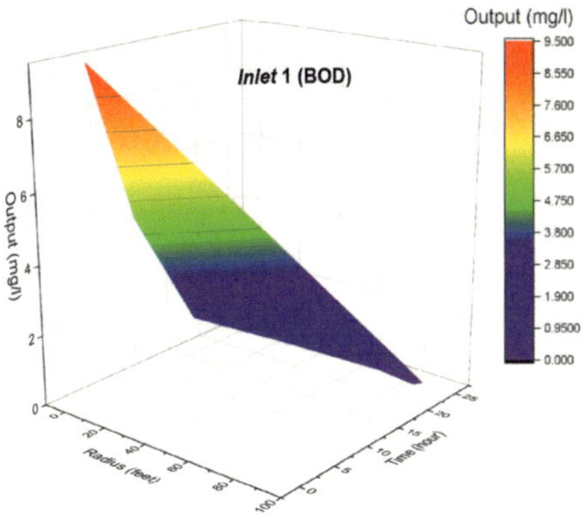

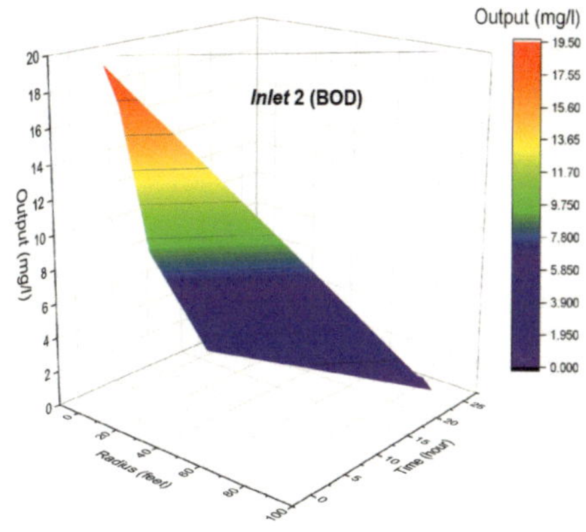

From the prediction analysis, it is observed that the pollutant concentration decreases as the radius increases and reaches zero concentration at a certain distance. Hence, for the known quality of recharge water and site condition, it is possible to establish the pollutant travel distance.

6. Conclusion

Sewage management is a burning issue and there is a need for this should be addressed because of the disposal issue and nuisance most developing and underdeveloped countries face today. Each one of the counties has set up its own standards for disposal and degree of the treatment that needs to be provided based on the organic load in order to prevent environmental degradation. There is a lag in addressing the treated sewage disposal irrespective of the reuse for different purposes and needs. Since India is one of the countries where there is a large amount of fresh water available and the reuse of treated sewage has never been given importance and focus. But today, each continent is facing a drastic climate change and the global temperature is increasing, there is a need for addressing and promotig sustainable practice and conserving water. This will be one of the sustainability goal that can be reached by the United Nations by 2030.

Acknowledgements

I would like to thank the co-author Dr. Ganesha A for his support in preparing the report and reviewing the report. Also, I would like to thank Nealesh Dalal, managing trustee, JD school of design for giving me an opportunity to utilize the resources available in campus.

Conflict of interest

"The authors declare no conflict of interest."

Author details

Maipady Ramu Dhanraj[1*] and A. Ganesha[2]

1 JD School of Design, Bangalore, India

2 International Center for Applied Science, MAHE, Manipal, India

*Address all correspondence to: dhanraj.s@jdinstitute.edu

References

[1] CPCB. Status of water supply and Wastewater Collection Treatment & Disposal in Class I Cities-1999, Control of Urban Pollution Series; 1999. CUPS/44/1999-2000

[2] Dillon P. Future management of aquifer recharge. Hydrogeology Journal. 2005;**13**(1):313-316

[3] Sanders LL. Manual of field hydrogeology. Prentice Hall; 1998

[4] United States. Environmental Protection Agency. Office of Wastewater Management. Municipal Support Division, National Risk Management Research Laboratory (US). Technology Transfer, and Support Division. Guidelines for water reuse. US Environmental Protection Agency; 2004

[5] Li D, Xinze W, Lina C, Jie W. The design and operation of subsurface wastewater infiltration systems for domestic wastewater. Water Environment Research. 2019;**91**(9):843-854

[6] Yang L, Yinghua L, Fei S, Haibo L. Metabolomics study of subsurface wastewater infiltration system under fluctuation of organic load. Current Microbiology. 2020;**77**(2):261-272

[7] Dhanraj AG. Pilot model study for artificial recharge of treated domestic sewage in unconfined aquifer. Indian Journal of Environmental Protection. 2021;**41**(6):698-702

Chapter 13

Options for the Disposal and Reuse of Wastewater Sludge, Associated Benefit, and Environmental Risk

Astha Kumari, Nityanand Singh Maurya, Abhishek Kumar, Rajanee Kant Yadav and Amit Kumar

Abstract

To protect human and environmental health, wastewater treatment is one of the important activities in urban and industrial areas. Urbanized increasing population with industrialization demands more amount of wastewater treatment. Despite wastewater treatment's positive impact on human and environmental health, it also produces sludge as a by-product of the process. Characteristics of the sludge mainly depend on the source of wastewater and the process applied for its treatment. Domestic sludge generally contains a large number of pathogenic bacteria carrying biodegradable compounds. Characteristics of industrial sludge vary greatly. It may contain biodegradable, non-biodegradable, toxic compounds, heavy metals, etc. The sludge may be in the form of liquid or semisolid with 0.25–12% solids. Thus, the handling and disposal/reuse of sludge may become a complex task due to its large volume and infectious and/or toxic nature. This chapter analyses the characterization and quantity estimate of the sludge produced during the application of various municipal and industrial wastewater treatment options. Current practices for the disposal and reuse options such as anaerobic digestion for biogas production, composting to utilize as a fertilizer, brick production, filler material, and bioplastic production will be reviewed and the suitability of each option in terms of benefit and risk will be critically analyzed.

Keywords: sewage sludge, biogas, bioplastic, anaerobic digestion, sludge reuses

1. Introduction

Wastewater contains a lot of solid and liquid waste discarded after use and defined in various combinations based on their sources of generation. The wastewaters from residential apartments, colonies, and institutions are termed domestic, wastes from agricultural fields, mining effluents and industrial are termed as industrial wastewater [1].

With the increase in population coupled with urbanization and industrialization, the quantity of wastewater is also increasing. To cater to the increasing demand for water for urban, industrial, and agricultural purposes, the need for its treatment, call for water reuse, and resource recovery are growing exponentially. Wastewater treatment is one of the important activities in urban and industrial areas in order to protect human and environmental health and to eliminate or neutralize these pollutants. Though wastewater treatment has a positive impact on human and environmental health, it also produces sludge as by-product of the process. The sludge may be in the form of a liquid or a semisolid, with 0.25 to 12% solids that come from households and industries [2]. This sewage sludge may cause various environmental and health problems such as nutrient leaching, loss of soil biodiversity, emission of GHGs, and pathogenic outbreak [2]. Challenges related to managing the sludge thus produced are cost related to handling and transportation, disposal methods, strict regulations, and environmental threats [3]. A clean and sustainable environment can only be built by introducing technologies that are environment friendly and economically feasible. Hence, the treatment of wastewater as per quality standards is no longer the only objective of WWTPs; they also need to focus on its discharge management units. Sewage sludge contains insecticides, cleansing agent, oils, fats, grease, thinners, paints, etc. The high fat content of the sludge may be utilized for biofuel productions such as biodiesel, biohydrogen, bioethanol, and bio-oils [4].

Therefore, the chapter will include the characterization and quantity estimate of the sludge produced during the application of various municipal and industrial wastewater treatment options. Current practices for disposal and reuse options such as anaerobic digestion for biogas production, composting to utilize as fertilizer, brick production, filler materials, or bioplastic production will be reviewed, and the suitability of each option in terms of benefit and risk will be critically analyzed.

2. Wastewater generation, sludge production, and treatment

2.1 Global scenario of wastewater and sludge generation

According to a research carried out by Qadir et al. [5], around 380 billion m^3 of wastewater is generated annually across the world. And if we consider the annual growth of population and development, the daily wastewater generation can hike up by 24% by the end of 2030 and 51% by 2050. Asian countries have contributed largely to wastewater generation with 42% (159 billion m^3) of wastewater globally and hence need attention [5]. Due to population growth and rapid urbanization, immediate attention is required to manage wastewater.

2.2 Indian scenario of wastewater and sludge generation

India is the second most populated (1.38 billion) country in the world with 900 million people living in rural areas and 483 million in urban areas (around 35%). In rural regions, the wastewater produced is approximately 39,604 MLD (72,368 MLD for the years 2020–2021; NITI [6]). Urban areas have higher waste generation due to higher water needs for flushing and sewage drainage as compared to rural areas.

According to a report published by the Central Pollution Control Board (CPCB), sewage generated is about 72,368 MLD with treatment capacity of only about 20,235 (≈28.0%) MLD. 52,133 MLD (≈72.0%) of domestic sewage from cities and towns is being disposed of without treatment, thus making it the biggest source of pollution of water bodies (NITI [6]). Total 1631 STPs are installed/proposed in India with a total

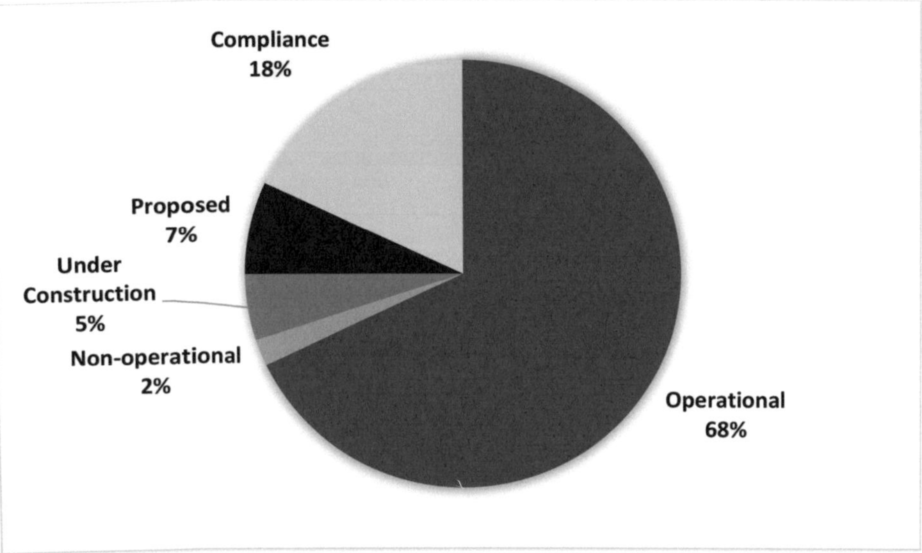

Figure 1.
Current status of STPs in India as per CPCB [7].

treatment capacity of 36,668 MLD. Out of this, 1093 (68%) STPs are operational, 5% are under construction, and 2% are non-operational. **Figure 1** describes the current status of STPs in India as per a report from CPCB [7]. Around 39,55,000 metric tonnes of dry sludge is generated every year after complete treatment of sewage [8].

2.3 Treatment technologies adopted in India

In India, different treatment technologies such as activated sludge process (ASP), up-flow anaerobic sludge digester (UASB), oxidation pond (OP), and advanced technologies like SBR and MBBR are adopted for the treatment of sewage [7]. **Figure 2** presents the capacity distribution of technologies adopted for domestic wastewater treatment in India.

2.4 Sewage classification and characteristics

Sewage can be broadly classified in three ways: domestic sewage, industrial sewage, and storm sewage. Domestic sewage is the water from houses and apartments, containing 99.9% water by weight and < 0.1% of a wide variety of dissolved and suspended impurities [1]. It contains high concentrations of organic matter and nutrients (phosphorus and nitrogen) [9]. Water discharged from industries having chemical compounds during various industrial processes is called "Industrial Sewage" [10].

Components of wastewater based on their source of origin are presented in **Table 1**.

The size and capacity of WWTPs are determined by assessing the total volume of sewage generated from nearby areas connected to sewer systems in terms of inflows and infiltration [5]. The degree of treatment depends upon environmental conditions and effluent discharge standards prescribed by government/local bodies. Stream standards include amount of DO, coliforms, turbidity, acidity, and heavy metal contents, intended to maintain the existing water quality of streams where it has to be disposed of [5].

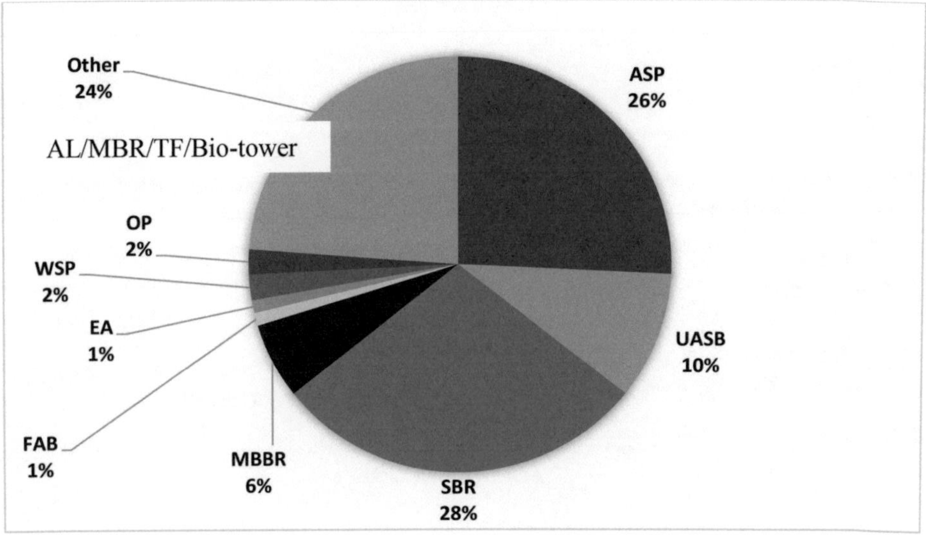

Figure 2.
Wastewater treatment technologies adopted for domestic wastewater treatment in India.

Nature of sludge	Sludge components	Environmental impact
Domestic	Pathogens	Waterborne/communicable disease
Domestic Industrial	Suspended solids	Anaerobic layer formation in aquatic environment
	Biodegradable organics	Oxygen depletion and biological degradation
	Nutrients	Eutrophication
	Dissolved inorganic solids	
Industrial	Refractory organics	Carcinogenic
	Heavy metals	Carcinogenic

Table 1.
Components of wastewater based on their source of origin.

"Sludge is a byproduct of STP having organic compounds, macro and micronutrients, trace elements including toxic metals, microorganisms, and micro pollutants" [11]. Micro/macro-nutrients are the source of plant nutrients, whereas organic constituents serve as soil conditioner [12]. Sewage sludge is neutral to slightly alkaline in nature with high organic matter and high concentrations of N, P, Ca, and Mg [9].

Characteristics of the sludge thus produced mainly depend on the source of wastewater and the process applied for its treatment. Domestic sludge generally contains a large number of pathogenic bacteria along with biodegradable compounds. Characteristics of sludge generated in different STP processes are presented in **Table 2**.

Raw (untreated) sewage has approximately 90 gram per day per capita suspended solids, and about 60% solids are removed in PST, leaving behind 4–5% solid, and the remaining suspended solids are either oxidized in a secondary tank or amalgamated in the biological mass [13]. The amount of solids thus produced depends on the sludge age, and the volume of sludge depends on its water content and the volume of the solids [4].

Sludge-Producing Processes	Physical Characteristics of Sludge	Digestion process involved
Primary Settling tank sludge	• Gray slimy liquid • Settleable solids around 50–60% of the total applied solids • Offensive odor	• Chemical conditioning followed by dewatering • Can be readily digested due to high volatile solid content (60–80%)
Chemical-Precipitation sludge	• Dark-colored slime with objectionable odor, but less from sludge obtained from PST	• Decomposition at a slower rate • Its density is increased by long residence time in storage
Activated sludge process	• Brown or dark flocculants • Brown—under-aeration in good condition without offensive odor • Dark—it may be approaching a septic condition.	• This can be thickened by floatation or centrifugation with or without chemical addition • It will digest readily alone or mixed with fresh sewage solids
Trickling filter	• Brownish, flocculants, and relatively inoffensive when fresh	• Sludge digests readily
Aerobically digested	• Brown to dark brown with a flocculent appearance	• Well-digested sludge
Anaerobically digested sludge	• Dark brown to black in color with lots of gasses	• Primary sludge produces two times CH_4 than activated sludge
Composted sludge	• Dark brown to black varying on the basis of bulking agents used. • Inoffensive odor	• Can be used as a soil conditioner by mixing with bulking agents such as saw dust and wood chips.

Table 2.
Characteristics of sludge generated in different unit operations and processes in an STP [11].

2.5 Sludge generation and treatment

The STPs are designed based on the influent characteristics, primarily total suspended solids (TSS), biochemical oxygen demand (BOD), and fecal coliform (FC). The preliminary treatment process along with the primary sedimentation tank is referred as primary treatment. It is designed to remove 60–70% of the suspended solids (organic and inorganic) and 30–40% of BOD (organic) associated with it. The sludge from PST is gray and slimy with offensive odor and can be digested by employing simpler operations [11].

In secondary treatment, microorganisms decompose the organic matter and more than 85% of both suspended solids and BOD is removed. The treated effluent from WWTPs usually contains BOD suitable for disposal. Aerobic systems such as stabilization ponds, ASP, SBR, and MBBR and anaerobic systems such as anaerobic ponds and UASB are secondary treatment processes. Sludge from the activated sludge process approaches the septic conditions rapidly and can be digested alone and/or along with primary sludge [11].

Tertiary treatment includes chemical precipitation and membrane technologies. Tertiary treatment methods can remove >99% BOD but are used in special cases due to their cost of operation. Chemical precipitation with metal salts ($FeCl_3$) and nitrification-denitrification is commonly used to remove phosphorus and nitrogen, respectively, from sewage in tertiary treatment methods [11].

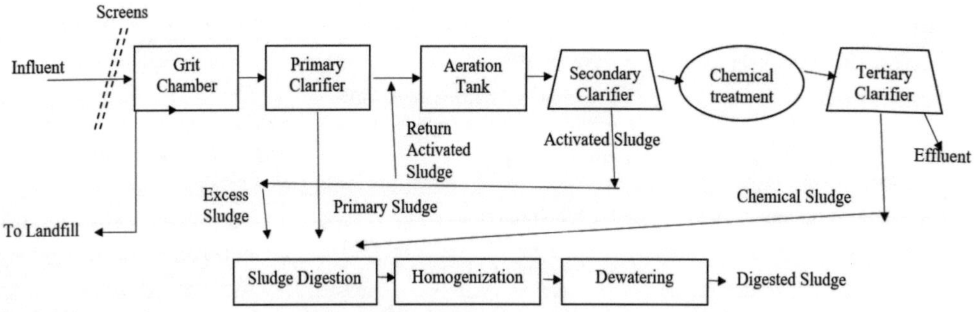

Figure 3.
Types of sewage sludge generated through various unit processes of sludge treatment [11].

"Primary sludge" is the TSS settled in PST, secondary sludge is the mixed liquor settled in SST by gravity, "return sludge" is a part of secondary sludge going into aeration tank, and "excess sludge" is wasted sludge from SST [4]. Sewage after chemical precipitation produces chemical sludge [14]. In biological treatment, 1–2% of BOD/COD is converted into solids, making sewage sludge [15].

The sludge from PST and SST is passed through a sludge-thickening process in sludge thickeners. If the thickened sludge is put through the digestion process anaerobically to produce CH_4, it may be termed anaerobic digester, and if digested aerobically, it is called aerobic digestion [2]. The digested sludge will have to be dewatered using sludge drying beds (centrifuge/filter press/natural solar drying beds) [11]. **Figure 3** explains the various types of sewage sludge generated through various unit processes of sludge treatment taken from Metcalf and Eddy [11].

Anaerobic digestion has various advantages over other methods. It leads to recovery of methane, nutrients, and dying-off pathogens due to relatively long detention periods. Use of larger closed tanks may lead to increase in capital cost. Further sensitivity of micro-organisms involved in anaerobic digestion toward small environmental changes is the major con of using this facility. The residue liquid from the system has very high oxygen demand, suspended solids, and high concentration of nitrogen. Anaerobically digested sludge produces about twice as much methane gas as does waste-activated sludge [8].

In aerobic digestion, biological degradation of organic matter takes place in the presence of oxygen. In this process, microorganisms (sludge) are oxidized to CO_2, H_2O, and ammonia. Aerobically digested sewage can be dewatered easily on drying beds. The pH of the system is required to be maintained as pH drop may occur when ammonia is oxidized to nitrate and the alkalinity of the sewage is insufficient [3]. Long-term aeration of the waste-activated sludge creates a bulking material difficult to thicken [9].

Treated wastewater can be viewed as a resource for energy, nutrients, and water, which is a much undervalued resource in India [16]. However, the main challenge toward generating a common statement about waste management is quantifying it in terms of volumes of wastewater generated, collected, treated, and reused at different scales.

3. Sludge management

The methods used to treat and dispose of sludge are sludge thickening, sludge digestion or stabilization, conditioning, dewatering, drying, incineration, and ultimate disposal, as described in **Figure 4**.

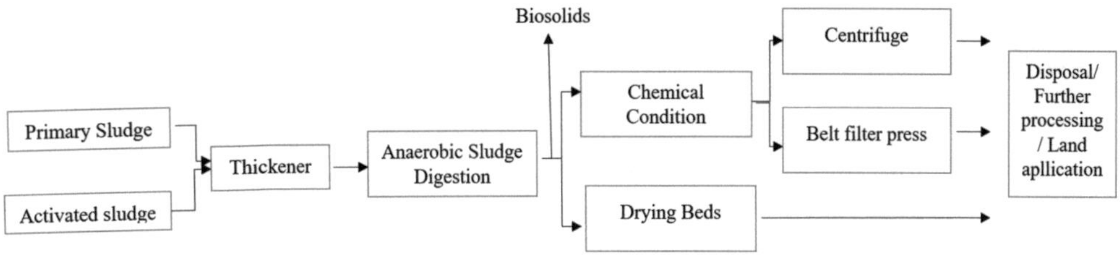

Figure 4.
Process of treatment and disposal of sewage sludge.

Anaerobic digestion is regarded as a major and essential part of a modern WWTP [4]. Three types of anaerobic digesters are being used: single-stage process (single-chamber bioconversion of sludge), double-stage process (separate acidogenic and methanogenic chambers), and temperature-phased anaerobic digestion (combination of a thermophilic unit prior to a mesophilic unit) [4]. **Table 3** provides details about sludge treatment methods along with processes involved and their impact on sludge mass and volume.

Thermal drying and incineration can reduce the volume of sludge by carbonizing organic constituents in sludge. But thermal drying is very costly due to its high energy requirement. In incineration, organic matter gets destroyed, the heavy metals get mixed with ash, and its efficiency is based on the degree of dewatering to reduce

Unit processes	Unit operations	Impact on sludge
Sludge Thickening	Gravity	Reduction in sludge volume
	Floatation	
	Centrifugation	
Stabilization of sludge	Alkaline	Stabilized sludge
	Anaerobic digestion	Stabilized sludge with reduced mass
	Aerobic digestion	
	Thermal aerobic digestion	
	Composting	Stabilization of sludge followed by recovery of useful products
Conditioning of sludge	Chemical	Improvement in dewatering conditions
Dewatering of sewage sludge	Centrifuge	Reduction in sludge volume
	Sludge drying beds	
	Lagoons	
Incineration	Multiple hearth incineration	Resource recovery and reduced volume
Lands Application of bio solids	Land application	Beneficial use + disposal
	Dedicated land application	Disposal + land reclamation
	Landfilling	Disposal

Table 3.
Sludge treatment methods along with processes involved and their impact on sludge [11].

moisture content that is applied on the sludge prior to incineration. Sewage sludge ash has very high P_2O_5 content compared to commercial superphosphate [12].

The ash can be used as a raw material for the manufacture of construction materials, namely, bricks, tiles, pavers, and cement. The energy required to heat the sludge for moisture removal can be achieved using oil, natural gas, coal, and even electricity [3]. To use the ash as a phosphate fertilizer, it is needed to be extracted and also checked for heavy metal content to be within safe limits as per standards.

4. Reuse practices commonly adopted

The sludge generated from the different WWTP units, namely, PST, SST, and others, has become a problem for mankind due to the unavailability of land for disposal, high population growth, and very fast urbanization/industrialization. Therefore, it is indeed a requirement to develop long-term solutions by recycling and reusing the sludge thus produced to achieve a zero-waste strategy.

4.1 Biogas

Anaerobic digestion of sludge produces fuel-rich biogas as a by-product, which can be utilized to meet the energy and fuel demand, making WWTPs self-efficient. Biogas is a combination of methane (50–75%), carbon dioxide (25–50%), and other gases [17]. It can be collected from an anaerobic digester tank and converted to electrical energy and heat energy [12].

Biogas can also be upgraded to a relatively larger fraction of methane. It can be used for domestic purpose, electricity generation, and transportation in place of compressed natural gas (CNG). CO_2 and H_2S produced in the process can be recovered by using activated bio-char as adsorbents or biologically using CO_2-fixing microalgae or sulfur-reducing bacteria (SRB) [9]. Bio-electro-chemical systems may also be employed for upgradation of biogas through the electro-methanogenesis process, in which CO_2 gets converted to CH_4 [18].

In a global context, around 10,100–14,000 TWh (1 TWh = 10^8 KWh) biogas can be produced using currently available substrate ranges, and the energy thus produced by consuming almost all resources gives 6–9% of the total energy consumption globally and reduces 23–32% of the world's coal consumption [9].

4.2 Bio-hydrogen

Bio-hydrogen is an intermediate product of anaerobic digestion (AD), having a higher calorific value than biogas. It is considered green energy as its combustion only generates water without GHG emissions [19]. This process is known as dark fermentation by avoiding methanogenic activity and controlling the operating parameters of the AD system, namely, pre-treatment of inoculum, short HRT, and acidic pH [20]. Until now, the achieved yield of hydrogen production is very less (<6%). Very few studies have been conducted to study yield improvement by pre-treating substrates using calcium peroxide and nitrous acid. It can also be improved by the co-fermentation of sludge with other materials that can reduce the C/N ratio of the sludge [21].

4.3 Bioplastic

Bioplastic is made from the union of microorganisms, with macromolecules, namely, starch, cellulose, and protein [22]. It is biodegradable in nature and thus not a threat to humans. It is even used in producing postoperative sutures (medical surgical equipment). The high investment cost of these products has made them uncommon for regular use and manufacturing. The bio-synthesis of micro-organisms as an energy storage component produces compounds such as polyhydroxyalkanoates (PHA) [22].

The bioplastic derived from only biomass materials is often called all-bioplastic and others are part-bioplastics. "All-bioplastics" are "protein plastics" (soybean fiber, cellulosic, and algae-based resin), and "part-bioplastics" mainly contain "starch bioplastic," modified with starch and cellulose [23]. All-bioplastics are also called biodegradable bioplastic, and they have poor water and moisture resistance [18]. PHA is a biodegradable plastic, with all good features except high cost of the raw material [24]. Bioplastics are broadly used in making packaging products such as shopping and trash bags, bottles, labels, packaging films, cushioning, fibers of synthetic clothes, children's toys, and home interior furnishings and décor items [22].

Conventional plastic has petroleum residues as the raw material, which is going to be exhausted someday, but the source of bioplastic is organic biomass, which is inexhaustible. Bioplastic is biodegradable in nature and can be broken down as water and carbon dioxide. It has very low CO_2 emission, thereby reducing the temperature of Earth [22]. As it is completely biomass-based plastic derived from starch, cellulose, and protein, it does not contain any organic toxic substances. High cost, lack of technology and market, and lesser customer awareness have undervalued its use on a larger platform.

The global bioplastic production capacity was 2.11 million tonnes in 2018 and is expected to exceed 2.6 million tonnes in 2023 [22]. The market price of per kilogram of bioplastic PHA is approximately six times higher compared to petroleum-based plastic [5]. However, replacing petroleum-derived plastics with bioplastics does not necessarily solve the plastic waste issue. To make the bioplastic use an effective solution, it is need to study its recycling, reuse, and the carbon footprint gathered in throughout life cycle.

4.4 Bio-fertilizers

The sludge from anaerobic digestion process is used to obtain biogas and waste is left in the form of slurry, termed "digestate" [25]. This digestate may be used as a fertilizer for plants as a source of macro/micro nutrients. Bio-fertilizer produced from the digested sludge may become a substitute of chemical fertilizers. Bio-fertilizer improves the fertility of soil and provides the option for waste disposal and resource recovery, thus solving environmental issues associated with waste [25].

The fertilizer of sewage sludge gives rise to the problem of bio-accumulation of heavy metals in agricultural soil in the topsoil and can be transferred to the food chain in a magnified way [18]. Because the higher doses of sludge application on ground have higher heavy metal concentration instead of comparatively lower doses, its intermittent uses with additional analysis of its exposures will be a great way to deal with its negative impacts.

4.5 Syngas

Syngas is different from biogas as biogas is formed during the biological degradation of organic mass in anaerobic conditions (CO_2 + CH_4), whereas syngas is composed of carbon CO, CO_2, and H_2 when coal or biomass is gasified [26]. In thermochemical treatment, sludge is fed into a reactor, where it is partially oxidized at 300–900°C (pyrolysis), and syngas is produced along with tar and other products. Various useful products can be derived from this syngas, namely, fertilizers, synthetics, and fuels [17]. Despite the high cost of production and complexity in operating procedure, this technology has ranked among the most advanced technologies to convert biomass to energy due to a large yield. Gasifying agents such as air, steam, and oxygen were used to produce different types of syngases. Air is the most commonly used gasifying agent [19].

4.6 Compost

Composting is an efficient and cost-effective method for treating and reusing sewage sludge post-digestion and used as soil amendments. The compost properties are controlled and modified by using bulking agents such as high moisture content, lesser porosity, and low C/N ratio [25]. Composting could reduce polycyclic aromatic hydrocarbons (PAHs), but biodegradation processes of sludge can form toxic intermediary products, causing soil toxicity, leading to environmental stress and reduction in soil microbial activity [18]. In this process, the organic matter is turned into a stabilized product, which can be applied as a form of returning organic matter to soils, which acts as a carbon sink. The safety and efficacy of sewage sludge composting should be monitored carefully in terms of microbial indicators such as community structure, diversity, and composition.

4.7 Bio-oil

Sewage sludge can be recycled as a jet fuel (hydrocarbons C_8-C_{16}) by pre-conditioning and processing through pyrolysis at temperatures 450–700°C to produce a bio-oil [27]. This is a two-stage process of hydrodeoxygenation and hydrocracking in a batch reactor under high pressure (autoclave). This fuel may meet the jet fuel specifications in terms of calorific value, viscosity, density, and freeze point; however, it fails in terms of smoke release, flash point, and total acid number [27]. The conversion of sewage sludge into jet fuel can be a sustainable pathway for energy production and a promising route for sewage sludge management [28].

4.8 Construction materials

Use of dried sludge as a clay substitute to produce an engineering quality brick can be a suitable option of sludge reuse. The proportion of sludge in the mixture and the firing temperature are the two key factors affecting the quality of bricks [10].

Low organic matter sewage sludge is also used in manufacturing concrete mix alterations [29]. According to various researchers, strength is inversely proportional to sludge content when greater than 10% mixing is done; higher the sludge content, greater the strength loss. Though its use in the manufacture of construction materials solves a very small portion of the problem, but this method is assumed to be safe for human health and environment [1]. The by-products obtained from sludge recycling and processing are summarized in **Table 4**.

By-product of the sludge treatment process		Remarks/significance/process	References
Energy source	**Biogas (60% CH_4)**	**1 m^3 biogas = 23.3 M/m^3 energy value**	**[9]**
Cooking fuel	Biogas	Source of cooking fuel.	[9]
Transport fuel	CBM and LBM	Upgraded from Biogas	[12]
Liquid anaerobic digestate		Struvite precipitation (with phosphate and ammonium ion recovery)	[9]
Dried digested sludge	Bio-fertilizer	Land spreading within the safe limits.	[25]
Electricity and heat		Energy is recovered from incineration of dried raw or digested sludge	[17]
Production *via* syngas	Bio-methane	Steam catalytic/gasification and pyrolysis of dried sludge to produces syngas feedstock for producing bio-methanol	[12]
Phosphorus recovery	Biological Phosphorus	Obtained thermo-chemically from ash or tar obtained while gasification	[1]
Biofuels	Bio Oil Biodiesel Bioethanol	Fast pyrolysis Transesterification of dried sludge Fermentation	[2, 12, 18]
Carbonaceous materials	Bio-char Hydro-char	Pyrolysis hydrothermal carbonization	[12]
Green hydrogen	Biosyngas	Gasification/pyrolysis + bio-methane	[26]

Table 4.
By-products obtained from sludge recycling and processing.

5. Limitations and risks associated

Limitations and challenges while dealing with reuse practices are maintaining the quality standards with precise monitoring in order to reduce the pollution risk. In this regard, the source and impact of contamination needs to be checked regularly by employing risk assessment studies. In this study, environmental systems, exposure pathways, and the recipients of the pollution loads including human populations should be considered and analyzed for exposure. When the bio-solids are released to the soil, they do not need to meet the water quality standards. **Figure 5** explains the potential risks imposed on humans and the ecological environment on utilizing sewage sludge as a resource.

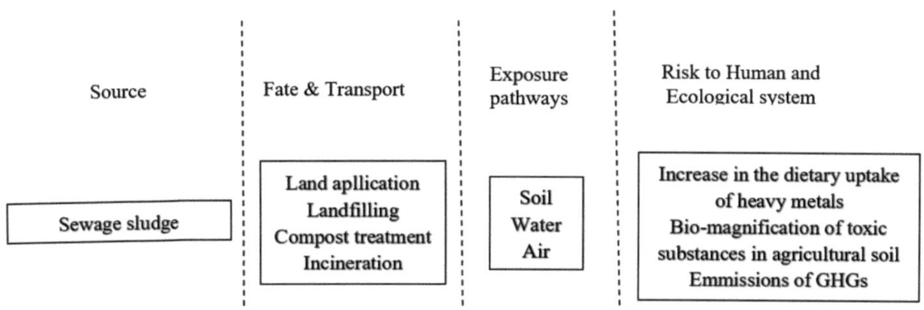

Figure 5.
Potential risk imposed on humans and the ecological environment on utilizing sewage sludge as a resource.

6. Conclusions

Wastewater sludge reuse and recycling prior to disposal serves as the basis of sludge management. The energy and resources that can be obtained using sewage sludge could be a long-term solution for disposal, resource recovery, and future energy needs. The application of these steps on a practical ground will be a fascinating approach for ensuring the long-term sustainability and lowering the overall carbon footprint. This chapter provided a brief discussion of the possible options for the treatment of sewage sludge and its reuse options after smaller amendments on soil as fertilizer, compost, concrete production, biogas, and bioplastics. Wastewater treatment plants generate a certain level of carbon footprint and wastes; therefore, it is necessary to reduce the carbon footprint and manage the waste in order to keep the world safe.

Author details

Astha Kumari[1], Nityanand Singh Maurya[2*], Abhishek Kumar[3], Rajanee Kant Yadav[2] and Amit Kumar[2]

1 Department of Civil Engineering, BCE Bakhtiyarpur, Patna, India

2 Department of Civil Engineering, National Institute of Technology, Patna, India

3 Department of Civil Engineering, Government Engineering College, Bihar, India

*Address all correspondence to: nsmaurya@nitp.ac.in

References

[1] Vilakazi S, Onyari E, Nkwonta O, Bwapwa JK. Reuse of Domestic Sewage Sludge to Achieve a Zero Waste Strategy & Improve Concrete Strength & Durability - a Review. Vol. 43. Elsevier B.V; 2023

[2] Laura F, Tamara A, Müller A, Hiroshan H, Christina D, Serena C. Selecting sustainable sewage sludge reuse options through a systematic assessment framework: Methodology and case study in Latin America. Journal of Cleaner Production. 2020;**242**:118389. DOI: 10.1016/j.jclepro.2019.118389

[3] Liu G, Yang Z, Chen B, Zhang J, Liu X, Zhang Y, et al. Scenarios for sewage sludge reduction and reuse in clinker production towards regional eco-industrial development: A comparative Emergy-based assessment. Journal of Cleaner Production. 2015;**103**:371-383. DOI: 10.1016/j.jclepro.2014.09.003

[4] Guangyin Z, Youcai Z. Sewage sludge generation and characteristics. In: Pollution Control and Resource Recovery. Elsevier; 2017. pp. 1-11

[5] Qadir M, Drechsel P, Cisneros BJ, Kim Y, Pramanik A, Mehta P, et al. Global and regional potential of wastewater as a water, nutrient and energy source. Natural Resources Forum. 2020;**44**(1):40-51. DOI: 10.1111/1477-8947.12187

[6] NITI Aayog. Urban Wastewater Scenario in India. NITI Aayog; 2022

[7] CPCB. "National Inventory of Sewage Treatment Plants March 2021." (March):183, CPCB, India. 2021

[8] Owusu-Twum MY, Sharara MA. Sludge Management in Anaerobic Swine Lagoons: A review. Journal of Environmental Management. 2020;**271**

[9] Werkneh AA, Gebru SB. Development of ecological sanitation approaches for integrated recovery of biogas, nutrients and clean water from domestic wastewater. Resources, Environment and Sustainability. 2022;**100095**. DOI: 10.1016/j.resenv.2022.100095

[10] Weng C-H, Lin D-F, Chiang P-C. Utilization of sludge as brick materials. Advances in Environmental Research. 2003;**7**:679-685

[11] Metcalf, and Eddy. Wastewater Engineering: Treatment and Reuse. Vol. 179. Metcalf, and Eddy; 2004

[12] Shanmugam K, Gadhamshetty V, Tysklind M, Bhattacharyya D, Upadhyayula VKK. A sustainable performance assessment framework for circular management of municipal wastewater treatment plants. Journal of Cleaner Production. 2022;**339**. DOI: 10.1016/j.jclepro.2022.130657

[13] Yan S, Bala Subramanian S, Tyagi RD, Surampalli RY. Wastewater sludge Charecteristics. In: Sustainable Sludge Management: Production of Value Added Products. Reston, Virginia, USA: American Society of Civil Engineers; 2009. pp. 6-36

[14] Monzó J, Payá J, Borrachero MV, Girbés I. Reuse of sewage sludge ashes (SSA) in cement mixtures: The effect of SSA on the workability of cement mortars. Waste Management. 2003;**23**(4):373-381. DOI: 10.1016/S0956-053X(03)00034-5

[15] Singh RP, Agrawal M. Potential benefits and risks of land application

of sewage sludge. Waste Management. 2008;**28**(2):347-358. DOI: 10.1016/j.wasman.2006.12.010

[16] Minhas PS, Saha JK, Dotaniya ML, Sarkar A, Saha M. Wastewater irrigation in India: Current status, impacts and response options. Science of the Total Environment. 2022;**808**:152001. DOI: 10.1016/J.SCITOTENV.2021.152001

[17] Nkuna SG, Olwal TO, Daniel SP, Chowdhury. Assessment of thermochemical Technologies for Wastewater Sludge-to-Energy: An advance MCDM model. Cleaner Engineering and Technology. 2022;**9**. DOI: 10.1016/j.clet.2022.100519

[18] Raheem A, Sikarwar VS, He J, Dastyar W, Dionysiou DD, Wang W, et al. Opportunities and challenges in sustainable treatment and resource reuse of sewage sludge: A review. Chemical Engineering Journal. 2018;**337**:616-641

[19] Mun TY, Kim JW, Kim JS. Air gasification of dried sewage sludge in a two-stage gasifier: Part 1. The effects and reusability of additives on the removal of tar and hydrogen production. International Journal of Hydrogen Energy. 2013;**38**:5226-5234

[20] Pandey AK, Pilli S, Bhunia P, Tyagi RD, Surampalli RY, Zhang TC, et al. Dark fermentation: Production and utilization of volatile fatty acid from different wastes- a review. Chemosphere. 2022;**288**:132444. DOI: 10.1016/J.CHEMOSPHERE.2021.132444

[21] Kumar LR, Yellapu SK, Tyagi RD, Drogui P. Cost, energy and GHG emission assessment for microbial biodiesel production through valorization of municipal sludge and crude glycerol. Bioresource Technology. 2020;**297**:122404. DOI: 10.1016/J.BIORTECH.2019.122404

[22] Liu F, Li J, Zhang XL. Bioplastic production from wastewater sludge and application. In: IOP Conference Series: Earth and Environmental Science. Vol. 344. Institute of Physics Publishing; 2019

[23] Mohapatra DP, Brar SK, Tyagi RD, Surampalli RY. Physico-chemical pre-treatment and biotransformation of wastewater and wastewater sludge - fate of Bisphenol a. Chemosphere. 2010;**78**(8):923-941. DOI: 10.1016/j.chemosphere.2009.12.053

[24] Yadav B, Anita Talan RD, Tyagi, and Patrick Drogui. Concomitant production of value-added products with Polyhydroxyalkanoate (PHA) synthesis: A review. Bioresource Technology. 2021;**337**(June):125419. DOI: 10.1016/j.biortech.2021.125419

[25] Ezemagu IG, Ejimofor MI, Menkiti MC, Diyoke C. Biofertilizer production via composting of Digestate obtained from anaerobic digestion of post biocoagulation sludge blended with saw dust: Physiochemical characterization and kinetic study. Environmental Challenges. 2021;**5**:100288. DOI: 10.1016/J.ENVC.2021.100288

[26] Liu H, Hongyun H, Luo G, Li A, Minghou X, Yao H. Enhancement of hydrogen production in steam gasification of sewage sludge by reusing the calcium in lime-conditioned sludge. International Journal of Hydrogen Energy. 2013;**38**(3):1332-1341. DOI: 10.1016/j.ijhydene.2012.11.072

[27] Bashir MA, Lima S, Jahangiri H, Majewski AJ, Hofmann M, Hornung A, et al. A step change towards sustainable aviation fuel from sewage sludge. Journal of Analytical and Applied Pyrolysis. 2022;**163**:105498. DOI: 10.1016/j.jaap.2022.105498

[28] Kumar LR, Yellapu SK, Tyagi RD, Drogui P. Biodiesel production from microbial lipid obtained by intermittent feeding of municipal sludge and treated crude glycerol. Systems Microbiology and Biomanufacturing. 2021;**1**(3):344-355. DOI: 10.1007/s43393-021-00030-2

[29] Lu JX, Zhou Y, He P, Wang S, Shen P, Poon CS. Sustainable reuse of waste glass and incinerated sewage sludge ash in insulating building products: Functional and durability assessment. Journal of Cleaner Production. 2019;**236**:117635. DOI: 10.1016/j.jclepro.2019.117635

Chapter 14

Development of the Phosphorus Recovery System (PRS) Utilizing Ultrasonic Wave in Incinerated Sewage Sludge Ash (ISSA)

Ye Duk Choi, Jun Yeon Lee, Zoo Ho Jang, Jung Gone Joung, Kyu Mun Han, Mok Young Lee and Kweon Jung

Abstract

This study was performed to develop a Phosphorus Recovery System(PRS) for the recovery of phosphorus from incinerated sewage sludge ash using struvite precipitation. Fly ash generated at the Seonam Sewage Treatment Plant(SSTP) has a high P_2O_5 content (13.9%). We developed a PRS consisting of an ultrasonic extractor, solid-liquid separator, mixing tank, and phosphorus recovery tank. The ultrasonic extractor had a 28 kHz vibrator for high speed and efficiency, which could perform the extraction in one-quarter of the time required in the conventional stirring method. Results of tests on the ultrasonic extractor showed that up to 0.044 g of P per gram of ash could be extracted with 1 N NaOH at an L/S ratio of 10 mL/g and an ultrasonic output of 500 Wh for 0.5 hr. The PRS is needed to improve the operation method and economic analysis to commercialize the technology and its application through further studies.

Keywords: phosphorus, recovery, ash, sludge, ultrasonic

1. Introduction

Incineration ash generated from sewage treatment facilities is known to contain a large amount of phosphorus, about 10% [1, 2]. Republic of Korea relies entirely on imports of phosphorus, which is referred to as the 3rd element of fertilizer, in the form of phosphate rock. Phosphate rock is expected to run out of resources within the next 50 to 100 years because of the increase in world population and demand for fertilizers [1, 3, 4].

Recently, due to the prohibition of dumping of sewage sludge at sea, the rate of onshore disposal is greatly increasing. Phosphorus recovery from incineration ash has many advantages in terms of improving economic efficiency and minimizing volume of t reaction tank through high concentration of phosphorus concentrated in incineration ash. Most of the conventional phosphorus recovery technologies eluted phosphorus from sewage sludge under anaerobic and high-temperature conditions, but commercialization of the technology was not easy due to high water content and low phosphorus content. Water content of sludge in incineration facility is more than 80%, and percentage of organic matter is higher than 60%.

Looking at the previous research results, phosphorus recovery is mostly in the form of MAP (Struvite, $MgNH_4\,PO_4\,6H_2O$ or $MgNH_4\,PO_4\,H_2O$) and HAP (Hydroxyapaptite, $Ca_5(PO_4)_3OH$) [5]. The principle of precipitation or crystallization is applied. Struvite can be expected to have a fertilizer effect because it contains essential elements necessary for plant growth, and excessive input is possible because it does not dissolve instantly due to the slow-release rate of nitrogen and phosphorus [6]. On the other hand, Struvite decomposes when heated and has high solubility in acidic conditions.

2. Materials and methods

Currently, the S sewage treatment facility incinerates about 150 ton/day of sewage waste by putting it into a fluidized bed incinerator at 800°C or higher. As an incineration by-product, incineration ash is generated at approximately 10 ton/day, and incineration ash is largely classified into fly ash and bottom ash. Particle size analysis was performed using a particle size analyzer (Sympatec GmbH, HELOS/HAER), and the chemical composition was analyzed with an X- ray fluorescence spectrometer (Rigaku, NEX-CG) and a scanning electron microscope (Shimadzu, SS-550). First, the L/S ratio (Leaching solution/Solid ratio, mL/g) was fixed at 10, which is the same as the experimental conditions suggested by other researchers [7]. fly ash 200 g and eluate 2 L was put into a standard Jar(2 L), and after stirring, solid-liquid separation was performed at 2000 rpm for 20 min using a centrifuge. The supernatant was analyzed. Heavy metals were prepared by pretreating the sample. Analyzed by ICP-OES (Varian, VISTA PRO) and $PO_4{}^{3-}$ and $SO_4{}^{2-}$ Anions such as were analyzed using IC(Metrohm, CH/Advanced IC) .

The phosphorus recovery process largely consists of two steps: phosphorus extraction and crystallization. First, the phosphorus extraction method is classified into an acid extraction method and an alkali extraction method, and the phosphorus crystallization step is MAP and HA depending on the input cation.

Phosphorus Recovery System(PRS), which was developed for the first time in Korea by applying the above phosphorus recovery principle, is made of acrylic

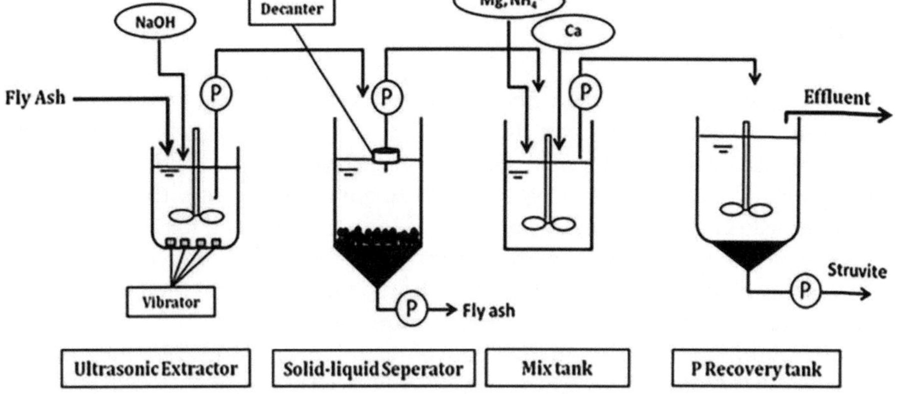

Figure 1.
Configuration of the PRS.

material except for the ultrasonic extraction tank, and the ultrasonic extractor, Solid-Liquid Seperator, Mixing Tank, and P-Recovery Tank (**Figure 1**) .

Ultrasonic extraction tank that elutes phosphorus contained in the incineration ash is made of stainless material to prevent corrosion, and the fly ash and An agitator was installed to mix the eluate (**Figure 2**). In addition, by installing a cover on the top, evaporation of moisture due to temperature change during operation is minimized. The internal volume of the reactor is 48 L (0.4 m × 0.4 m × 0.3 m), and a total of 20 vibrators (4 × 5) with a frequency of 28 kHz are installed on the floor (**Figure 3**). The maximum ultrasonic power is 700 W, and the fly ash and of the eluate When mixing The L/S ratio (leaching solution/solid ratio, mL/g) was set to 10.

The solid-liquid separation tank has an inner diameter at the rear end of ultrasonic extraction tank. It is a cylinder with a height of 190 mm and a height of 500 mm and has an internal volume of 14 L. It separates the leached mixture of incineration ash. The remaining incineration ash is transferred to the hopper at the bottom by a pump, and the supernatant is transferred to the mixing tank through the Decanter located on the surface. It is a cylinder with a height of 150 mm and a height of 205 mm, and the

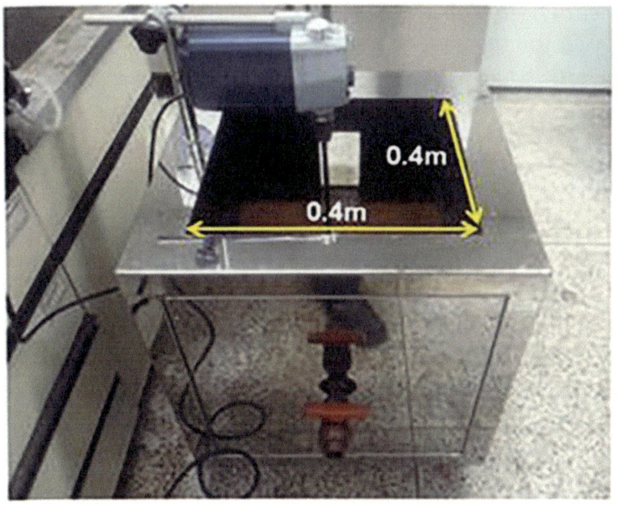

Figure 2.
Photo of ultrasonic extractor.

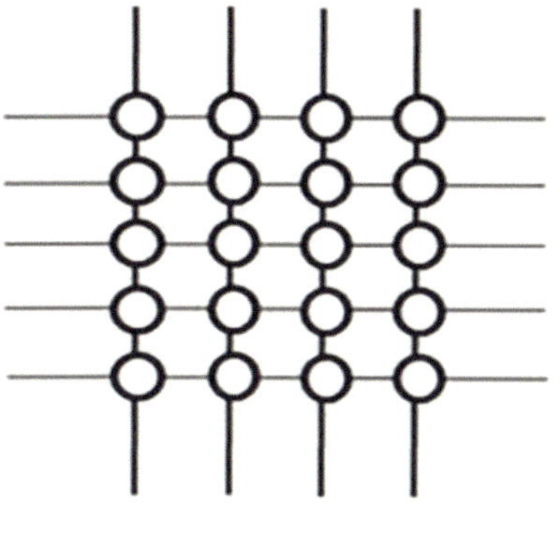

Figure 3.
Arrangement of vibrators.

Component	Volume(L)	Stirring velocity(rpm)	Remark
Ultrasonic Extractor	48.0	200	Vibrator(28 kHz)
Solid-Liquid Seperator	14.0	—	Decanter
Mixing Tank	3.6	150	—
P-Recovery tank	6.4	—	—

Table 1.
Specification of the PRS.

internal volume is 3.6 L, and an agitator is installed in the center for complete mixing. During MAP crystallization, 2 M $MgCl^{2+}$ $_{and}$ 2 M NH_4Cl were used as Mg^{2+} and NH_4^+ sources, respectively and $CaCl_2$ $_{was}$ used as Ca^{2+} source during HAP crystallization. Phosphorus (P) recovery tank inner diameter It is a cylinder with a height of 190 mm and a height of 250 mm, and the internal volume is 6.4 L. In addition, it was designed so that phosphorus crystal (Struvite) precipitated at the bottom was transferred to the storage tank through a pump. The specifications of the P(phosphorus) recovery system (PRS) are shown in **Table 1**.

The recovered precipitate was analyzed using an X- ray diffractometer(XRD, Shimadzu, JP/XRD-7000) and an X- ray fluorescence analyzer (XRF, Rigaku, NEX-CG).

3. Result

3.1 Fly ash and bottom ash

3.1.1 Particle size

Ash generated from sewage treatment facilities is largely divided into fly ash and bottom ash, and in particular, fly ash accounts for more than 97%. Fly ash and bottom ash generated from the incinerator were collected and the distribution by particle size was analyzed with a particle size analyzer.

Figures 4 and 5 shows the particle size distribution of fly ash and bottom ash, respectively. X_{50} is Fly ash and bottom ash were 31.02 μm and 42.61 μm, respectively,

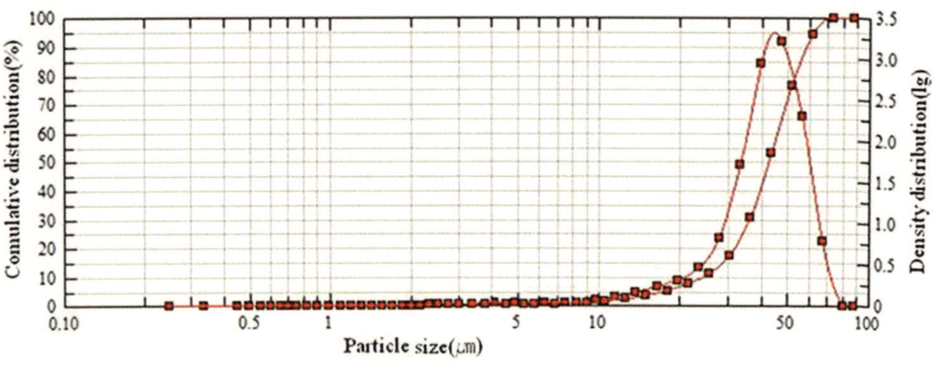

Figure 4.
Grain size distribution of fly ash.

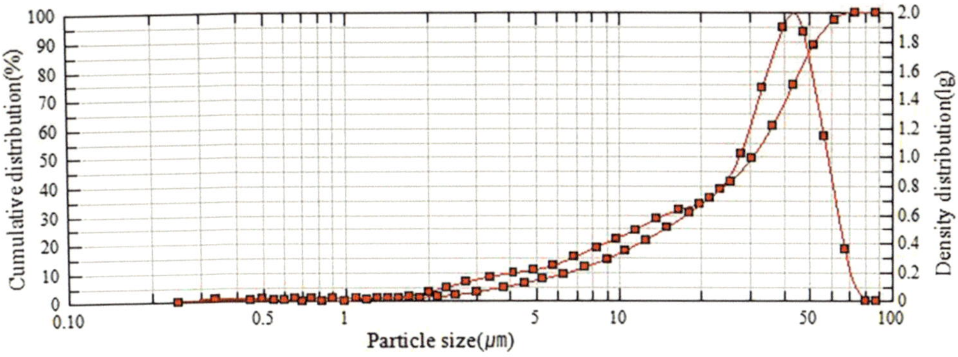

Figure 5.
Grain size distribution of bottom ash.

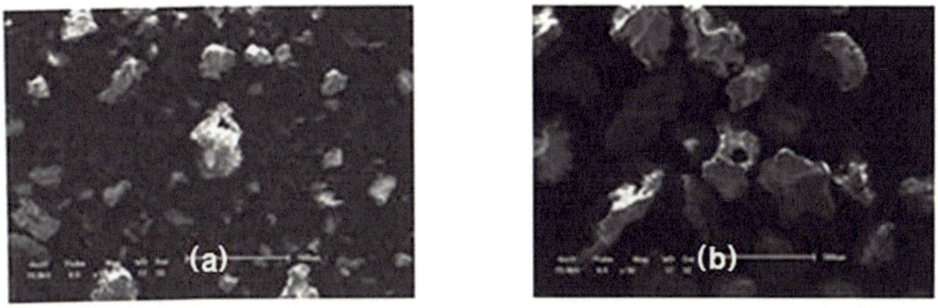

Figure 6.
SEM photographs(×50) of ash; a) Fly ash, b)bottom ash.

and a large number of relatively small particles were distributed in the fly ash. In addition, as a result of specific surface area(Sv) analysis, fly ash and bottom ash were 0.46 m^2/cm^3 respectively and 0.17 m^2/cm^3, the specific surface area of the fly ash was more than 2.5 times larger. Considering the experimental results of Roy [8] and the amount of incinerated ash, fly ash is judged to be more suitable for P elution. Phosphorus (P) contained in incineration ash is mostly in the form of Whitlockite and is known to be thermally stable and not volatilized even at high temperatures of 800 to 900°C [9].

Figure 6 is a photograph (magnification ratio: 50 times) of the surface of fly ash and incineration ash using a Scanning Electron Microscope (SEM), and it was confirmed that a large number of relatively large particles were distributed in the bottom ash.

3.1.2 Chemical composition

Incineration ash, a by-product of incineration, is mostly emitted in the form of metal oxides along with combustion gases. **Table 2** shows the test results of fly ash and bottom ash analyzed using X- ray fluorescence spectrometry (XRF). First of all, the P_2O_5 of the fly ash was 13.65 ± 2.35%, indicating that it contained more P than the bottom ash (10.23 ± 2.38%). Merino et al. [10] reported that P_2O_5 of sewage sludge incineration ash in Spain was 14.2%, and Donatello et al. [11] reported P_2O_5 of incineration ash as 14.8%. Fly ash is also believed to contain similar levels of

Element	Fly ash	Bottom ash
P^2O^5	13.65 ± 2.35	10.23 ± 2.38
Al_2O_3	16.90 ± 1.19	11.50 ± 1.90
SiO_2	24.15 ± 2.95	13.75 ± 2.45
Fe_2O_3	19.40 ± 2.60	42.45 ± 9.45
CaO	16.45 ± 1.55	12.73 ± 5.38
MgO	3.40 ± 0.27	3.83 ± 0.61
K_2O	3.31 ± 0.35	2.49 ± 1.34
ZnO	0.64 ± 0.31	1.04 ± 0.71
MnO	0.24 ± 0.04	0.29 ± 0.23
TiO_2	1.32 ± 0.20	0.95 ± 0.53
Na_2O	1.12 ± 0.19	0.58 ± 0.58
CuO	0.60 ± 0.33	0.72 ± 0.72

Table 2.
Chemical composition of ash by XRF. (unit: %).

phosphorus. In addition, Al_2O_3 of fly ash was 16.90 ± 1.19%, and Al_2O_3 of bottom ash was 11.50 ± 1.90%, indicating that fly ash contained more Al. According to literature, Donatello et al. [11] reported 13.1% Al_2O_3 in incineration ash, and Adam et al. [12] reported 10.8% Al_2O_3 in incineration ash. P contained in incineration ash is mostly a Whitlockite type($Ca_3(PO_4)_2$) orthophosphate.

Incineration ash As a result of analysis by X- ray fluorescence spectrometry (XRF), the $P_2 O_5$ of the fly ash was 13.65 ± 2.35% and contained more phosphorus (P) than the bottom ash (10.23 ± 2.38%). In addition, MgO, a component of MAP precipitation, was 3.40 ± 0.27% and 3.83 ± 0.61% for fly ash and bottom ash, respectively, and there was no significant difference.

SiO_2(24.15 ± 2.95%) was the oxide that occupied the largest proportion in fly ash, which was attributed to the large amount of soil contained in sewage sludge. SiO_2 in incineration ash reported by other researchers Looking at the content, Huacheng Xu et al. [13] and Donatello et al. [11] reported that SiO_2 was the most present in incineration ash, with 43.1 ~ 49.1% and 31.6%, respectively. On the other hand, $Fe_2 O_3$ in the bottom ash accounted for the largest proportion at 42.45 ± 9.45% due to the iron-based coagulant mainly used in the return water treatment process of the S sewage treatment facility. Comparing the CaO content, fly ash was 16.45 ± 1.55% and bottom ash was 12.73 ± 5.38%, similar to the 15.7% CaO content reported by Adam et al. [12].

3.2 Jar-test

3.2.1 Eluted phosphorus

P elution efficiency was compared by adding sulfuric acid and sodium hydroxide to the fly ash. For the Jar-test, 0.4 N and 1 N sulfuric acid solutions and sodium hydroxide solutions were prepared as extraction liquids, and the stirring time was It was set to 120 min. **Figure** 7 shows P eluted from 1 N sulfuric acid solution and sodium hydroxide solution.

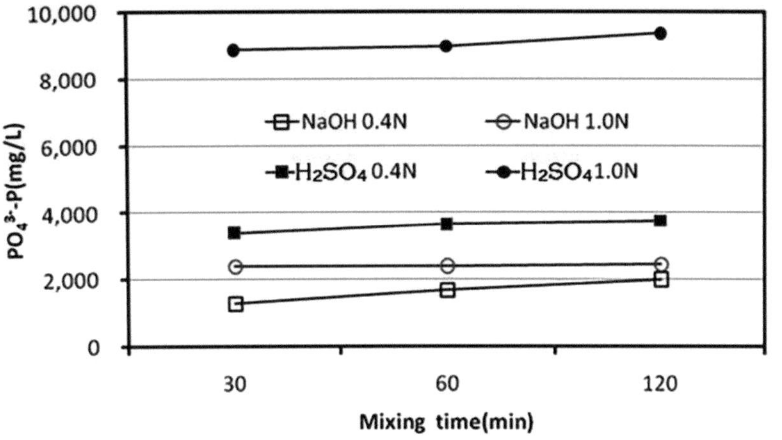

Figure 7.
Comparison of H_2SO_4 and NaOH as extraction liquid.

1 N sulfuric acid and 1 N sodium hydroxide were used, the eluted P was 9362.2 mg/L and 2462.7 mg/L, respectively, and the elution efficiency of sulfuric acid was about 3.8 times higher than that of sodium hydroxide. Through the above experimental results, when using sulfuric acid The maximum recovery of phosphorus that can be eluted is It was more than 0.094 g of PO_4^{3-}-P/g ash. P eluted when 0.4 N sulfuric acid and 0.4 N sodium hydroxide were used under the same conditions were 3747.4 mg/L and 1986.1 mg/L, respectively, and P eluted when 1 N sulfuric acid was used as the eluent. P increased 2.5 times in proportion to 0.4 N sulfuric acid. Biswas, B et al. reported that when sodium hydroxide is used for phosphorus elution from incineration ash, about 40% of phosphorus elution is possible compared to acid [14].

3.2.2 Stirring speed

Figure 8 shows P eluted at stirring speeds of 150 rpm and 200 rpm when 1 N sulfuric acid solution was used as the eluent. In order to select the optimal stirring speed for elution of incineration ash, It was fixed at 120 min. Stirring speed PO_4^{3-}-P

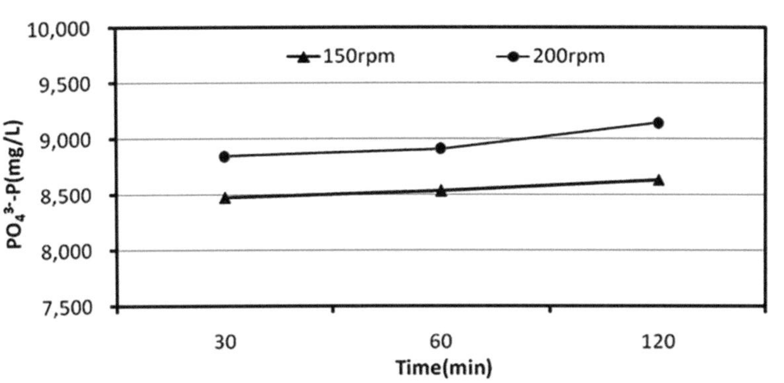

Figure 8.
Variation of PO_4^{3-}-P with different stirring speed.

eluted at 200 rpm was 9141.0 mg/L, showing stirring speed It was slightly higher than P(8,634.9 mg/L) eluted at 150 rpm.

Based on the above experimental results, the phosphorus elution efficiency was determined by the stirring speed. At 200 rpm, Agitation speed through the superiority of contact area and the number of collisions It is judged to be better than 150 rpm.

3.2.3 Heavy metal

When using 1 N sulfuric acid solution Al, Ca and Mg concentrations are shown in **Figure 9**. After 120 minutes, the Al concentration was 2790.1 mg/L and Mg concentration was 1781.4 mg/L. Al and Mg concentrations increased with stirring time, but Ca concentration continuously decreased from 2404.1 mg/L(5 min) to 1211.3 mg/L(120 min).

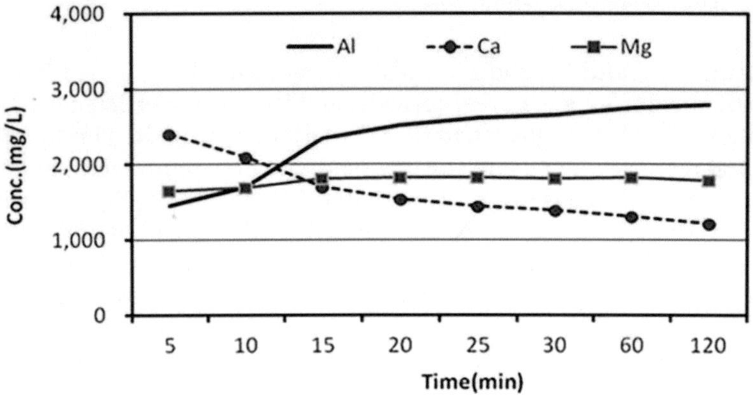

Figure 9.
Variations of heavy metals with H_2SO_4.

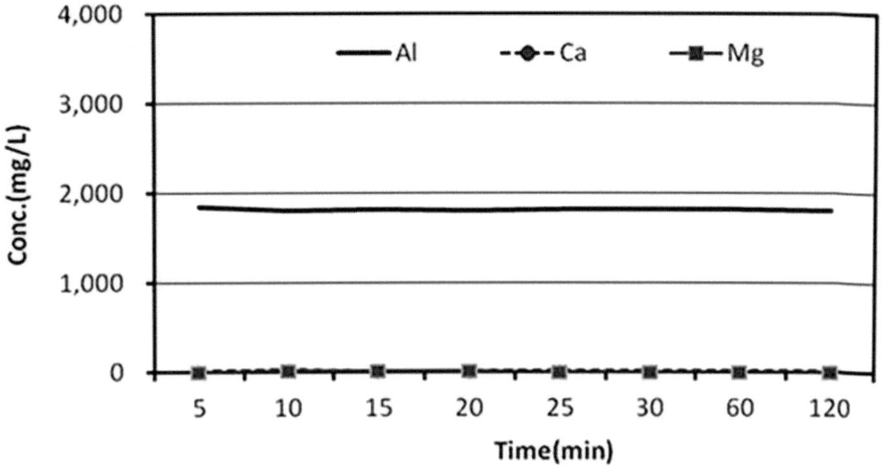

Figure 10.
Variations of heavy metals with NaOH.

This phenomenon is thought to be due to chemical precipitation of sulfate and calcium contained in the eluate as calcium sulfate ($CaSO_4$). In addition, K and Na concentrations were produced below 400 mg/L, and As, Cr, Cd, Pb and Se were detected below the limit of quantification.

On the other hand, when using 1 N sodium hydroxide solution Al, Ca and Mg concentrations are shown in **Figure 10**. The Al concentration appeared constant at a level of 1800 mg/L, and Mg and Ca concentrations were maintained at a low concentration of 20 mg/L or less compared to when sulfuric acid was used as an eluent. Schaum et al. argued that phosphorus extraction efficiency is as low as 30% or less, but heavy metal elution is low when phosphorus is separated by applying alkali extraction method from sewage sludge incineration ash [15]. When sulfuric acid is used, phosphorus elution efficiency is excellent, but a large amount of other heavy metals are generated as by-products, and phosphorus crystallization. It is necessary to select an effective phosphorus recovery method in consideration of the process of adjusting above pH 9.5 and recovery cost.

3.3 Operating result of PRS

3.3.1 Phosphorus

Figure 11 shows $PO_4{}^{3-}$-P according to ultrasonic power in 1 N sodium hydroxide. Same as 12. L: S ratio of 10 without ultrasonic irradiation When stirred for 30 min, $PO_4{}^{3-}$-P was 2117.7 mg/L, and when ultrasonic waves were irradiated at 100 W, $PO_4{}^{3-}$-P was 3194.8 mg/L at same stirring time during ultrasonic irradiation.

It was found that the elution concentration of phosphorus increased by more than 50%. when ultrasonic power was increased to 500 W, $PO_4{}^{3-}$-P was 4104.6 mg/L, indicating that phosphorus elution amount Improved over 94%. This is judged to be the result of an increase in the mass transfer rate of phosphorus adsorbed on fly ash or bonded in the form of a compound by the shear force of ultrasonic waves. The reason why the phosphorus elution concentration is high according to ultrasonic output after

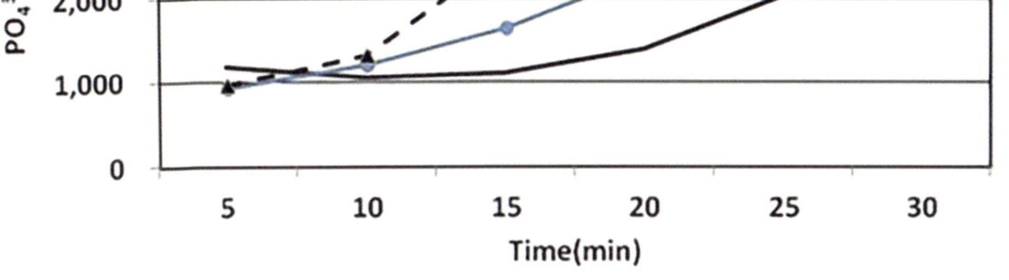

Figure 11.
Variations of $PO_4{}^{3-}$-P at different ultrasonic power with NaOH.

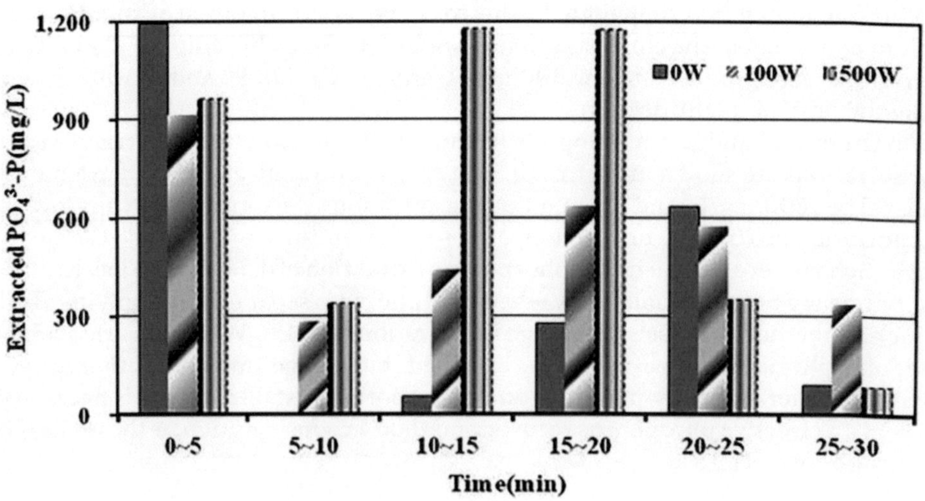

Figure 12.
Extracted PO_4^{3-}-P at different ultrasonic power with NaOH.

ultrasonic irradiation time of 10 min is that the P adsorbed on the surface of the fly ash is easily separated in the beginning, but P bonded in the form of P compound is thought to be because elution is possible through a strong shear force such as ultrasound and long exposure.

The amount of PO_4^{3-}-P dissolution by time according to the ultrasonic output was **Figure 12**. In the 0–5 min section, P adsorbed on the surface of the fly ash is easily separated, and it is found that a lot of P(phosphorus) over 900 mg/L (0.9 g P/100 g Ash) is eluted. On the other hand, ultrasonic waves were irradiated at 500 W, phosphorus elution amount It was significantly higher than 1100 mg/L (1.1 g P/100 g Ash) in the intervals of 10 ~ 15 min and 15 ~ 20 min.

3.3.2 Dissolution rate

The dissolution rate of PO_4^{3-}-P per unit time according to the ultrasonic power in 1 N sodium hydroxide is shown in **Figure 13**. When stirring for 5 min without ultrasonic irradiation, PO_4^{3-}-P dissolution rate was 238.4 mg/L-min, and when the ultrasonic wave was irradiated at 100 W, PO_4^{3-}-P dissolution rate was 183.2 mg/L-min. Appeared as On the other hand, when ultrasonic waves were not irradiated (0 W), the dissolution rate decreased to less than 100 mg/L-min, but when ultrasonic waves were irradiated at 100 W, it was maintained constant at 100 mg/L-min.

Compared to when ultrasonic waves were not irradiated, P elution concentrations increased by 2.1 times and 2.7 times when ultrasonic power was irradiated with 100 W and 500 W, respectively. The effect was found to be large. Based on these results, it is judged that the phosphorus elution effect can be maximized if the operating conditions are derived in consideration of the phosphorus elution amount and the ultrasonic irradiation cost when sodium hydroxide is used for phosphorus recovery from fly ash.

3.3.3 Heavy metal

The concentration of heavy metals eluted from 1 N sodium hydroxide according to the ultrasonic power is shown in **Figure 14**. Without ultrasound irradiation When

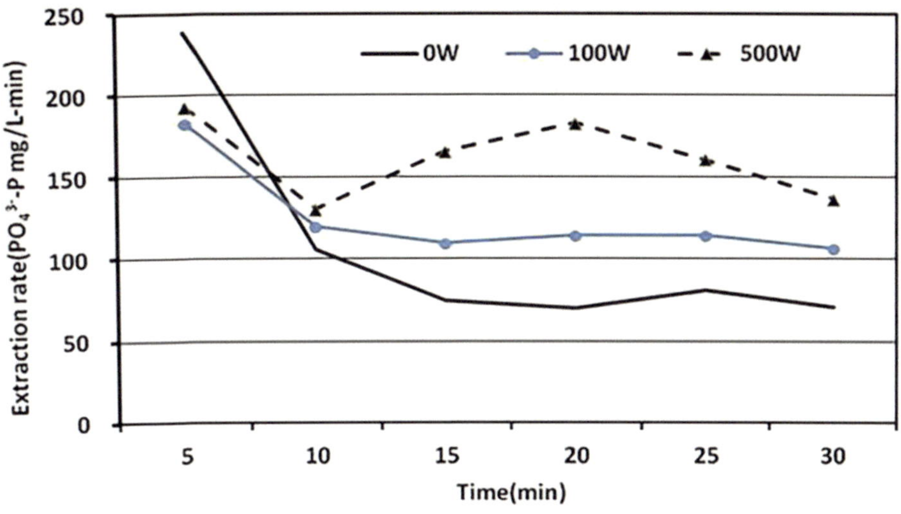

Figure 13.
Variations of extraction rate at different ultrasonic power with NaOH.

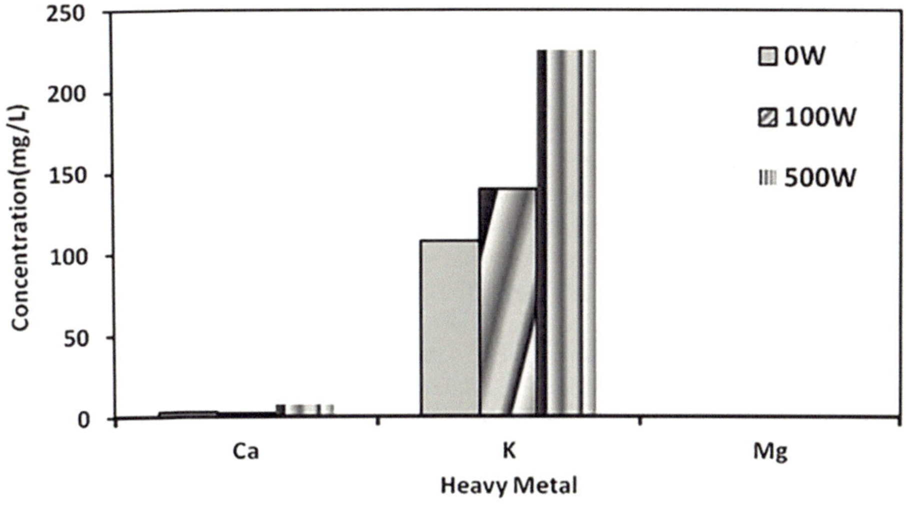

Figure 14.
Variations of heavy metals at different ultrasonic power with NaOH.

stirring for 5 min (0 W), Ca concentration was 3.5 mg/L, and when ultrasonic waves were irradiated at 100 W, the Ca concentration was 2.6 mg/L. Also, the Mg concentration was measured without ultrasonic irradiation.

When stirred for 5 min and when irradiated with ultrasonic waves at 100 W, 0.2 mg/L and non-detection were found, respectively, indicating that the Ca and Mg concentrations were insignificantly affected by ultrasonic irradiation. On the other hand, the K concentration was measured without ultrasonic irradiation. Concentrations were 108.5 mg/L and 224.9 mg/L, respectively, when stirring and when ultrasonic waves were irradiated at 500 W, confirming that the K concentration increased as the ultrasonic power increased.

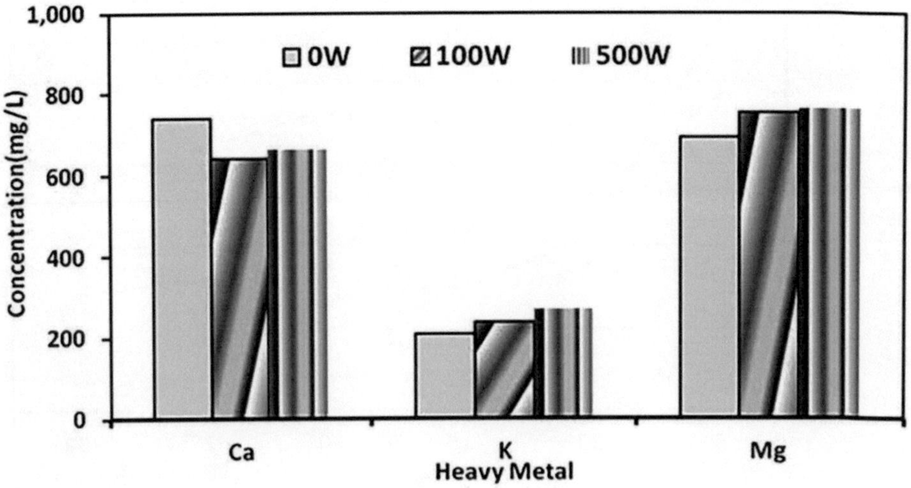

Figure 15.
Variations of heavy metals at different ultrasonic power with H_2SO_4.

The heavy metal concentration according to the ultrasonic power in 1 N sulfuric acid is shown in **Figure 15**. Mg concentration was measured without ultrasonic irradiation. When stirring and ultrasonic irradiation at 100 W were 691.8 mg/L and 754.9 mg/L, respectively, effect of ultrasonic irradiation was not significant. On the other hand K without ultrasonic irradiation When stirred and when irradiated with 500 W of ultrasonic waves, it was 209.7 mg/L and 269.0 mg/L, respectively, and the K concentration slightly increased with the increase in ultrasonic power. On the other hand, Ca and Mg concentrations showed a significant increase in heavy metal elution concentration in sulfuric acid eluate compared to sodium hydroxide.

When sulfuric acid is used as eluent, Cu concentration more than 10 times higher than that of sodium hydroxide is detected. It is expected that the purity of struvite will be lowered or secondary contamination will occur. On the other hand, when sodium hydroxide was used, As, Ba, Cd, and Ni were not detected, and Cr, Pb, Se, and Zn concentrations were detected at low concentrations of less than 1 mg/L. When sulfuric acid is used as an eluent, phosphorus elution efficiency is high, but heavy metals such as Cu, Zn, and As are eluted at high concentrations, causing other side reactions.

3.3.4 Optimization of operating condition

In order to determine the elution conditions of P in fly ash, an experiment was conducted by setting the stirring time and irradiation time to 3 hr. On the other hand, without ultrasound Stirring time when only stirring at 120 min, PO_4^{3-}-P was 4360.3 mg/L, compared to P(4,402.0 mg/L) eluted at 30 min at 500 W ultrasonic irradiation. It could be drastically reduced to 1/4. Considering this fact, fly ash When irradiated with 500 W ultrasound for 30 min in 1 N sodium hydroxide, P elution of up to 0.044 g P/g ash is possible. Compared to experimental results of recovering more than 0.094 g of P/g ash in 1 N sulfuric acid, about 46.8% or more of phosphorus

Element	Content(%)	Element	Content(%)
Mg	47.84	Si	3.9
P	22.6	Fe	1.2
Al	17.5	K	0.5
S	4.0	Zn	0.1

Table 3.
Main chemical composition of precipitated struvite by XRF.

contained in incineration ash can be recovered by using ultrasonic waves in 1 N sodium hydroxide. Based on the above experimental results, the optimal operating condition of ultrasonic elution tank is to minimize the concentration of harmful heavy metals and recovery cost when irradiating for 30 min at ultrasonic output of 500 W using 1 N sodium hydroxide at L/S ratio of 10.

3.4 Phosphorus crystallization analysis

Table 3 shows the results of X-ray Fluorescence Spectrometry (XRF) analysis of recovered struvite. Main components Mg and P were 47.84% and 22.6%, respectively, and Si and Fe were 3.9% and 1.2%. Struvite recovered through the above analysis results contains a large amount of Mg and P at more than 70%, so it is expected that it can be used as a fertilizer when cultivating crops. In addition it was confirmed through X-ray Diffractometer (XRD) analysis that Mg and P are the main components constituting phosphorus crystallization, considering the position of the main peak appearing in struvite.

4. Conclusion

1. Fly ash included in the incineration ash generated from the S sewage treatment facility was suitable for phosphorus recovery due to its large amount, specific surface area and high P_2O_5 content.

2. Phosphorus Recovery System(PRS) consisting of an ultrasonic extraction tank, a solid-liquid separation tank, a mixing tank, and a P recovery tank was developed to recover phosphorus from sewage sludge incineration ash.

3. As a result of operated ultrasonic extraction tank, compared to the existing agitation method, P(phosphorus) Separation time can be drastically reduced by 1/4, and phosphorus elution of up to 0.044 g P/g ash is possible when irradiating 500 W ultrasonic waves for 30 min in L/S ratio 10 and 1 N NaOH solution.

4. Phosphorus from incineration ash recognized as waste It is expected that the recovery will contribute to the expectation of import substitution effect of phosphate rock facing the crisis of resource depletion.

4.1 Discussion

When recovering phosphorus from incineration ash, the biggest problem is considered to be economic feasibility. It is still difficult to secure economic feasibility because the recovery cost is higher than that of phosphate rock. However, natural resources are finite and resource recovery technology needs to be developed to prepare for resource depletion. In particular, when recovering phosphorus from incineration ash, it is possible under acidic and basic conditions.

Although acidic conditions are favorable for phosphorus extraction efficiency from incineration ash, there is a disadvantage in that the purity of phosphorus crystallization is low due to the elution of many other heavy metals. In addition, the process is complicated because a separate pH control is required under acidic conditions. On the other hand, in basic conditions, phosphorus extraction efficiency is low, but phosphorus crystallization with high purity can be obtained because the elution concentration of heavy metals is low. In the future, considering these problems, it is judged that the phosphorus recovery method should be applied according to the site conditions of the sewage treatment plant and the e of phosphorus crystal (MAP or HAP).

Author details

Ye Duk Choi[1]*, Jun Yeon Lee[2], Zoo Ho Jang[2], Jung Gone Joung[2], Kyu Mun Han[1], Mok Young Lee[1] and Kweon Jung[1]

1 Seoul Metropolitan Government Research Institute of Public Health and Environment, Seoul, Republic of Korea

2 Seonam Sewage Treatment Center, Seoul, Republic of Korea

*Address all correspondence to: ydchoi9@seoul.go.kr

References

[1] Franz M. Phosphate fertilizer from sewage sludge ash. Waste Management. 2008;**28**(10):1809-1818

[2] Schipper W, Klapwijk A, Potjer B, Rulkens W, Temmink B, Kiestra F, et al. Phosphate recycling in the phosphorus industry. Environmental Technology. 2001;**22**(11):1337-1345

[3] Cordell D, Drangert J, White S, White S. The story of phosphorus: Global food security and food for thought. Global environmental. Change. 2009;**19**:292-305

[4] Steen I. Phosphorus availability in the 21st century: Management of a non-renewable resource. British Sulfur Publishing. 1998;**217**:25-31

[5] Zhou K, Barjenbruch M, Kabbe K, Inial G, Remy C. Phosphorus recovery from municipal and fertilizer wastewater: China's potential and perspective. Journal of Environmental Sciences. 2017;**52**:151-159

[6] Tilman D, Cassman KG, Matson PA, Naylor R, Polasky S. Agricultural sustainability and intensive production practices. Nature. 2002;**418**:671-677

[7] Petzet S, Peplinski B, Cornel P. On wet chemical phosphorus recovery from sewage sludge ash by acidic or alkaline leaching and an optimized combination of both. Water Research. 2012;**46**:3769-3780

[8] Roy DM, Luke K, Diamond S. Characterization of fly ash and its reactions in concrete. Materials Research Society Symposium Proceedings. 1984;**43**:3-11

[9] Donatello S, Cheeseman CR. Recycling and recovery routes for incinerated sewage sludge ash (ISSA): A review. Waste Management. 2013;**33**:2328-2340

[10] Merino I, Arévalo LF, Romero F. Preparation and characterization of ceramic products by thermal treatment of sewage sludge ashes mixed with different additives. Waste Management. 2007;**27**(12):1829-1844

[11] Donatello S, Tong D, Cheeseman CR. Production of technical grade phosphoric acid from incinerator sewage sludge ash (ISSA). Waste Management. 2010;**30**(8-9):1634-1642

[12] Adam C, Peplinski M, Michaelis GK, Simon FG. Thermochemical treatment of sewage sludge ashes for phosphorus recovery. Waste Management. 2009;**29**(3):1122-1128

[13] Xu H, He P, Gu W, Wang G, Shao L. Recovery of phosphorus as struvite from sewage sludge ash. Journal of Environmental Sciences. 2012;**24**(8):1533-1538

[14] Biswas B, Inoue K, Harada H, Ohto K, Kawakita H. Leaching of phosphorus from incinerated sewage sludge ash by means of acid extraction followed by adsorption on orange waste gel. Journal of Environment Science. 2009;**21**(12):1753-1760

[15] Schaum C, Cornel P, Jardin N. Possibilities for a phosphorus recovery from sewage sludge ash. In: Proceeding of Management of Residues Emanating from Water and Wastewater Treatment. Johannesburg, South Africa. 2005